VOLUME THREE HUNDRED AND FORTY NINE

INTERNATIONAL REVIEW OF CELL AND MOLECULAR BIOLOGY

Immunobiology of Dendritic Cells Part B

INTERNATIONAL REVIEW OF CELL AND MOLECULAR BIOLOGY

VOLUME THREE HUNDRED AND FORTY NINE

INTERNATIONAL REVIEW OF CELL AND MOLECULAR BIOLOGY

Immunobiology of Dendritic Cells Part B

Edited by

CLAIRE LHUILLIER
Weill Cornell Medical College, New York, NY, United States

LORENZO GALLUZZI
Weill Cornell Medical College, New York, NY, United States

ELSEVIER

ACADEMIC PRESS
An imprint of Elsevier

Academic Press is an imprint of Elsevier
50 Hampshire Street, 5th Floor, Cambridge, MA 02139, United States
525 B Street, Suite 1650, San Diego, CA 92101, United States
The Boulevard, Langford Lane, Kidlington, Oxford OX5 1GB, United Kingdom
125 London Wall, London, EC2Y 5AS, United Kingdom

First edition 2019

ISBN: 978-0-12-818357-1
ISSN: 1937-6448

For information on all Academic Press publications
visit our website at https://www.elsevier.com/books-and-journals

Publisher: Zoe Kruze
Acquisition Editor: Ashlie M. Jackman
Editorial Project Manager: Leticia M. Lima
Production Project Manager: Denny Mansingh
Cover Designer: Greg Harris

Typeset by SPi Global, India

Contents

Contributors *ix*
Preface: Dendritic cells: Master regulators of innate and adaptive immunity *xi*

1. Origin and development of classical dendritic cells **1**
Pierre Guermonprez, Yohan Gerber-Ferder, Kristine Vaivode, Pierre Bourdely, and Julie Helft

1. Introduction 2
2. Classical dendritic cell subsets 3
3. Growth factors dependence of classical dendritic cell subsets 10
4. Bone marrow progenitors for classical dendritic cells 15
5. Circulating precursors for classical dendritic cells 19
6. Peripheral niches 21
7. Transcriptional regulation of classical dendritic cell development 24
8. Conclusion and open questions 33

Acknowledgment 33
References 33
Further reading 54

2. Transcriptional control of dendritic cell development and functions **55**
Lukas Amon, Christian H.K. Lehmann, Anna Baranska, Janina Schoen, Lukas Heger, and Diana Dudziak

1. Introduction 56
2. The unique functions of DCs 58
3. Overview about antigen uptake, processing and presentation by DCs 59
4. Phenotypical characterization of cDCs and pDCs 62
5. Identification and functional specialization of the cDC1 subset 63
6. Identification and functional specialization of the cDC2 subset 65
7. cDC1 and cDC2 and the induction and maintenance of tolerance 68
8. Identification and functional specialization of pDCs 69
9. Origin, identification and functional specialization of Langerhans cells 77
10. Identification and functional specialization of MoDCs 79

11. Phenotypical characterization, functional specialization, and presence of DC subsets in human tissues 80
12. The development of early DC progenitors within the bone marrow 82
13. Crucial factors for DC development: Flt3 and Flt3L, GM-CSF and GM-CSFR 86
14. Transcriptional regulation of DC differentiation and function 87
15. Impact of post-transcriptional regulation by microRNAs 104
16. Epigenetic regulation of DCs 113
17. Conclusions 115
Acknowledgments 115
Statement 115
References 116

3. The impact of endoplasmic reticulum stress responses in dendritic cell immunobiology 153

Camilla Salvagno and Juan R. Cubillos-Ruiz

1. A brief introduction to dendritic cell biology 154
2. ER stress responses in DCs 156
3. Concluding remarks 169
Acknowledgments 170
Competing interest 170
References 170

4. The versatile plasmacytoid dendritic cell: Function, heterogeneity, and plasticity 177

Rebecca Leylek and Juliana Idoyaga

1. Introduction 178
2. Functional diversity of plasmacytoid dendritic cells 178
3. Sources of pDC functional diversity: Developmentally encoded subsets versus environmentally mediated plasticity 185
4. Concluding remarks 197
Acknowledgments 198
References 199

5. Dendritic cell extracellular vesicles **213**

Joanna Kowal and Mercedes Tkach

1. Introduction 214
2. Molecular identity of dendritic cell-derived extracellular vesicles 216
3. Diversity of extracellular vesicles from dendritic cells 221
4. The journey of an extracellular vesicle: From its biogenesis to its final destination in the target cell 222
5. Role of dendritic cell-derived extracellular vesicles as modulators of immune responses 227
6. Clinical application of dendritic cell-derived extracellular vesicles 235
7. Concluding remarks and perspectives 236

Acknowledgments 237

References 238

6. *Ex vivo* dendritic cell generation—A critical comparison of current approaches **251**

Patrick Han, Douglas Hanlon, Olga Sobolev, Rabib Chaudhury, and Richard L. Edelson

1. Cytokine signals 253
2. Physiological signals for DC differentiation 260
3. Gene engineering tools 267
4. Clinical trials of DC-based vaccines 281
5. Concluding remarks 289

Conflict of interest 290

References 290

Contributors

Lukas Amon
Laboratory of Dendritic Cell Biology, Department of Dermatology, University Hospital Erlangen, Friedrich-Alexander University of Erlangen-Nürnberg, Erlangen, Germany

Anna Baranska
Laboratory of Dendritic Cell Biology, Department of Dermatology, University Hospital Erlangen, Friedrich-Alexander University of Erlangen-Nürnberg, Erlangen, Germany

Pierre Bourdely
King's College London, Centre for Inflammation Biology and Cancer Immunology, The Peter Gorer Department of Immmunobiology, London, United Kingdom

Rabib Chaudhury
Department of Chemical and Environmental Engineering, School of Engineering and Applied Science, Yale University, New Haven, CT, United States

Juan R. Cubillos-Ruiz
Department of Obstetrics and Gynecology; Sandra and Edward Meyer Cancer Center, Weill Cornell Medicine; Weill Cornell Graduate School of Medical Sciences, Cornell University, New York, NY, United States

Diana Dudziak
Laboratory of Dendritic Cell Biology, Department of Dermatology, University Hospital Erlangen, Friedrich-Alexander University of Erlangen-Nürnberg, Erlangen, Germany

Richard L. Edelson
Department of Dermatology, School of Medicine, Yale University, New Haven, CT, United States

Yohan Gerber-Ferder
Institut Curie, PSL Research University, INSERM U932, SiRIC «Translational Immunotherapy Team»; Université de Paris, Immunity and Cancer Department, INSERM U932, Institut Curie, Paris, France

Pierre Guermonprez
King's College London, Centre for Inflammation Biology and Cancer Immunology, The Peter Gorer Department of Immmunobiology, London, United Kingdom; Université de Paris, CNRS ERL8252, INSERM1149, Centre for Inflammation Research, Paris, France

Patrick Han
Department of Chemical and Environmental Engineering, School of Engineering and Applied Science, Yale University, New Haven, CT, United States

Douglas Hanlon
Department of Dermatology, School of Medicine, Yale University, New Haven, CT, United States

Lukas Heger
Laboratory of Dendritic Cell Biology, Department of Dermatology, University Hospital Erlangen, Friedrich-Alexander University of Erlangen-Nürnberg, Erlangen, Germany

Julie Helft
Institut Curie, PSL Research University, INSERM U932, SiRIC «Translational Immunotherapy Team»; Université de Paris, Immunity and Cancer Department, INSERM U932, Institut Curie, Paris, France

Juliana Idoyaga
Department of Microbiology and Immunology, and Immunology Program, Stanford University School of Medicine, Stanford, CA, United States

Joanna Kowal
Department of Oncology, Ludwig Institute for Cancer Research, University of Lausanne, Lausanne, Switzerland

Christian H.K. Lehmann
Laboratory of Dendritic Cell Biology, Department of Dermatology, University Hospital Erlangen, Friedrich-Alexander University of Erlangen-Nürnberg, Erlangen, Germany

Rebecca Leylek
Department of Microbiology and Immunology, and Immunology Program, Stanford University School of Medicine, Stanford, CA, United States

Camilla Salvagno
Department of Obstetrics and Gynecology; Sandra and Edward Meyer Cancer Center, Weill Cornell Medicine, New York, NY, United States

Janina Schoen
Laboratory of Dendritic Cell Biology, Department of Dermatology, University Hospital Erlangen, Friedrich-Alexander University of Erlangen-Nürnberg, Erlangen, Germany

Olga Sobolev
Department of Dermatology, School of Medicine, Yale University, New Haven, CT, United States

Mercedes Tkach
Institut Curie, PSL Research University, INSERM U932, Paris, France

Kristine Vaivode
King's College London, Centre for Inflammation Biology and Cancer Immunology, The Peter Gorer Department of Immmunobiology, London, United Kingdom

Preface: Dendritic cells: Master regulators of innate and adaptive immunity

Claire Lhuillier[a,*], Lorenzo Galluzzi[a,b,c,d,*]
[a]Department of Radiation Oncology, Weill Cornell Medical College, New York, NY, United States
[b]Sandra and Edward Meyer Cancer Center, New York, NY, United States
[c]Department of Dermatology, Yale School of Medicine, New Haven, CT, United States
[d]Université de Paris, Paris, France
[*]Corresponding authors: e-mail address: cfl2002@med.cornell.edu; deadoc80@gmail.com

Dendritic cells (DCs) are a population of myeloid cells that play a key role in the initiation of adaptive immune responses, a core immunological activity that reflects their proficiency at presenting extracellular antigens to naïve T cells and earned them the name of "sentinels of the immune system" (Reis e Sousa, 2006). These professional antigen-presenting cells were originally discovered by the Canadian physician Ralph Steinman in 1973 (Steinman, 1991; Steinman and Cohn, 1973), who was the first to observe cells with a peculiar tree-like morphology in mouse peripheral lymphoid organs and named them dendritic cells after the Greek word *dendreon* (tree). A few years later, Steinman and colleagues demonstrated that DCs are potent inducers of the so-called "mixed leukocyte reaction" (a rather outdated term referring to the experimental assessment of T cell priming) in mice (Steinman and Witmer, 1978) as they highlighted multiple phenotypic and functional properties of DCs (Nussenzweig and Steinman, 1982; Steinman, 1991). In 2011, Ralph Steinman was awarded the Nobel Prize in Physiology or Medicine "for his discovery of the dendritic cell and its role in adaptive immunity," jointly with the American immunologist Bruce Beutler and the Luxembourger biologist Jules Hoffmann "for their discoveries concerning the activation of innate immunity" (Lemaitre et al., 1996; Poltorak et al., 1998).

The phenotypic and functional heterogeneity of DCs became apparent early after their discovery (Maldonado-Lopez et al., 1999; Pulendran et al., 1999; Vremec et al., 1992). Indeed, despite a shared ability for antigen uptake, processing and presentation to naïve T cells, DCs exhibit a wide spectrum of phenotypic and functional properties that—coupled to their anatomical location—have a major impact on their capacity to regulate immune responses (Alcantara-Hernandez et al., 2017; Hashimoto et al., 2011). Both human and mouse DCs can be broadly classified into two major lineages based on morphology: myeloid DCs (mDCs), also known as classical or conventional DCs (cDCs), which (prior to activation) resemble circulating monocytes, and plasmacytoid DCs (pDCs), which display some morphological resemblance to plasma cells (Merad et al., 2013). Two other subsets of DCs have been described: monocyte-derived (or inflammatory) DCs, which are believed to originate from $CD16^+$ monocytes, and Langerhans cells (LCs), a skin-resident population of DCs with potent immunosuppressive activity (Collin and Bigley, 2018; Price et al., 2015). cDCs originate from a common myeloid precursor and can be classified into two phenotypically distinct subsets: (1) cDC1, which express CD8α or CD103 in mice, or CD141 in humans and (2) cDC2, which express CD11b in mice, or CD1c in humans. pDCs have been suggested to originate from a poorly defined lymphoid precursor and are characterized by the coexpression of CD45R/B220 and Ly6C in mice or CD123 and CD304 in humans. DCs have also been classified based on the differential expression of key transcription factors such as basic leucine zipper transcription factor ATF-like 3 (BATF3), which is known for its role in the development of $CD8\alpha^+$ DCs in lymphoid tissues and $CD103^+CD11b^-$ DCs in the periphery (Collin and Bigley, 2018; Hildner et al., 2008; Schlitzer et al., 2013). Recently, the use of high-dimensional analysis techniques (e.g., single-cell RNA-sequencing, mass cytometry) has opened new avenues for studying the complexity of DCs, casting doubts on the precision of the aforementioned taxonomy (See et al., 2017; Villani et al., 2017).

The ability of DCs to induce robust immune responses is highly dependent on their maturation status (Truxova et al., 2018). Through the expression of pattern recognition receptors such as Toll-like receptors (TLRs), C-type lectin receptors, Nod-like receptors (NLRs), and others (Akira et al., 2006; Galluzzi et al., 2018a; Takeuchi and Akira, 2010), immature DCs can indeed decode microenvironmental signals of danger, including microbe-associated molecular patterns (MAMPs) as well as pathogen-associated molecular patterns (DAMPs), and undergo a phenotypic and functional shift that underlies their ultimate immunological activity (Galluzzi et al., 2017). Thus, while immature DCs display intense endocytic and phagocytic capacity but are poor at antigen presentation, their mature counterparts express high amounts of MHC class II molecules, costimulatory molecules (e.g., CD80, CD86, CD40), and chemokine receptors (e.g., CCR7) allowing them to migrate to lymphoid organs and stimulate naïve T cells (Banchereau et al., 2000; Reis e Sousa, 2006). Importantly, depending on subset, phenotype, and maturation status, DCs may induce tolerogenic or immunogenic responses downstream of the polarization of different $CD4^+$ helper T cell populations (e.g., T_H1, T_H2, T_H9,

T_H17, T_H22) or the functional interaction with $CD8^+$ cytotoxic T lymphocytes (CTLs) or $CD4^+CD25^+FOXP3^+$ regulatory T (T_{REG}) cells (Devi and Anandasabapathy, 2017; Tanchot et al., 2013). The precise mechanisms whereby different DCs subsets orchestrate T cell responses remain to be fully elucidated.

From a functional perspective, the cDC1 subset is specialized at cross-presenting microenvironmental antigens on MHC class I molecules to $CD8^+$ T cells in the context of type I interferon (IFN) secretion (Bachem et al., 2010; den Haan et al., 2000), an activity that is critical for anticancer immune responses (Galluzzi et al., 2018b; Zitvogel et al., 2015). Conversely, the cDC2s subset efficiently presents exogenous antigens on MHC class II molecules to $CD4^+$ T cells (Collin and Bigley, 2018; Desch et al., 2014; Segura and Amigorena, 2015). However, cDC2 DCs are more heterogeneous than their cDC1 counterparts and some $CD8\alpha^-$ DCs have been shown to engage in cross-presentation upon activation (Collin and Bigley, 2018; Desch et al., 2014; Segura and Amigorena, 2015). Finally, pDCs are extraordinarily potent at producing type I and type III IFNs upon viral infection, de facto constituting a major player in innate antiviral immunity (Dubois et al., 2019; McNab et al., 2015; Swiecki and Colonna, 2015). Multiple DC subtypes can also interact with natural killer (NK) cells, a cross-talk that can result in the pruning of the DC compartment and mutual activation (Barry et al., 2018; Ferlazzo and Moretta, 2014). Of note, the immunomodulatory functions of DCs are not only mediated by cell-to-cell interaction and cytokine secretion, but also by the release of extracellular vesicles (Pitt et al., 2014).

Despite their fundamental role in early responses against viral pathogens, DCs can also contribute to the etiology of various human disorders. For instance, human immunodeficiency virus type 1 (HIV-1) highjacks DC functions in support of viral replication, distal dissemination, and cell-to-cell transmission (Wacleche et al., 2018). Moreover, some DCs subsets express high levels of the immunosuppressive enzyme indoleamine 2,3-dioxygenase 1 (IDO1), which can be detrimental in some pathophysiological settings including chronic infections and cancer (Harden and Egilmez, 2012). Finally, Langerhans cell histiocytosis is a rare myeloproliferative disorder mostly affecting children that is driven by uncontrolled LC proliferation (Allen et al., 2018; Chakraborty et al., 2003).

Not surprisingly, DCs have attracted considerable attention as potential targets for the development of novel therapeutic interventions against a variety of human disorders. On the one hand, infusion of tolerogenic DCs as induced by specific pharmacological agents, growth factors, or cytokines may be beneficial in the context of allergic disorders, autoimmune diseases, and transplant rejection. Conversely, vaccines based on autologous DCs loaded with tumor-associated antigens in the context of robust immunostimulatory signals have demonstrate significant anticancer activity in both preclinical and clinical settings (Garg et al., 2017; Kantoff et al., 2010; Medler et al., 2019; Palucka and Banchereau, 2012; Tacken et al., 2007; Vacchelli et al., 2013). Intriguingly, one therapeutic option for prostate carcinoma approved by the US Food and Drug Administration (FDA), consists in a DC-enriched mixture of autologous myeloid cells engineered to express a tumor-associated antigens fused to colony stimulating factor 1 (CSF1, best known as GM-CSF) (Cheever and Higano, 2011).

In *Immunobiology of Dendritic Cells* (a special issue of the *International Review of Cell and Molecular Biology* series) worldwide recognized experts in the field provide an overview of ontological, biological, and functional DCs features in the context of human pathophysiology. Each of the 12 chapters of *Immunobiology of Dendritic Cells* critically discusses the key role of DCs at the interface between innate and adaptive immunity from a unique mechanistic or therapeutic perspective. The book (which consists of two parts, part A and part B) is equally addressed to expert researchers who may be interested in acquiring in-depth insights into the key role of DCs at the interface between innate and adaptive immunity, and to neophytes of this expanding field of translational investigation.

Acknowledgments

C.L. is supported by the American Association for Cancer Research triple-negative breast cancer fellowship. L.G. is supported by a Breakthrough Level 2 grant from the US DoD, Breast Cancer Research Program (BRCP) [#BC180476P1], by a startup grant from the Department of Radiation Oncology at Weill Cornell Medicine (New York, USA), by industrial collaborations with Lytix (Oslo, Norway) and Phosplatin (New York, USA), and by donations from Phosplatin (New York, USA), the Luke Heller TECPR2 Foundation (Boston, USA) and Sotio a.s. (Prague, Czech Republic).

References

Akira, S., et al., 2006. Pathogen recognition and innate immunity. Cell 124, 783–801.
Alcantara-Hernandez, M. et al. (2017) High-dimensional phenotypic mapping of human dendritic cells reveals interindividual variation and tissue specialization. Immunity 47, 1037–1050 e1036.
Allen, C.E., et al., 2018. Langerhans-cell histiocytosis. N. Engl. J. Med. 379, 856–868.
Bachem, A., et al., 2010. Superior antigen cross-presentation and XCR1 expression define human CD11c+CD141+ cells as homologues of mouse CD8+ dendritic cells. J. Exp. Med. 207, 1273–1281.
Banchereau, J., et al., 2000. Immunobiology of dendritic cells. Annu. Rev. Immunol. 18, 767–811.

Barry, K.C., et al., 2018. A natural killer-dendritic cell axis defines checkpoint therapy-responsive tumor microenvironments. Nat. Med. 24, 1178–1191.
Chakraborty, M., et al., 2003. Irradiation of tumor cells up-regulates Fas and enhances CTL lytic activity and CTL adoptive immunotherapy. J. Immunol. 170, 6338–6347.
Cheever, M.A., Higano, C.S., 2011. PROVENGE (Sipuleucel-T) in prostate cancer: the first FDA-approved therapeutic cancer vaccine. Clin. Cancer Res. 17, 3520–3526.
Collin, M., Bigley, V., 2018. Human dendritic cell subsets: an update. Immunology 154, 3–20.
den Haan, J.M., et al., 2000. CD8(+) but not CD8(−) dendritic cells cross-prime cytotoxic T cells in vivo. J. Exp. Med. 192, 1685–1696.
Desch, A.N., et al., 2014. Dendritic cell subsets require cis-activation for cytotoxic CD8 T-cell induction. Nat. Commun. 5, 4674.
Devi, K.S., Anandasabapathy, N., 2017. The origin of DCs and capacity for immunologic tolerance in central and peripheral tissues. Semin. Immunopathol. 39, 137–152.
Dubois, H., et al., 2019. Nucleic acid induced interferon and inflammasome responses in regulating host defense to gastrointestinal viruses. Int. Rev. Cell Mol. Biol. 345, 137–171.
Ferlazzo, G., Moretta, L., 2014. Dendritic cell editing by natural killer cells. Crit. Rev. Oncog. 19, 67–75.
Galluzzi, L., et al., 2017. Immunogenic cell death in cancer and infectious disease. Nat. Rev. Immunol. 17, 97–111.
Galluzzi, L., et al., 2018a. SnapShot: CGAS-STING signaling. Cell 173, 276.e271.
Galluzzi, L., et al., 2018b. The hallmarks of successful anticancer immunotherapy. Sci. Transl. Med. 10.
Garg, A.D., et al., 2017. Trial watch: dendritic cell-based anticancer immunotherapy. Oncoimmunology 6, e1328341.
Harden, J.L., Egilmez, N.K., 2012. Indoleamine 2,3-dioxygenase and dendritic cell tolerogenicity. Immunol. Invest. 41, 738–764.
Hashimoto, D., et al., 2011. Dendritic cell and macrophage heterogeneity in vivo. Immunity 35, 323–335.
Hildner, K., et al., 2008. Batf3 deficiency reveals a critical role for CD8alpha+ dendritic cells in cytotoxic T cell immunity. Science 322, 1097–1100.
Kantoff, P.W., et al., 2010. Sipuleucel-T immunotherapy for castration-resistant prostate cancer. N. Engl. J. Med. 363, 411–422.
Lemaitre, B., et al., 1996. The dorsoventral regulatory gene cassette spatzle/Toll/cactus controls the potent antifungal response in Drosophila adults. Cell 86, 973–983.
Maldonado-Lopez, R., et al., 1999. CD8alpha+ and CD8alpha- subclasses of dendritic cells direct the development of distinct T helper cells in vivo. J. Exp. Med. 189, 587–592.
McNab, F., et al., 2015. Type I interferons in infectious disease. Nat. Rev. Immunol. 15, 87–103.
Medler, T., et al., 2019. Activating the nucleic acid-sensing machinery for anticancer immunity. Int. Rev. Cell Mol. Biol. 344, 173–214.
Merad, M., et al., 2013. The dendritic cell lineage: ontogeny and function of dendritic cells and their subsets in the steady state and the inflamed setting. Annu. Rev. Immunol. 31, 563–604.
Nussenzweig, M.C., Steinman, R.M., 1982. The cell surface of mouse lymphoid dendritic cells. Immunol. Today 3, 65–68.
Palucka, K., Banchereau, J., 2012. Cancer immunotherapy via dendritic cells. Nat. Rev. Cancer 12, 265–277.
Pitt, J.M., et al., 2014. Dendritic cell-derived exosomes as immunotherapies in the fight against cancer. J. Immunol. 193, 1006–1011.

Poltorak, A., et al., 1998. Defective LPS signaling in C3H/HeJ and C57BL/10ScCr mice: mutations in Tlr4 gene. Science 282, 2085–2088.
Price, J.G., et al., 2015. CDKN1A regulates Langerhans cell survival and promotes Treg cell generation upon exposure to ionizing irradiation. Nat. Immunol. 16, 1060–1068.
Pulendran, B., et al., 1999. Distinct dendritic cell subsets differentially regulate the class of immune response in vivo. Proc. Natl. Acad. Sci. U. S. A. 96, 1036–1041.
Reis e Sousa, C., 2006. Dendritic cells in a mature age. Nat. Rev. Immunol. 6, 476–483.
Schlitzer, A., et al., 2013. IRF4 transcription factor-dependent CD11b+ dendritic cells in human and mouse control mucosal IL-17 cytokine responses. Immunity 38, 970–983.
See, P., et al., 2017. Mapping the human DC lineage through the integration of high-dimensional techniques. Science 356.
Segura, E., Amigorena, S., 2015. Cross-presentation in mouse and human dendritic cells. Adv. Immunol. 127, 1–31.
Steinman, R.M., 1991. The dendritic cell system and its role in immunogenicity. Annu. Rev. Immunol. 9, 271–296.
Steinman, R.M., Cohn, Z.A., 1973. Identification of a novel cell type in peripheral lymphoid organs of mice. I. Morphology, quantitation, tissue distribution. J. Exp. Med. 137, 1142–1162.
Steinman, R.M., Witmer, M.D., 1978. Lymphoid dendritic cells are potent stimulators of the primary mixed leukocyte reaction in mice. Proc. Natl. Acad. Sci. U. S. A. 75, 5132–5136.
Swiecki, M., Colonna, M., 2015. The multifaceted biology of plasmacytoid dendritic cells. Nat. Rev. Immunol. 15, 471–485.
Tacken, P.J., et al., 2007. Dendritic-cell immunotherapy: from ex vivo loading to in vivo targeting. Nat. Rev. Immunol. 7, 790–802.
Takeuchi, O., Akira, S., 2010. Pattern recognition receptors and inflammation. Cell 140, 805–820.
Tanchot, C., et al., 2013. Tumor-infiltrating regulatory T cells: phenotype, role, mechanism of expansion in situ and clinical significance. Cancer Microenviron. 6, 147–157.
Truxova, I., et al., 2018. Mature dendritic cells correlate with favorable immune infiltrate and improved prognosis in ovarian carcinoma patients. J. Immunother. Cancer 6, 139.
Vacchelli, E., et al., 2013. Trial watch: dendritic cell-based interventions for cancer therapy. Oncoimmunology 2, e25771.
Villani, A.C., et al., 2017. Single-cell RNA-seq reveals new types of human blood dendritic cells, monocytes, and progenitors. Science 356 (6335), eaah4573.
Vremec, D., et al., 1992. The surface phenotype of dendritic cells purified from mouse thymus and spleen: investigation of the CD8 expression by a subpopulation of dendritic cells. J. Exp. Med. 176, 47–58.
Wacleche, V.S., et al., 2018. The biology of monocytes and dendritic cells: contribution to HIV pathogenesis. Viruses 10.
Zitvogel, L., et al., 2015. Type I interferons in anticancer immunity. Nat. Rev. Immunol. 15, 405–414.

CHAPTER ONE

Origin and development of classical dendritic cells

Pierre Guermonprez[a,b,*], Yohan Gerber-Ferder[c,d], Kristine Vaivode[a], Pierre Bourdely[a], Julie Helft[c,d,*]
[a]King's College London, Centre for Inflammation Biology and Cancer Immunology, The Peter Gorer Department of Immmunobiology, London, United Kingdom
[b]Université de Paris, CNRS ERL8252, INSERM1149, Centre for Inflammation Research, Paris, France
[c]Institut Curie, PSL Research University, INSERM U932, SiRIC «Translational Immunotherapy Team», Paris, France
[d]Université de Paris, Immunity and Cancer Department, INSERM U932, Institut Curie, Paris, France
*Corresponding authors: e-mail address: pierre.guermonprez@cnrs.fr; julie.helft@inserm.fr

Contents

1. Introduction 2
2. Classical dendritic cell subsets 3
 2.1 Type 1 cDCs 5
 2.2 Type 2 cDCs 7
 2.3 Human cDC subsets 9
3. Growth factors dependence of classical dendritic cell subsets 10
 3.1 FLT3L 10
 3.2 GM-CSF 12
4. Bone marrow progenitors for classical dendritic cells 15
 4.1 Progenitor populations and developmental hierarchies 15
 4.2 cDC specification at the single progenitor cell level 18
5. Circulating precursors for classical dendritic cells 19
 5.1 Mouse pre-cDCs 19
 5.2 Human pre-cDCs 19
6. Peripheral niches 21
 6.1 Infiltration 21
 6.2 Proliferation 22
 6.3 Differentiation 22
 6.4 Positioning 23
7. Transcriptional regulation of classical dendritic cell development 24
 7.1 Transcription factors controlling the development of the cDC lineage 24
 7.2 Transcription factors controlling the development of cDC subsets 27
8. Conclusion and open questions 33
Acknowledgment 33
References 33
Further reading 54

International Review of Cell and Molecular Biology, Volume 349
ISSN 1937-6448
https://doi.org/10.1016/bs.ircmb.2019.08.002

Abstract

Classical dendritic cells (cDCs) are mononuclear phagocytes of hematopoietic origin specialized in the induction and regulation of adaptive immunity. Initially defined by their unique T cell activation potential, it became quickly apparent that cDCs would be difficult to distinguish from other phagocyte lineages, by solely relying on marker-based approaches. Today, cDCs definition increasingly embed their unique ontogenetic features. A growing consensus defines cDCs on multiple criteria including: (1) dependency on the fms-like tyrosine kinase 3 ligand hematopoietic growth factor, (2) development from the common DC bone marrow progenitor, (3) constitutive expression of the transcription factor ZBTB46 and (4) the ability to induce, after adequate stimulation, the activation of naïve T lymphocytes. cDCs are a heterogeneous cell population that contains two main subsets, named type 1 and type 2 cDCs, arising from divergent ontogenetic pathways and populating multiple lymphoid and non-lymphoid tissues. Here, we present recent knowledge on the cellular and molecular pathways controlling the specification and commitment of cDC subsets from murine and human hematopoietic stem cells.

1. Introduction

Classical dendritic cells (cDCs) are professional antigen presenting cells of hematopoietic origin that reside in virtually all lymphoid and non-lymphoid tissues. cDCs were originally identified in mouse spleen by Ralph Steinman and Zanvil Cohn (Steinman, 1973). In a series of seminal studies Steinman and Cohn described the morphology and the immune function of cDCs (Steinman and Cohn, 1974; Steinman et al., 1975). They found that splenic cDCs were short-lived and continuously replenished from bone marrow progenitors (Steinman et al., 1974) and were 100 times more efficient than other leukocyte populations (including macrophages) to stimulate allogenic T cells (Steinman and Witmer, 1978). In the past 40 years, cDCs have been shown to be sentinels of the immune system by their unique ability to bridge innate sensing and adaptive immunity. In peripheral tissues, cDCs can sense and capture pathogens and cellular debris. This triggers a maturation process leading to the migration of cDCs to the T cell zones of secondary lymphoid organs, where they present antigen-derived peptides to T lymphocytes and induce T cell-mediated immune responses (Banchereau and Steinman, 1998). During homeostasis, immature cDCs also migrate from peripheral tissues to lymphoid organs where they induce tolerance by presenting self-antigens to T lymphocytes (Bonifaz et al., 2002; Cella et al., 1997; Hawiger et al., 2001; Inaba et al., 2000). Beyond antigen presentation, cDCs exert innate immune functions, including secreting regulatory and inflammatory mediators (Del Fresno et al., 2018; Janela et al., 2019; Lauterbach et al., 2010; Mashayekhi et al., 2011; Spranger et al., 2017). Therefore, cDCs exert a plethora of immune functions and are key regulators of innate and adaptive immunity.

cDCs were first described to constitutively express the CD11c integrin and high levels of major histocompatibility complex (MHC) class II. While CD11c expression is still considered to be a surrogate marker for cDC identification, it is now well established that other immune cell lineages, such as inflammatory monocytes and some tissue-resident macrophages, also express CD11c. This phenotypic overlap has led to many confusions in the identification and functional assessment of tissue cDCs. Using CD11c as the sole marker to identify cDCs has also undermined their cellular diversity. cDCs are indeed heterogeneous and comprise various subtypes, which are phenotypically and functionally distinct. Recent technical advances have led to a precise and refined definition of cDCs, an essential pre-requisite for their functional assessment.

Understanding the cellular origin and the developmental biology of cDCs have been key steps in defining cDCs as an independant cellular lineage distinct from monocytes and macrophages (see Fig. 1). Comparative transcriptomic studies between cDC subsets, monocytes and macrophages isolated from mouse tissues have allowed the identification of a specific cDC-gene signature, and have pinpointed important molecular players of cDC development and functions (Crozat et al., 2010b; Miller et al., 2012). In addition, the lineage relationship between cDC subsets across tissues has been established with the identification of cDC-committed progenitors and the transcription factors (TFs) and growth factors, which control their development. This knowledge further allows the discrimination of cDCs from monocyte-derived cells or tissue-resident macrophages (Guilliams et al., 2014).

Here we review the latest knowledge on cDC origin and development both in mouse and human.

2. Classical dendritic cell subsets

cDCs arise from DC-committed hematopoietic precursors (pre-DCs), which circulate in the blood and continuously seed non-lymphoid and lymphoid organs (Breton et al., 2015; Liu et al., 2009; See et al., 2017). Most organs comprise at least two distinct cDC subsets: the IRF8^{+}XCR1^{+} type

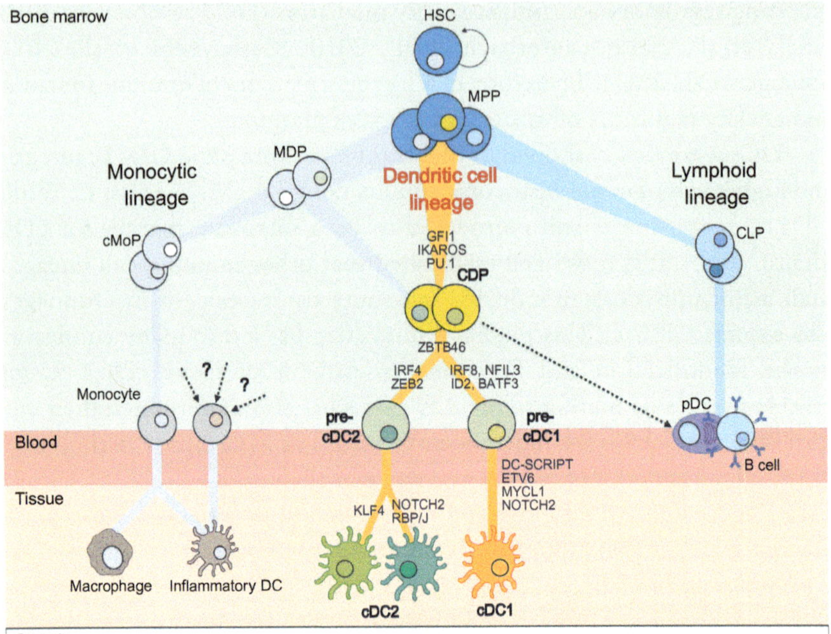

Growth Factors	M-CSF	GM-CSF	FLT3L / GM-CSF	FLT3L / IL-7

Function	Th2, ILC2 allergens parasites	Th17, ILC3 extracellular bacteria fungi	T CD8+, ILC1 intracellular bacteria viruses

Fig. 1 The illustration depicts the current view on cDC development from bone marrow progenitors toward circulating blood precursors and tissue-resident cDC subsets. Dotted lines indicate developmental transitions that exist in some conditions but that may not be the principal developmental pathway occuring in steady state (CDP to pDC), or ill-understood developmental pathways (inflammatory DC). Developmental intermediates are characterized by the expression of specific transcription factors (TFs) like ZBTB46 in pre-cDCs. Other TFs also control developmental transitions. TFs are indicated in capital letters. Below, growth factors involved in each lineage development and peripheral survival of tissue-resident subsets are indicated. Finally, known immunological function of the cDC subsets are highligthed with a color code corresponding to the cDC subsets. (*HSC*: hematopoietic stem cells; *MPP*: multipotent progenitor; *MDP*: macrophage-DC progenitor; *cMoP*: common monocyte precursor; *CDP*: common DC progenitor, *pre-cDC*: precursor of cDC; *CLP*: common lymphoid progenitor; *pDC*: plasmacytoid DC).

1 DCs (cDC1) and the IRF4$^+$SIRPα^+ type 2 DCs (cDC2), each performing specific immune function (del Rio et al., 2007; Ginhoux et al., 2007; Murphy et al., 2016; Poulin et al., 2007; Vremec et al., 2000). The division of labor between cDC subsets is currently one of the major interest in the DC-research field. However, while cDC1s are a well-defined cell

population expressing a unique set of markers (Edelson et al., 2010), some of which are evolutionarily conserved (Crozat et al., 2010b; Guilliams et al., 2016), cDC2s are a more heterogeneous population and remain to be further characterized.

Recent transcriptomic profiling of cDC subsets across tissues and species has provided a more thorough classification based on the expression of specific gene signatures, expression of growth factor receptors and TFs (Durai and Murphy, 2016; Gautier et al., 2012; Miller et al., 2012). Transcriptomic and developmental studies have established the following criteria for cDCs identification: (1) cDC-development and survival in tissues depend on the growth factor FLT3L; (2) cDCs stem from the bone marrow common DC progenitor (CDP) but not from monocytes; (3) cDCs constitutively express the TF ZBTB46 under steady state conditions. Using comparative transcriptomics between cDCs and other myeloid cell subsets, two laboratories have concomitantly discovered the TF ZBTB46 as being specifically expressed in mature cDCs (Meredith et al., 2012a; Satpathy et al., 2012). *Zbtb46*^{gfp} reporter mouse showed that ZBTB46 is expressed in bone marrow pre-cDCs (Grajales-Reyes et al., 2015). The exact functional role of ZBTB46 in cDCs is yet to be elucidated, as *Zbtb46*-deficient mice do not have any major immune deficiency (Meredith et al., 2012b). However, expression of ZBTB46 is believed to be specific of the cDC lineage and to enables its identification (Meredith et al., 2012a; Satpathy et al., 2012).

These three criteria are particularly important for the identification of cDC2s, which phenotypic features overlap with monocyte-derived cells and tissue resident macrophages. This framework, combined with the use of mouse genetics and of high dimensional techniques, allows the precise identification of cDC subsets across various organs and species.

2.1 Type 1 cDCs

cDC1s specifically and uniformly express the chemokine receptor XCR1 and the C-type lectin CLEC9A, regardless of the organs in which they reside (Kroczek and Henn, 2012; Poulin et al., 2012; Robbins et al., 2008). Yet the expression of other cDC1 surface markers are tissue-dependent. For example, splenic cDC1s express CD8α, while dermal cDC1s express langerin (CD207) and the integrin ITGAE (CD103) (Ginhoux et al., 2007; Henri et al., 2010). *In vitro*, cDC1-like cells generated in bone marrow cultures supplemented with FLT3L growth factor express high levels of CD24 (Naik et al., 2005).

cDC1 development depends on two specific TFs, BATF3 and IRF8. Genetic deletion of *Batf3* or point mutation in the *Irf8* gene (*Irf8*^{R294C}) abrogate cDC1-development in all organs. Mouse models deficient for *Batf3* or *Irf8* have allowed to probe the immune function of cDC1s *in vivo* (Edelson et al., 2010; Hildner et al., 2008; Tailor et al., 2008). cDC1s play a non-redundant role in the induction of type 1 immunity, particularly for the priming of anti-viral and anti-tumor cytotoxic CD8^{+} T cells (Bedoui et al., 2009; Hildner et al., 2008; Helft et al., 2012), and in the control of intracellular parasites and bacteria (Mashayekhi et al., 2011; Yamazaki et al., 2013). Indeed, various functional properties of cDC1s enable the efficient activation of CD8^{+} T cells:

(1) a molecular machinery for cross-presenting exogenously acquired cell-associated antigens on MHC-I to CD8^{+} T cell (Bougnères et al., 2009; Helft et al., 2012; Kretzer et al., 2016; Sancho et al., 2009; Savina et al., 2009; Theisen et al., 2018). Some genes favoring cross-presentation like *Plin2* (Bougnères et al., 2009), *Clec9a* (Sancho et al., 2009), *Rab43* (Kretzer et al., 2016) or *Wdfy4* (Theisen et al., 2018) are over-expressed in cDC1s as compared to cDC2s or other mononuclear phagocytes.

(2) the specific expression of the chemokine receptor XCR1, which binds to XCL1, a unique ligand secreted by recently activated CD8^{+} T cells. The XCR1/XCL1 axis established between cDC1s and T cells enhances both the expansion and survival of antigen specific CD8^{+} T cells. (Dorner et al., 2009). Recently, intra-tumor NK cells have been shown to be a relevant source of XCL1, participating in the recruitment of cDC1s in tumors and enhancing anti-tumor CD8^{+} T cell responses (Böttcher et al., 2018).

(3) the ability to produce CXCL9, a chemokine required to attract effectors and memory CXCR3^{+}CD8^{+} T cells to the site of injury (Alexandre et al., 2016; Spranger et al., 2017).

These functional properties make cDC1s essential players for the initiation of cytotoxic CD8^{+} T cell immune responses toward intracellular pathogens and tumors (Hildner et al., 2008).

cDC1s also play a role in the regulation of CD4^{+} Th1 cell responses. Besides antigenic presentation by MHC-I, cDC1s are also able to efficiently present antigenic peptide on MHC-II to CD4^{+} T cells (Helft et al., 2012; Kastenmüller et al., 2013). However, MHC-II presentation by cDC1s is not strictly required to prime CD4^{+} T cells, since mice lacking the cDC1 subset (*Batf3*-deficient mice or *langerin-DTR* mice) can efficiently mount CD4^{+} T cell responses to West-Nile virus infection (Hildner et al., 2008) or

Leishmania major infection (Brewig et al., 2009). Nonetheless, cDC1s are responsible for bridging $CD4^+$ and $CD8^+$ T cells during an anti-viral immune response, enabling efficient anti-viral $CD8^+$ T cell response during systemic virus infection (Eickhoff et al., 2015; Hor et al., 2015). cDC1s are also crucial for the induction of Th1 immunity against *Candida albicans* (Brewig et al., 2009). Dermal cDC1s are required for priming pathogenic $CD4^+$ T cells involved in experimental autoimmune encephalomyelitis (King et al., 2010). Furthermore, cDC1s are key players in the immunity against *Toxoplasma Gondii* (Mashayekhi et al., 2011) and *Leishmania major* parasites (Ashok et al., 2014; Martínez-López et al., 2015), by producing the IL-12 cytokine in the earliest stages of infection. TLR11 expression in cDC1s is essential for *Toxoplasma Gondii*-induced IL-12 production *via* profilin/TLR11 interaction (Yarovinsky et al., 2005). Interestingly, constitutive IL-12 production by migratory cDC1s also inhibits Th2 responses against helminth infection (Everts et al., 2016), suggesting that IL-12 production by cDC1s is a regulator of Th1 *versus* Th2 immunity.

Two recent studies have uncovered innate immune functions for cDC1s, independent of T cell priming. Production of VEGFα by murine cDC1s (Janela et al., 2019) or inhibition of CXCL2 production by cDC1s (Del Fresno et al., 2018) regulate neutrophil recruitment to various infectious sites.

cDC1s have a unique molecular machinery enabling specific recognition and interaction with pathogens. For example cDC1s are the major source of type III IFN lambda upon TLR3 stimulation by double stranded RNA injection *in vivo* (Lauterbach et al., 2010) or in response to hepatitis C virus *in vitro* (Yoshio et al., 2013). Splenic cDC1s are the primary site of entry and proliferation of $GRAM^+$ bacteria *Listeria Monocytogenes* (Edelson et al., 2011a; Neuenhahn et al., 2006). In addition, cDC1s are poorly infectable by influenza virus and HIV (Helft et al., 2012; Silvin et al., 2017). Finally, cDC1s are very efficient presenter of malaria antigens by MHC-I to $CD8^+$ T cells (Guermonprez et al., 2013; Lundie et al., 2008). These specific host-pathogen attributes underscore the crucial role of cDC1s in innate immunity.

2.2 Type 2 cDCs

cDC2s share phenotypic features with monocyte-derived cells and tissue resident macrophages, including constitutive expression of the CD11b integrin and the inhibitory receptor SIRPα. They are therefore less easily identified than cDC1s in peripheral tissues, particularly during inflammation where monocytes extravasate in tissues and differentiate in CD11c$^+$MHC-II$^+$CD11b$^+$SIRPα$^+$ phagocytes (Greter et al., 2012; Menezes et al., 2016; Serbina et al., 2003). The identification of TFs selectively impairing cDC2 development or function independently of monocyte and macrophages has been key to better assess cDC2s function *in vivo*. The first TF shown to regulate cDC2s homeostasis is IRF4 (Suzuki et al., 2004). IRF4 affects cDC2s terminal differentiation (Persson et al., 2013; Schlitzer et al., 2013), migration (Bajaña et al., 2012) and antigenic presentation (Vander Lugt et al., 2014).

cDC2s are less efficient than cDC1s to perform cross-presentation of exogenous antigens on MHC-I and to cross-prime CD8$^+$ T cells *in vivo*. Consequently, cDC2s cannot compensate the lack of cDC1s in *Batf3*-deficient mice for cross-priming of anti-viral and anti-tumor CD8$^+$ T cells (Hildner et al., 2008). The inability of cDC2s to efficiently prime CD8$^+$ T cell responses is partly due to their endocytic molecular machinery, which is more adapted for antigen processing in the MHC-II pathway (Dudziak et al., 2007). cDC2s have reduced phagosomal ROS production and increased acidification (Savina et al., 2009) together with distinct lipid storage impacting on phago-lysosomal function (Bougnères et al., 2009). They also express higher levels of TFEB, a TF which enhances lysosomal activation and antigen degradation, two processes known to inhibit cross-presentation and enhance MHC-II presentation (Samie and Cresswell, 2015).

Two other transcription regulators, NOTCH2 (Lewis et al., 2011) and KLF4 (Tussiwand et al., 2015), further subdivide IRF4$^+$ cDC2s in two populations, which exert non-redundant immune functions.

2.2.1 NOTCH2-dependent cDC2 subset

The NOTCH2$^+$ cDC2s play a prominent role in shaping innate and CD4$^+$ T cell-mediated immunity at mucosal sites. This was demonstrated in mice lacking the NOTCH2 receptor in CD11c expressing cell (using the *CD11c-cre* deleter strain for DC-specific deletion of conditional *Notch2* allele, called the DC-*Notch2*$^{\Delta}$ mouse) (Lewis et al., 2011). DC-*Notch2*$^{\Delta}$ mice lack the intestinal CD103$^+$CD11b$^+$ cDC2 subset and show a reduced IL-17 production by CD4$^+$ T cells in the lamina propria (Lewis et al., 2011). This is in line with the non-redundant role of CD103$^+$CD11b$^+$ intestinal FLT3L-dependent cDC2s in producing IL-23 and promoting IL-22 production by innate lymphoid cell (Kinnebrew et al., 2012; Satpathy et al., 2013). This IL-23/IL-22 axis has been shown to be essential to control *Citrobacter*

Rodentium infection and IL-23 is also involved in Th17 differentiation (Ivanov et al., 2008). In the lung, IRF4-dependent $CD24^+CD11b^+$ cDC2s were also shown to produce IL-23 and to control Th17 responses to *Aspergillus fumigatus* fungii infection (Schlitzer et al., 2013). $CD103^+CD11b^+$ intestinal cDC2s also express high levels of CCR7 and are the first DCs to migrate and transport *Salmonella Typhimurium* to lymph nodes after oral infection (Bogunovic et al., 2009). Altogether, these studies demonstrate that intestinal NOTCH2-dependent cDC2s shape innate and adaptive immunity to pathogenic bacteria.

More recently, a study demonstrated that NOTCH2-dependent splenic $ESAM^+$ cDC2s are essential for the activation of follicular helper T cells, the formation of germinal center and the production of high affinity antibodies against sheep red blood cell and heat-inactivated *Listeria monocytogenes* (Briseño et al., 2018). This again demonstrates the prominent role of NOTCH2-dependent cDC2s in regulating $CD4^+$ T cell immunity.

2.2.2 KLF4-dependent cDC2 subset

Mice deficient in *Irf4* in CD11c expressing cells ($Cd11c^{cre}xIrf4^{flox/flox}$named DC-$Irf4^{\Delta}$) were shown to have defective Th2 responses towards *Nippostrongylus brasiliensis* parasitic infection or immunization with papain adjuvant (Gao et al., 2013). Interestingly, DC-$Notch2^{\Delta}$ mice do not phenocopy DC-$Irf4^{\Delta}$ mice since they conserve a normal ability to prime Th2 responses. This hinted at the existence of a yet unknown $IRF4^+$ cDC2 subset specialized for Th2 priming. Accordingly, *Klf4* deficiency in CD11c expressing cells ($Cd11c^{cre}xKlf4^{flox}$) was later shown to ablate a subset of $IRF4^+$ cDC2s different from NOTCH2-dependent cDC2s (Tussiwand et al., 2015). This study identified KLF4-dependent cDC2s as key player in Th2 responses upon various type of immunization (house dust mite, *Schistosoma* egg antigen) or parasitic infection (*S. mansoni*).

2.3 Human cDC subsets

cDC1 and cDC2 subsets homologs have been first identified in peripheral blood in human (Dzionek et al., 2000). Comparative transcriptomic studies have shown that mouse and human cDC subsets in blood, spleen, gut and lungs have conserved expression programs (Robbins et al., 2008), including the expression of XCR1, CLEC9A and IRF8 by human cDC1s (Poulin et al., 2010) and of *SIRPα* and *IRF4* by human cDC2s (Persson et al., 2013; Schlitzer et al., 2013). Comparative functional studies have also confirmed a relatively preserved division of labor between human cDC subsets, with cDC1s being more efficient at cross-presenting exogenously acquired antigens (Balan et al., 2014; Haniffa et al., 2012) and cDC2s more prone at inducing Th17 and Th2 responses (Persson et al., 2013; Schlitzer et al., 2013). Human intestinal cDC2s also express $CD103^+$ together with SIRPα and share various common features with mouse NOTCH2-dependent cDC2s, including Th17 polarization (Watchmaker et al., 2014). These studies are further reviewed in a following book chapter.

3. Growth factors dependence of classical dendritic cell subsets

3.1 FLT3L

FMS-like tyrosine kinase (FLT3, previously Flk2) is a class III receptor tyrosine kinase and FLT3L is its ligand. Ligation of FLT3 by FLT3L induces FLT3 homo-dimerization supporting receptor auto-phosphorylation. This process activates multiple signaling pathways enabling STAT3 and STAT5 phosphorylation, PI3kinase and MAP kinases activation. Overall, these signals control the survival, proliferation and differentiation of FLT3-expressing cells.

In mice, FLT3 expression is one of the hallmarks of early lymphoid-primed multipotent progenitors (LMPP) (Adolfsson et al., 2001, 2005; Christensen and Weissman, 2001). The PU.1 TF promotes FLT3 expression and supports cDC commitment (Carotta et al., 2010). FLT3 expression is maintained in a subset of myeloid-committed (CMP) and lymphoid-committed (CLP) progenitors (D'Amico and Wu, 2003). Further analysis has identified high levels of FLT3 expression on cDC-committed progenitors (Liu et al., 2009; Naik et al., 2007; Onai et al., 2007). Finally, among mature hematopoietic lineages, cDCs are characterized by high levels of FLT3 expression, particularly in the cDC1 subset (Miller et al., 2012).

Pioneer work by Pulendran et al. has shown that administration of supraphysiological levels of recombinant FLT3L dramatically expand cDC numbers in mice (Karsunky et al., 2003; Pulendran et al., 1997). This interesting property can be harnessed in immunotherapy for the mobilization of precursors of cDC1s within tumors (Salmon et al., 2016). The crucial role of FLT3L on cDC development was first demonstrated by analyzing *Flt3l*$^{-/-}$ mice in which cDC populations are almost entirely absent (McKenna et al., 2000). Later on, it was found that mice lacking the FLT3 receptor (*Flt3*$^{-/-}$) only partially phenocopy *Flt3l*$^{-/-}$ mice. Indeed, unlike *Flt3l*$^{-/-}$ mice, *Flt3*$^{-/-}$ mice are only partly deficient in lymphoid organ cDCs of both cDC1 and cDC2 subtypes (Ginhoux et al., 2009; Guermonprez et al., 2013; Waskow et al., 2008). By contrast, cDC1s in non-lymphoid tissues like dermis and lung are deeply affected in both *Flt3*$^{-/-}$ and *Flt3l*$^{-/-}$ mice (Ginhoux et al., 2009). These findings demonstrate that tissue and lymphoid organ resident cDCs differ in their growth factor requirements. In addition, these findings have fueled the speculation for the existence of an alternative receptor for FLT3L, which might compensate for the lack of expression of FLT3. This hypothesis was recently invalidated by a study showing that *Flt3*$^{-/-}$ mice are actually unresponsive to administration of supra-physiological levels of FLT3L (Durai et al., 2018). Instead, it was found that *Flt3*$^{-/-}$ mice compensate for the loss of FLT3 by up-regulating the responsiveness to other class III receptor tyrosine kinases CSF1-R/M-CSFR and c-KIT whose contribution is enhanced in the context of *Flt3*$^{-/-}$ mice (Durai et al., 2018). As a paradoxical consequence, inactivation of FLT3 receptor in the context of FLT3L deficiency increases cDC pools in periphery. These findings underline the fine tuning and coordination of multiple signaling pathways controlling cDC development.

Alterations in the efficiency of FLT3 signaling can be created in the mouse system by inhibiting the PTEN phosphatase, a negative regulator of mTOR activation downstream PI3kinase (Sathaliyawala et al., 2010). cDC-specific gene inactivation of PTEN leads to the expansion of cDC1s and can be counteracted by mTOR inhibition provided by rapamycin delivery. Similarly, inactivation of the LAMTOR protein prevents FLT3 internalization in endolysosomes (Scheffler et al., 2014). This leads to the upregulation of FLT3 at the cell surface membrane, and in increased FLT3 signaling and mTOR-dependent accumulation of cDC1s in aging mice (Scheffler et al., 2014).

Initially established for murine cDCs, the crucial role of FLT3L on DC development has been confirmed for human cDCs. First, immunodeficient mice reconstituted with human CD34^{+} hematopoietic stem and progenitor cells develop multiple human lineages but are very scarcely populated by human cDCs (Ding et al., 2014; Guermonprez et al., 2013). Administration of recombinant FLT3L is sufficient to trigger a massive expansion of human cDCs in blood and spleen of humanized mice (Ding et al., 2014; Guermonprez et al., 2013; Traggiai et al., 2004). Finally, administration of FLT3L to healthy volunteers induces a dramatic increase in circulating pre-cDCs and differentiated cDC1s and cDC2s (Anandasabapathy et al., 2015; Breton et al., 2015). Regimen of 1–75 μg/kg per day were administrated during 1–5 days did not induce severe adverse effects (Anandasabapathy et al., 2014).

Of note, FLT3L induced the mobilization of $CD34^+$ hematopoietic stem and progenitor cells in the blood stream and also expanded circulating monocytes and granulocytes populations but to a lesser extent than cDCs (Anandasabapathy et al., 2015; Breton et al., 2015). These results are consistent with FLT3 expression in human hematopoietic stem cells and in human and murine early multipotent hematopoietic progenitors (Adolfsson et al., 2001; Beaudin et al., 2014; Christensen and Weissman, 2001; Doulatov et al., 2010).

From a practical perspective, murine FLT3L has been used to generate cDC1s and cDC2s in culture from total murine bone marrow (Naik et al., 2005). Interestingly, provision of NOTCH ligands DLL1 by stromal cells to these cultures leads to the terminal maturation of $CD24^+$ cDC1-like cells, which acquire CD8α, DEC205 and CCR7 expression like their splenic primary counterparts (Kirkling et al., 2018). Also, human FLT3L is incorporated in most protocols aiming at generating human cDC subsets from $CD34^+$ hematopoietic stem and progenitor cells (Balan et al., 2018; Chicha et al., 2004; Kirkling et al., 2018; Lee et al., 2015a,b, 2017; Poulin et al., 2010; Proietto et al., 2012).

3.2 GM-CSF

Granulocyte Macrophage Colony Stimulating factor (GM-CSF/CSF2) is a monomeric, soluble cytokine engaging the GM-CSF receptor. The GM-CSF receptor comprises the cytokine-specific alpha chain and the common beta chain (shared with IL-3 and IL-5 receptors), which activates PI3Kinase, mTOR and STAT5 phosphorylation (Becher et al., 2016; Hamilton and Anderson, 2004; Martinez-Moczygemba and Huston, 2003). GM-CSF and GM-CSFR deficiencies in mice lead to the complete absence of lung alveolar macrophages (Shibata et al., 2001). cDCs from lymphoid organs are not massively affected by the lack of GM-CSF while they are fully depleted in *Flt3*-deficient mice (Kingston et al., 2009; Vremec et al., 1997). By contrast to lymphoid organs, cDCs are largely impacted in multiple non-lymphoid organs in *Gm-csf*-deficient mice (Greter et al., 2012; Mortha et al., 2014). Of note, there has been some controversy on the extent of the depletion in the cDC1 compartment. Indeed, King et al. initially reported that dermal $CD103^+$ cDC1s were absent from *Gm-csfr*-deficient mice (King et al., 2010). A thorough investigation of tissue cDCs concluded that GM-CSF supports the differentiation of tissue $CD103^+$ cDC1s (Greter et al., 2012). These data were challenged by a further study evidencing a crucial role of GM-CSF for the acquisition of the CD103 integrin expression in CD11b^{low} cDC1s (Edelson et al., 2011b; Sathe et al., 2011). It was therefore suggested that cDC1s from non-lymphoid tissues were largely independent from GM-CSF and that GM-CSF was instead required for their acquisition of the CD103^{+} cDC1-tissue phenotype (Edelson et al., 2011b). In line with these findings, GM-CSF stimulates the expression of CD103 in FLT3L-dependent cultures of total bone marrow in mice (Edelson et al., 2011b; Mayer et al., 2014). GM-CSF is also required to support the terminal phenotypic maturation of human cDC1s generated from CD34^{+} hematopoietic stem and progenitor cells in FLT3L, IL-7 and TPO *in vitro* culture: in this context, GM-CSF stimulates the expression of XCR1 and CADM1, which are two lineage-specific markers of human cDC1s (Balan et al., 2018). In summary, GM-CSF signaling is considered as a promoter of the terminal maturation of tissue cDC1s, supporting the acquisition of the full phenotypic identity of tissue cDC1s at late stage of their development.

Regarding cDC2s, the role of GM-CSF signaling is complex. Pulendran et al. clearly evidenced that administration of recombinant GM-CSF triggered the expansion of splenic CD11b^{+}MHC-II^{high} DCs (Pulendran et al., 1999; Pulendran et al., 2004). Similar findings were reported in mononuclear phagocytes infiltrating B16 melanoma tumors over-expressing GM-CSF (Dranoff et al., 1993; Mach et al., 2000). Also, recall T cell-dependent responses against methylated BSA induce CD11b^{+}MHC-II^{+} cells relying on GM-CSF for their differentiation (Cook et al., 2004). While Pulendran et al. noticed the phenotypic similarity between splenic CD11b^{+}MHC-II^{+} DCs and CD11b^{+} resident cDC2s, subsequent studies from Naik et al. showed that GM-CSF-dependent CD11b^{+}, unlike cDC2s, do not arise from pre-cDCs (Naik et al., 2006). These findings are in line with their reliance on the CCR2 chemokine receptor, which is a distinctive feature of monocytes as compared to cDCs (Menezes et al., 2016). More recent in-depth characterization of *in vivo* induced CD11b^{+}MHC-II^{+} phagocytes in the melanoma model showed that they actually encompass at least two subsets differing for their expression of DC-SIGN/CD209a (Menezes et al., 2016). DC-SIGN/CD209a^{+} phagocytes rely, like cDC2s, on homozygous expression of the TF PU.1 (Menezes et al., 2016). By contrast CD11b^{+}MHC-II^{+}DC-SIGN/CD209a^{-} phagocytes are not compromised in PU.1 haplo-insufficient mice. *In vivo* fate mapping of the progeny of ZBTB46^{+} cDC-precursors (using the *Zbtb46*^{cre} driver mouse and a conditional fluorescent reporter allele) showed that CD209a^{-}GM-CSF-induced

$CD11b^+$ mononuclear phagocytes are largely distinct from the cDC lineage. However, a minor fraction of $CD209a^+$ positive phagocytes can be stained in *Zbtb46*^{cre} fate mapping experiments and *Zbtb46*^{gfp} knock-in reporter mice (Menezes et al., 2016). Also, IL-4 synergizes with GM-CSF to produce $ZBTB46^+$ DC-like cells from monocytes sorted from *Zbtb46*^{gfp} reporter mouse (Satpathy et al., 2012). Collectively these studies show how knowledge on the developmental pathways and their regulation enables to distinguish cDCs from other mononuclear phagocytes in the context of GM-CSF-dependent inflammation.

GM-CSF has been instrumental in multiple attempts to model DC differentiation *in vitro*, in both murine and human settings. In the mouse model, Inaba and Steinman have pioneered the use of GM-CSF to generate DCs from total murine bone marrow cultures (Inaba et al., 1993a,b). Of note, these methodologies have been developed before the identification of FLT3L as a critical factor for cDC homeostasis in mice (McKenna et al., 2000). A recent study has analyzed in-depth the diversity of mononuclear phagocytes generated in those murine bone marrow cultures (Helft et al., 2015, 2016). It appears that GM-CSF cultures encompass both $CD11c^+MHC\text{-}II^{int}M\text{-}CSFR^{high}$ *bona fide* macrophages and a heterogeneous population of cDC-like $CD11c^+MHC\text{-}II^{high}$ cells. $CD11c^+MHC\text{-}II^{high}$ cells are overall dependent on the TF IRF4 (Gao et al., 2013), which controls the expression of MHC-II α and β chain together with MHC-II-related genes (the invariant chain Ii and H2-DM) as well as lysosomal proteases (Vander Lugt et al., 2014). IRF4 requires PU.1 heterodimerization to engage Ets/ISRE common elements EICE and, as a consequence, GM-CSF-induced $CD11c^+MHC\text{-}II^{high}$ cells also require homozygous expression of PU.1 for their differentiation (Carotta et al., 2010; Menezes et al., 2016). $CD11c^+MHC\text{-}II^{high}$ cells are heterogeneous and can be split in $FLT3^+M\text{-}CSFR^-$ and $FLT3^-M\text{-}CSFR^{low}$ populations (Helft et al., 2015, 2016). Transcriptional analysis of the $FLT3^+$ fraction show that they resemble migratory cDCs. $FLT3^+$ cells exhibit the cDC-defining capability to prime naïve T cells (Helft et al., 2015, 2016) and derive from cDC-committed progenitor (CDP), unlike the $CD11c^+MHC\text{-}II^{int}M\text{-}CSFR^{high}$ fraction, showing that they correspond to *bona fide* cDCs (Helft et al., 2015, 2016).

In human culture systems, the use of GM-CSF together with IL-4 has been pioneered by Sallusto and Lanzavecchia who showed that GM-CSF + IL-4 could induce the differentiation of DC-like cells from $CD14^+$ monocytes, called "mo-DCs" (Sallusto and Lanzavecchia, 1994).

In the mouse system, IL-4 modulates the cross-presentation ability of GM-CSF-dependent monocyte-derived DCs (Briseño et al., 2016). Recent mechanistic studies shed some light on the molecular processes by which human monocytes acquire a DC-like phenotype. Sander et al. identified NCOR2, nuclear receptor co-repressor 2, as an important transcriptional regulator of IL-4 dependent DC differentiation (Sander et al., 2017). Also, previous reports highlighted the need for high levels of PU.1 in supporting mo-DCs differentiation (Bakri et al., 2005). Human GM-CSF+IL4 mo-DCs differ from *bona fide* cDCs found in human blood by both their surface, transcriptional phenotypes and cellular origin (Balan et al., 2014; Sander et al., 2017). However, gene set enrichment analysis (GSEA) suggest that GM-CSF+IL4 mo-DCs might represent an *in vitro* proxy for $CD14^{+}CD1c^{+}$ inflammatory DCs (Segura et al., 2013) mobilized in multiple inflammatory settings. Nowadays, GM-CSF is included together with FLT3L and other cytokines (SCF, TPO, *e.g.*) in protocols enabling the differentiation of *bona fide* human cDCs from $CD34^{+}$ bone marrow or cord blood progenitors (Balan et al., 2018; Breton et al., 2015; Kirkling et al., 2018; Lee et al., 2017; Poulin et al., 2010; Proietto et al., 2012).

In conclusion, GM-CSF can stimulate the differentiation of both cDC2s and other cells distinct from the cDC lineage such as "inflammatory" DCs or monocyte-derived macrophages. In addition, GM-CSF also shapes the terminal differentiation of cDCs, including cDC1s.

4. Bone marrow progenitors for classical dendritic cells

4.1 Progenitor populations and developmental hierarchies

cDCs develop from adult hematopoietic stem cells residing in the bone marrow (see Fig. 1). Since cDCs are short-lived (Kamath et al., 2000), the cDC pool is constantly replenished from bone marrow progenitors.

The classical view of murine and human hematopoiesis involves a progressive and ordered loss of developmental potential. This definition is based on the identification and prospective isolation of progenitor populations with restricted developmental potential. For instance, the first hematopoietic tree embranchment splits early progenitors into progenitors restricted for myeloid and lymphoid fates (Akashi et al., 2000; Kondo et al., 1997). Early studies of cDC ontogeny aimed at positioning the road taken by cDCs on this well-ordered map. The characterization of murine cDC subsets has led to the identification of lymphoid-associated markers, such as CD8α expressed by splenic cDC1s and CD4 expressed by splenic cDC2s (Vremec et al., 2000). This supported the hypotheses that some subsets of cDCs may arise from lymphoid-committed progenitors while other arise from myeloid-commited progenitors. However, Traver et al. later showed in mice that either Common Myeloid Progenitor (CMP) or Common Lymphoid progenitor (CLP) give rise to both CD8α^+ and CD8α^- cDCs upon adoptive transfer into irradiated hosts (Traver et al., 2000). Accordingly, it was found that stimulation of the FLT3 receptor expressed by both CMP or CLP stimulates cDC differentiation and expansion (D'Amico and Wu, 2003; Karsunky et al., 2003). Later on, these data were challenged by genetically tracing the progeny of CLPs using *Il-7r* as a driver gene for the cre recombinase and a conditional fluorescent reporter line (Schlenner et al., 2010). Here, cDC ontogeny was physiologically assessed in homeostasic conditions without adoptive transfer of progenitors into irradiated host. This study showed an efficient genetic labeling of all lymphocytes lineages (B, T and NK). By contrast, cDCs were not labeled, supporting the idea that murine cDCs belong to the myeloid branch of hematopoiesis.

cDCs ontogeny has been proposed to occur after the loss of granulocytic potential. Fogg et al. have described in mice a bipotent macrophage/DC progenitors (MDP) endowed with the potential to generate virtually all mononuclear phagocytes, including monocytes and cDCs, but devoid of granulocytic potential (Auffray et al., 2009; Fogg et al., 2006). The existence of such bipotent intermediates has been challenged later on in mouse models where MDPs at the clonal level were found to produce macrophages, granulocytes but very few cDCs or pDCs (Sathe et al., 2014).

Pioneer work by Liu and Nussenzweig have analyzed cDC commitment in human progenitors. They identified the transition between GMDP, a progenitor endowed with a broad granulocyte and monocyte potential, and the mononuclear phagocyte-restricted MDP (Lee et al., 2015a,b). Although total MDP population give rise to both cDCs and CD14$^+$ monocyte, a clonal analysis of the MDP developmental potential at the single cell level provides a different picture. Only a very small proportion (around 10%) of individual MDP actually carries a mixed potential for DCs and monocytes/macrophages, a criterion functionally defining this population (Lee et al., 2015a,b). In mice, monocyte commitment would occur from MDP *via* the generation of monocyte-restricted progenitors such as the murine FLT3$^-$M-CSFR$^+$ cMoP endowed with a developmental potential restricted to CCR2-dependent Ly6C$^+$ monocytes and not cDCs (Hettinger et al., 2013). Using CLEC12A and FCGRIA/CD64 expression, Kawamura et al. have identified a human functional equivalent of the mouse cMoP, which give rise to CD14^{high} monocytes and not cDCs (Kawamura et al., 2017). In conclusion, MDP can give rise to monocyte-commited cMoP and cDCs. However, it is unclear if a sizeable proportion of individual MDPs carries both potential.

cDC develop from MDPs through a developmental intermediate, the "Common Dendritic cells Precursor" (CDP), generating cDCs and plasmacytoid DCs (pDCs) but not monocytes (Naik et al., 2007; Onai et al., 2007). CDPs express both FLT3 and M-CSFR and intermediate levels of c-KIT. Unlike CLPs, CDPs do not express IL-7R (Naik et al., 2007; Onai et al., 2007). Schraml et al. have noticed that CDP expressed CLEC9A, the cDC1-associated marker (Schraml et al., 2013). This led to the development of a genetic fate mapping strategy based on the *Clec9a*^{cre} deleter line crossed to a conditional fluorescent reporter allele. It was found that the majority of cDC2s, in addition to cDC1s, are genetically labeled with this strategy, thus arguing in favor of a substantial contribution of a common CDP progenitor for both subsets *in vivo* (Schraml et al., 2013).

On a peripheral note to the central topic of this review (cDCs), it is important to mention that early studies proposed that murine type I IFN producing pDCs would also arise from CDP (Naik et al., 2007; Onai et al., 2007). It was later shown that M-CSFR expression is associated to FLT3^{+} cDC-biased but not FLT3^{+} pDC-biased progenitors (Onai et al., 2013). However, *in vivo* fate mapping results do not support this notion and rather suggest that pDCs arise from IL-7R^{+} lymphoid-biased progenitors (Schlenner et al., 2010). By contrast, cDCs are poorly labeled in this system (10 to 15%) suggesting a minor contribution of IL-7R^{+} progenitors to cDC ontogeny at steady state (Schlenner et al., 2010). Recent studies also identified a subset of murine progenitors linked to the CLP involved in the generation of pDCs (Rodrigues et al., 2018). Similar findings have also been reported in humans (Karsunky et al., 2003) in which IL-7 is a major cytokine used to differentiate pDCs (Schotte et al., 2010). Therefore, distinct developmental pathways might generate distinct set of mature pDCs susbets, including one major subset of pDCs whose ontogeny is linked to lymphocytes (Reizis, 2019). Furthermore, recent studies of human pDCs have refined their phenotypic definition and showed that CD123^{+}BDCA2^{+} cell fraction contained both AXL^{-}SIGLEC6^{-} *bona fide* pDCs and CD123^{+}BDCA2^{+}AXL^{+}SIGLEC6^{+} defined as pre-cDCs or a subset on its own (See et al., 2017; Villani et al., 2017). Importantly, AXL^{-}SIGLEC6^{-} *bona fide* pDCs were found to be relatively inefficient at T cell activation, thereby questioning their very classification as DCs (See et al., 2017).

4.2 cDC specification at the single progenitor cell level

Recent progresses challenge the classical view that cDCs arise through a stepwise and ordered commitment. The classical model implies that functionally homogeneous progenitor populations of decreased developmental potential can be isolated. This is challenged by the development of single cell transcriptomics and barcoding approaches, which have revealed high levels of cellular heterogeneity of hematopoietic progenitor populations previously thought to be homogeneous (Giladi et al., 2018; Karamitros et al., 2018; Nestorowa et al., 2016; Paul et al., 2016; Perié et al., 2015; Velten et al., 2017). Accordingly, CDPs were found to be heterogeneous in single cell cultures with some individual CDPs biased toward the production of cDC1s or cDC2s only (Naik et al., 2007; Onai et al., 2007).

Using a genetic barcoding methodology, Naik et al. have characterised the fate of the early lympho-myeloid progenitor population (LMPPs) *in vivo*. Surprisingly, results indicate that a large fraction of individual LMPPs are endowed with a cDC potential. A sizeable fraction of individual LMPPs give rise to cDCs upon transfer in irradiated mice (Naik et al., 2013). In human, single cell culture analysis of the developmental potential of early hematopoietic progenitors also revealed that multipotent lympho-myeloid progenitors, equivalent of the murine LMPP, efficiently give rise to cDCs, cDC1s in particular (Helft et al., 2017). *In vitro*, Lin et al. have developped a unique methology to image and track the dynamics of single progenitor developmental trajectories giving rise to cDCs (Lin et al., 2018). cDCs and pDCs trajectories were found to be distinct and already programmed intrinsically at the single LMPP level (Lin et al., 2018). Thus, even if fate mapping experiments in mice exclude a major contribution of lymphoid-committed progenitors expressing IL-7R (Schlenner et al., 2010), these studies show that early progenitors like LMPPs, still endowed with a lymphoid potential, also carry a cDC potential (Naik et al., 2013).

Using mathematical modeling, Lee et al. proposed that cDC specification occurs through inherited specification events, followed by the establishment of a progressive commitment program activated by extrinsic cues (Lee et al., 2017). Environmental signals supporting DC fate would include recursive stimulation of the FLT3 receptor signaling for the re-enforcement of IRF8 expression needed for cDC1 development (see Section 7) (Grajales-Reyes et al., 2015; Kurotaki et al., 2019; Lee et al., 2017).

5. Circulating precursors for classical dendritic cells

5.1 Mouse pre-cDCs

cDCs derive from circulating precursors egressing the bone marrow to seed peripheral tissues. The identification of pre-cDCs was first established in the mouse model. Early work from the Ardavin laboratory established that a complex fraction of blood CD11c^{int}MHC-II$^{low/-}$ cells had the potential to generate cDC1s, cDC2s and pDCs but did not give rise to lymphocytes or other myeloid cells (del Hoyo et al., 2002). Further work established that the fraction of bone marrow and blood CD11c^{int}MHC-II$^{low/-}$ was heterogeneous for CD45RA expression and that IFNα secreting pDCs arose specifically from CD11b^{-}CD45RA^{+} circulating progenitors (Diao et al., 2004; O'Keeffe et al., 2003). Pre-cDCs were characterized in the bone marrow, blood and spleen as Lineage^{-}CD11c$^{+/int}$CD45RA^{low}FLT3^{high}CD43^{int}SIRPα^{int}CD4^{-}CD8α^{-} and display intracellular MHC-II expression but low or no expression of cell surface MHC-II (Diao et al., 2006; Liu et al., 2009; Naik et al., 2006). Importantly, cDC-committed CDP acts as an immediate precursor for pre-cDCs (but not for monocytes) (Liu et al., 2009). The CDP to pre-cDC transition is supported by FLT3 signaling and accompanied by a decrease in the proliferative capacity (Guermonprez et al., 2013; Liu et al., 2009).

Pre-cDCs are heterogeneous and commitment to generate cDC1s or cDC2s precedes egress from the bone marrow to the blood stream (Grajales-Reyes et al., 2015; Liu et al., 2009; Schlitzer et al., 2015). SiglecH^{-}Ly6C^{-} defines cDC1-commited pre-cDCs while SiglecH^{-}Ly6C^{+} pre-DCs are endowed with a cDC2-restriced potential (Schlitzer et al., 2015). In addition, pre-cDC2s are also defined by the expression of CD11c, FLT3 and M-CSFR (Grajales-Reyes et al., 2015; Menezes et al., 2016).

5.2 Human pre-cDCs

Several groups have recently reported the characterization of circulating progenitors acting as direct precursor for cDCs. The phenotyping methods used to characterize pre-cDCs are different (Breton et al., 2015, 2016; See et al., 2017; Villani et al., 2017). Nevertheless, all methods agree on the co-expression of CD45RA and CD123 on circulating pre-cDCs, a feature shared with bone marrow CDPs (Breton et al., 2015, 2016; See et al., 2017; Villani et al., 2017). Breton et al. define pre-cDCs as a population of Lineage$^-$CD14$^-$CD1c$^-$CD14$^-$CD34$^-$CD117$^+$CD45RA$^+$CD123^{int} M-CSFR$^-$FLT3$^+$ able to give rise to both cDC1s and cDC2s but not monocytes or pDCs upon culture with stromal cells and cDC-supporting cytokine cocktail (Breton et al., 2015, 2016). The identification of pre-cDCs was guided by the identification of FLT3-expressing cells in Lineage$^-$CD34$^-$CD117$^+$ cell fraction. A following study has shown that this population is actually heterogeneous and encompasses pre-cDC1s and pre-cDC2s that can be distinguished based on the differential expression of SIRPα on pre-cDC2s (Breton et al., 2016).

Using a combination of high dimensional cytometry and single cell RNA sequencing approaches, See et al. have identified bipotent pre-cDCs as well as committed pre-cDC1s and pre-cDC2s (See et al., 2017). Importantly, it was found that pre-cDCs overlap with the classical definition of pDCs (CD45RA$^+$CD123$^+$). The co-expression of AXL and SIGLEC6 defines the pre-cDC population within total CD45RA$^+$CD123$^+$. Also, pre-cDCs express high levels of CD33. In summary, CD33$^+$CD45RA$^+$ CD123$^+$FLT3$^+$ define early pre-cDCs identified by See et al. (2017). In addition to bipotent early pre-cDCs, See et al. also identified pre-cDCs committed to generate a single DC subset. Pre-cDC1s express CADM1 and intermediate levels of CD123 as well as high levels of the TFs IRF8 and ID2, which regulate cDC1 development (See et al., 2017). Pre-cDC2s express low levels of CD123 and stain positive for CD1c and SIRPα and express high levels of IRF4 and ZEB2, two TFs known to participate to cDC2 development (See et al., 2017).

A different picture was drawn by Villani and Hacohen who also described the contamination of classically defined pDCs by AXL$^+$SIGLEC6$^+$ cells, which, they propose, would be not be a DC precursor but rather a committed subset of DCs (termed as AS-DC, expressing various levels of CD123) (Villani et al., 2017). Along these lines, this study reports a weak activity of AS-DCs to give rise to cDC1s upon *ex vivo* cultures in cDC-supporting conditions, therefore excluding its precursor potential (Villani et al., 2017). Instead, Villani and Hacohen identify a rare CD45RA$^+$CD34^{int} CD100$^+$ population endowed with the potential to give rise to cDC1s and cDC2s (Villani et al., 2017). Regardless on the different interpretations of their developmental potential, both group agree on the functional activity of the different subsets of CD45RA$^+$CD123$^+$ cells. Indeed, it was reported that pre-cDCs/AS-DCs are endowed with a T cell stimulatory activity while pure pDCs are very inefficient at activating T cells (Colonna et al., 2004; See et al., 2017; Villani et al., 2017). Pre-cDCs/AS-DCs display an intermediate ability to secrete IL-12p70 upon TLR ligation, which is carried by its more mature CD123^{low} fraction (See et al., 2017; Villani et al., 2017). Conversely, efficient type I IFN production upon the triggering of TLR9 or TLR7 is an attribute of pDCs, which is not shared by pre-cDCs/AS-DCs (See et al., 2017; Villani et al., 2017). Despite its poor proliferative potential, pre-cDCs are dynamically regulated. For instance, administration of FLT3L to healthy volunteers dramatically expand the pool of circulating pre-cDC1s and pre-cDC2s as well as differentiated cDC1s and cDC2s (Breton et al., 2015).

6. Peripheral niches

6.1 Infiltration

As described above, pre-cDCs act as circulating precursors seeding the tissues. Experiments by Liu et al. (2007) have shown that pre-cDCs have a particularly short half-life in the bloodstream.

Pre-cDCs enter lymphoid organs through L-Selectin-dependent engagement of its ligand at the surface of high endothelial venules, a special feature of secondary and tertiary lymphoid organs (Liu et al., 2009). This pathway generates the so-called resident cDCs, which populate spleen and lymph nodes.

Pre-cDC infiltration within non-lymphoid organs has been recently analyzed. Both pre-cDC1s and pre-cDC2s express high levels of the CCR5 chemokine receptor (Cook et al., 2018). Pre-cDC1s have been shown to selectively express CXCR3 and to respond to the IFN-inducible CXCL9 and CXCL10 chemokines. By contrast, pre-cDC2s express CX3CR1 but not CXCR3 (Cook et al., 2018). Therefore, circulating pre-cDCs are readily equipped to respond to different inflammatory chemokines orchestrating their recruitment at inflammatory sites. Such an early imprinting supports the general notion that they fulfill different functions during immune response. CXCR3 is essential to mediate the recruitment of pre-cDC1s in melanoma tumors but not spleen or lymph nodes (Cook et al., 2018). Once recruited, pre-cDC1s terminally differentiate into cDC1s and gain or increase XCR1 expression (Bachem et al., 2010; Balan et al., 2014; Becker et al., 2014; Crozat et al., 2010a; Jongbloed et al., 2010; Poulin et al., 2010). XCR1, a chemokine receptor selectively expressed in murine and human cDC1s play an important role in the maintenance of recruited pre-cDC1s and their movements within lymphoid organs (Brewitz et al., 2017). NK cells and effector CD8$^+$ T cells secrete high amounts of XCR1 ligands

XCL1 and XCL2, together with CCR5 ligands including CCL3, CCL4 and CCL5 (Böttcher et al., 2018; Brewitz et al., 2017; Dorner et al., 2009). Brewitz et al. have shown that $CD8^+$ T cells release XCL1 to attract cDC1s within lymphoid organs at sites of interaction with pDCs and other infected DCs. The XCL1/XCR1 axis is essential for the efficiency of $CD8^+$ T cell priming by cDC1s (Dorner et al., 2009). Bottcher et al. have recently demonstrated that NK-dependent chemoattraction controls the intra-tumoral infiltration of cDC1s in an immunogenic melanoma cell line (Böttcher et al., 2018). Intra-tumor NK cells were also shown to be an important source of FLT3L and to deliver pro-survival signals to tumor-infiltrating cDC1s (Barry et al., 2018). In addition, Gajewsky et al. have demonstrated that inhibition of CCL4 by activation of the beta-catenin pathways in melanoma curtails the recruitment of intra-tumoral murine cDC1s in $Braf^{V600E}/Pten^{-/-}$ melanoma (Spranger et al., 2015).

6.2 Proliferation

The proliferative potential decreases along the cDC differentiation pathway. There is a noticeable drop of proliferative potential associated to the formation of circulating pre-cDCs from bone marrow CDP in both mice (Liu et al., 2007, 2009) and humans (Breton et al., 2015). However, the terminal differentiation of pre-cDCs in mature cDCs does not abrogate the proliferative potential of cDCs (Cabeza-Cabrerizo et al., 2019; Kabashima et al., 2005). Indeed, pioneer work by Cyster et al. have identified that a detectable fraction of differentiated cDCs continues cycling in the spleen (Kabashima et al., 2005). Signals delivered by the niche such as lymphotoxin beta expressed by stromal components of the marginal zone support the cycling activity of cDCs *via* ligation of the lymphotoxin beta receptor (Kabashima et al., 2005). These findings have been confirmed and extended to peripheral lymphoid and non-lymphoid organs (Cabeza-Cabrerizo et al., 2019).

6.3 Differentiation

Peripheral niches provide important signals for the differentiation of cDCs. Pioneering work by the Reizis laboratory had identified a requirement for NOTCH-RBPJ for the terminal differentiation of a subset murine splenic cDC2s (Caton et al., 2007). Caton et al. have shown that inactivation of RBPJ in CD11c expressing cells, afforded by the $Cd11c^{cre}$ line, impairs a subset of splenic cDC2s (Caton et al., 2007). In addition, *Notch2* inactivation in CD11c-expressing cells phenocopy the inactivation of a subset of cDC2s (Lewis et al., 2011).

Building on this observation, the same group has identified ESAM as a marker specifically distinguishing NOTCH2-dependent cDC2s from NOTCH2-independent cDC2s (Lewis et al., 2011). This finding is reminiscent of splenic marginal zone B cells, which also depend on NOTCH2 ligation by DLL4 ligand provided by the marginal zone niche for their terminal differentiation (Radtke et al., 2013).

The same requirement has also been evidenced in the gut where NOTCH2 controls the establishment of $CD103^{+}CD11b^{+}$ murine cDC2s (Satpathy et al., 2013). On a practical note, recent *in vitro* studies have shown that activation of the NOTCH pathway *via* stromal cells expressing DLL1 NOTCH ligand supports efficiently the differentiation of cDC1s arising from total murine bone marrow or human $CD34^{+}$ hematopoietic stem and progenitor cells (Balan et al., 2018; Kirkling et al., 2018). This finding is in line with previous reports of partial alteration in size and phenotype of cDC1s in mice deficient for *Notch2* in cDCs (Lewis et al., 2011; Satpathy et al., 2013).

6.4 Positioning

cDCs occupy specific niches within lymphoid organs enabling their efficient interaction with T cells (Eisenbarth, 2019). For instance, splenic cDC2s accumulate in the vicinity of the bridging channels of the marginal zone where interruptions in the marginal zone open connections between the red pulp and T cell zones. Gatto et al. have noticed that splenic cDC2s express higher levels of Epstein Barr Virus induced 2(EBI2)/GPR138, a G-protein coupled receptor activated by the 7a,25-OHC and 7a,27-OHC oxysterols (Gatto et al., 2013; Lu et al., 2017). Inactivation of *Ebi2* perturbs the homeostatic localization of cDC2s at homeostasis and also compromises the responsiveness toward adjuvanted antigens (Gatto et al., 2013). Both deficiencies in *Ch25h* (the 7a,25-OHC producing enzyme) and *Cyp27a1* (the 7a,27-OHC producing enzyme) lead to disturbed cDC2 positioning in the spleen (Gatto et al., 2013; Lu et al., 2017). Mis-localization of cDC2s caused by *Ebi2* inactivation prevents the activation of follicular helper T cells and the initiation of antibody responses (Gatto et al., 2013).

In contrast to cDC2s, some cDC1s reside in the T cell zone (Pack et al., 2008) while most of them are found at the border of marginal zones and red pulp (Alexandre et al., 2016; Calabro et al., 2016; Dorner et al., 2009; Idoyaga et al., 2009; Qiu et al., 2009). Localization of cDC1s make them particularly efficient at engulfing blood borne particles, such as *Plasmodium*-infected infected red blood cells (Lundie et al., 2008).

Using multifluorescent fate mapping under the *Clec9a*^{cre} driver line, it has been shown that cDCs differentiated from pre-cDCs in local patches of sister cells, mostly comprising the same cDC-subset (Cabeza-Cabrerizo et al., 2019). These organized structures of patches seen at homeostasis are disrupted during inflammation, which increases the influx of pre-cDCs from bone marrow (Cabeza-Cabrerizo et al., 2019).

7. Transcriptional regulation of classical dendritic cell development

cDCs develop from hematopoietic stem cells under the aegis of a unique set of TFs, including master regulators like PU.1 and cDC subset specific TFs like IRF8 and IRF4. These TFs enable specification of the cDC lineage and commitment toward cDC susbets. The hierarchy of the TFs driving the sequential developmental steps between multipotent progenitors, cDC-committed progenitors and cDC subsets is still under investigation together with its evolutionary conserved features.

7.1 Transcription factors controlling the development of the cDC lineage

cDC lineage development depends on instructive signals by FLT3L through its receptor FLT3 expressed by hematopoietic progenitors at different developmental stages (Adolfsson et al., 2005; D'Amico and Wu, 2003; McKenna et al., 2000). Various hematopoietic master regulators and TFs have been shown to regulate the expression or downstream signaling of FLT3 and therefore cDC lineage development.

7.1.1 STAT3

The FLT3L/FLT3 axis controls the development of all lymphoid lineages (B/T/NK) but cDC development selectively depends on STAT3 signaling downstream of FLT3 (Laouar et al., 2003). Indeed, *Stat3*-deficient mice have normal B/T/NK cell compartment (own to a normal hematopoietic stem cell compartment) but have diminished cDC progenitors and lack all cDC populations in the spleen. Conditional inactivation of *Stat3* in mature cDCs using the *Cd11c*^{cre} deleter line, has revealed an important regulatory role of STAT3 in keeping cDC activation and T cell activation at bay (Melillo et al., 2010). STAT3 activation seems to be selectively regulated in cDCs as compared to other immune lineages also expressing FLT3/STAT3 because cDCs specifically express the long non-coding RNA named lnc-DC (Wang et al., 2014b). Although most of the study was performed in human monocyte-derived DCs, the authors also reported that mouse bone marrow cells deficient for lnc-DC could not produce splenic cDCs in bone marrow chimera experiments nor cDCs *in vitro* in FLT3L-induced bone marrow culture.

7.1.2 GFI1

Phosphorylation and nuclear translocation of STAT3 are negatively regulated by SOCS3, the expression of which is controlled by the zinc-finger repressor molecule GFI1 (Rathinam et al., 2005). *Gfi1*-deficient mice show a decrease splenic cDC pool and defective cDC development in GM-CSF + IL-4 or Flt3L-derived bone marrow cultures (Rathinam et al., 2005).

7.1.3 PU.1

The ETS-family DNA-binding protein PU.1 is a hematopoietic TF controlling FLT3L induced-cDC development. PU.1 expression is continuously required from early multipotent progenitors to DC-committed progenitors for the development of cDCs (Anderson et al., 2000; Carotta et al., 2010; Guerriero et al., 2000). High levels of PU.1 expression in multipotent progenitors allows myeloid fate decision as opposed to lymphoid cell fate (DeKoter and Singh, 2000). PU.1 induces IRF8 expression in myeloid progenitors to allow cDC commitment (Schönheit et al., 2013). PU.1 also controls *FLT3* expression by hematopoietic progenitors and by mature cDCs in a dose-dependent manner (Carotta et al., 2010). Recently, Chopin et al. have shown that conditional ablation of PU.1 in cDC progenitors, using the $CD11c^{cre}$ deleter line, increases pDC production at the expense of cDC production (Chopin et al., 2019). PU.1 promotes the terminal differentiation and survival of cDC subsets through the induction of the ZFP366/DC-SCRIPT TF, which activates 767 genes (including *Xcr1, Clec9a, CD4, CD11b, Il12, Cd86, itgax/CD11c, Zbtb46*) in cDCs and represses the pDC transcriptional program (Chopin et al., 2019).

7.1.4 IKAROS

IKAROS is a DNA-binding zinc-finger protein controls lymphoid cell (Georgopoulos et al., 1994; Yoshida et al., 2006), pDC development (Allman et al., 2006) and cDC development in a dose-dependent manner. Mice expressing a dominant-negative form of *Ikaros* present a complete lack of cDC lineages, while *Ikaros* null mice present an ablation of cDC2s, pDCs and a reduction in cDC1s (Wu et al., 1997). In addition, mice expressing reduced levels of IKZF1 display a specific reduction of pDCs numbers while other subsets are conserved. Using an *Ikaros* reporter mouse, a study showed that IKAROS is required for FLT3 expression on LMPPs and for lymphoid lineage development (Yoshida et al., 2006). Although, cDC development was not assessed, this study suggests a role for IKAROS in controlling cDC responsiveness to instructive signals like FLT3L in hematopoietic progenitors. Recently Cytlak et al. have identified a reduction in pDCs in human patients carrying a human germline heterozygous mutation in IKZF1 and suffering from multiple and combined immunodeficiencies in T and B cell compartments (Cytlak et al., 2018). Interestingly, this reduction in pDCs is concomitant with an increase in cDC1s (Cytlak et al., 2018). Degradation of IKZF1 (and IKZF3) induced by the drug lenalinomide phenocopies the IKZF1 heterozygous status by increasing the cDC1/pDC ratio (Cytlak et al., 2018).

7.1.5 RUNX1/CBFβ

The TF RUNX1 and its co-factor CBFβ control the development of bone marrow hematopoietic progenitors expressing FLT3 including LMPPs, CLPs, MDPs and CDPs, and the development of all cDC subsets in the periphery (Satpathy et al., 2014). Of note, CBFβ deficiency also leads to the expansion of the granulocyte-macrophage progenitor, that can be counteracted by *Irf8* overexpression (Satpathy et al., 2014).

7.1.6 BCL6

BCL6, a transcriptional repressor controlling the development of follicular helper T cells (Nurieva et al., 2009), was also found to be required for the development of splenic cDC1s and cDC2s but not pDCs (Ohtsuka et al., 2011). BCL6 is not expressed in early hematopoietic progenitors but its expression is upregulated in the splenic pool of pre-cDCs (Zhang et al., 2014). BCL6 is also involved in the homeostatic proliferation of immature cDCs and is downregulated upon cDC maturation after adjuvant inoculation *in vivo* or upon treatment with LPS (Zhang et al., 2014).

7.1.7 GATA2

In humans, Collin and Bigley have identified the DC, monocyte, B and NK lymphocytes immunodeficiency (DCML), which is characterized by an absence of cDCs, monocytes, B and NK cells in peripheral blood mononuclear cells (Bigley et al., 2011). Whole exome sequencing showed that this syndrome is caused by heterozygous mutation in GATA2, a TF known to control stem cell homeostasis (Dickinson et al., 2011; 2014; Rodrigues et al., 2005). In mice, an inducible knock-out for *Gata2* revealed also its role in cDC and pDC development (Onodera et al., 2016).

7.2 Transcription factors controlling the development of cDC subsets

cDC subset diversification is controlled by various TFs regulating cell fate during the developmental transition between CDP and pre-cDCs in the bone marrow. These TFs, including IRF8, IRF4, NFIL3 and KLF4, often have mutually exclusive expression in pre-cDCs and cDC subsets. Other TFs, like BATF3 or ZFP366, are expressed by both cDC subsets but only control the development of one. Expression of some TFs is required only during the developmental transition while other TFs need to be continuously expressed to maintain the identity of terminally differentiated cells.

cDC1s and cDC2s specifically express IRF8 and IRF4, respectively (Suzuki et al., 2004). Expression of IRF8 and IRF4 in cDC subsets is conserved across tissues and across species and provides surrogate markers for univocal identification of the two branches of cDCs (Guilliams et al., 2016). IRF8 and IRF4 have roles both in the development and the terminal differentiation of cDCs.

7.2.1 Type 1 DC development

7.2.1.1 IRF8

The interferon responding TF IRF8 was the first TF to be identified as differentially affecting cDC1 development as compared to cDC2s (Schiavoni et al., 2002; Tamura et al., 2005). *Irf8*-deficient mice also present defects in other hematopoietic cell lineages, including monocytes/macrophages and pDCs, highlighting a central role of IRF8 in the regulation of hematopoiesis (Sichien et al., 2016; Tamura et al., 2000). It was later found that a spontaneous point mutation in the *Irf8* gene ($IRF8^{R294C}$) affects cDC1 development without affecting other immune lineages (Tailor et al., 2008). IRF8 acts at various stages of hematopoiesis by synergizing with other TFs (Ozato et al., 2007; Tailor et al., 2008). In particular, early IRF8 expression supports specification of LMPPs toward mononuclear phagocytes, cDCs in particular. The molecular basis for the early imprinting of the cDC program lies within PU.1-dependent IRF8 expression, which characterizes a substantial proportion of human and murine LMPPs (Helft et al., 2017; Kurotaki et al., 2019; Lee et al., 2017; Schönheit et al., 2013). PU.1 induces chromatin remodeling at the *Irf8* locus, enabling early cDC specification (Schönheit et al., 2013).

RUNX1/CBFβ, together with PU.1, also controls IRF8 expression in LMPPs and is essential for cDC development (Satpathy et al., 2014). As shown using genetically encoded reporters (Wang et al., 2014a), IRF8-expressing LMPPs give rise mostly to cDCs (cDC1s in particular) at the expense of other myeloid (monocytes and granulocytes) or lymphoid lineages (Lee et al., 2017; Kurotaki et al., 2019). Kurotaki et al. have shown that IRF8 expression in LMPPs controls chromatin remodeling at DC lineage genes before they become transcriptionally active in downstream progenitors (MDP or CDP) (Kurotaki et al., 2019). In parallel, IRF8 suppresses the onset of granulocyte differentiation in a cell autonomous fashion by binding and directly inhibiting C/EBPα (Becker et al., 2012; Kurotaki et al., 2014). IRF8 expression in MDP is induced by PU.1 and highest levels of IRF8 expression promote the cDC fate at the expense of macrophage differentiation (Schönheit et al., 2013). Finally, IRF8 is a major controller of cDC1 specification at the level of CDP. Here, BATF3 is required for *Irf8* auto-activation *via* a specific enhancer, which is essential for the developmental transition from CDP toward pre-cDC1s (Grajales-Reyes et al., 2015). High and sustained expression levels of IRF8 in pre-cDC1s is required for commitment to the cDC1 lineage (Grajales-Reyes et al., 2015).

In humans, biallelic mutations in the *Irf8* gene (K108E/K108E and R83C/R291Q) cause a severe immunodeficiency (Hambleton et al., 2011; Bigley et al., 2018; Salem et al., 2014). These mutations abrogate the development of cDC1s, cDC2s and pDCs, induce monocytopenia and an expansion of immature granulocytes. Mechanistically, the R291Q mutation is an ortologous of the R294C mutation found in mice (Tailor et al., 2008) and prevents functional dimerization with transcriptional partners, including PU.1 (Bigley et al., 2018; Hambleton et al., 2011; Salem et al., 2014). The R83C mutation is also impaired in activating Ets/IRF and ISRE sites. Finally, the K108E mutant is deficient in nuclear translocation (Hambleton et al., 2011; Bigley et al., 2018; Salem et al., 2014).

7.2.1.2 BATF3

BATF3 is part of the basic leucine zipper protein family and is expressed in both cDC1s and cDC2s. However, BATF3 only controls the development of cDC1s in homeostatic conditions (Hildner et al., 2008). Mice fully deficient in *Batf3*, therefore specifically lacking cDC1s, have been a key genetic model to probe the non-redundant immune functions of cDC1s *in vivo* and also helped establish lineage relationship between cDC1s across tissues

(Edelson et al., 2010; Ginhoux et al., 2009; Hildner et al., 2008). During infection by intracellular pathogens and in other inflammatory contexts, a BATF3-independent alternative cDC1 developmental pathway occurs *in vivo* (Seillet et al., 2013; Tussiwand et al., 2012). This alternative pathway for cDC1 development is due to physiological compensatory mechanisms operated by BATF or BATF2, two TFs of the same family able to bind IRF8 and stabilize its expression in pre-cDCs (Tussiwand et al., 2012). In human $CD34^+$ hematopoietic stem and progenitor cells, loss of BATF3 expression using shRNA knock-down was shown to inhibit cDC1 differentiation *in vitro* (Poulin et al., 2012).

7.2.1.3 ID2

ID2 is a DNA-binding protein inhibitor, which, like BATF3, is expressed in all cDC subsets but is only required for the development of cDC1s (Ginhoux et al., 2009; Hacker et al., 2003; Jackson et al., 2011). ID2 synergizes with BATF3 to promote cDC1 development (Jaiswal et al., 2013). However, without IRF8, BATF3 and ID2 are not sufficient to drive cDC1 development, placing IRF8 upstream of these TFs (Jaiswal et al., 2013; Theisen et al., 2018).

7.2.1.4 NFIL3

NFIL3 is another transcriptional repressor essential for the development of cDC1s (Kashiwada et al., 2011). NFIL3 also controls the development of NK cells and other innate lymphoid cells (Kashiwada et al., 2011; Xu et al., 2015). Absence of NFIL3 in pre-cDCs in the bone marrow reduces expression levels of BATF3, placing NFIL3 upstream of BAFT3. NFIL3 might also repress IRF4 expression, further supporting cDC1 commitment (Kashiwada et al., 2011).

7.2.1.5 ETV6

ETV6, an ETS-family TF involved in hematopoietic stem cell survival, was recently shown to control the terminal differentiation of murine cDC1s *in vivo* (Lau et al., 2018). In particular, *Etv6*-deficient cDC1s do not express CD8α in the spleen and have reduced ability to efficiently cross-prime anti-tumor $CD8^+$ T cells (Lau et al., 2018). *In vitro*, ETV6 facilitates the development of cDC1s in FLT3L-induced bone marrow cultures. Loss of ETV6 was shown to modify the chromatin landscape of cDC1s. In particular *Etv6*-deficient cDC1s express a set of IRF8-repressed genes and pDC specific-genes. The same study also showed that ETV6 directly interacts with IRF8 in cDC1s.

7.2.1.6 MYCL1

MYCL1, a member of the Myc TF family, was shown to be expressed in CDP, pre-cDCs and in immature and mature cDCs (Kc et al., 2014). Authors identified various IRF8 binding regions in the *Mycl1* gene suggesting that expression of *MYCL1* is controlled by IRF8. *Mycl1*-deficient mice showed normal cDC development in the spleen and reduced cDC1s in the lung and liver. MYCL1 regulates splenic cDC1 proliferation and GM-CSF-induced survival *in vivo*. In consequence, *Mycl1*-deficient mice present defect in $CD8^+$ T cell priming upon intracellular $GRAM^+$ bacteria *Listeria Monocytogenes* infection *in vivo*. Interestingly, *Mycl1*-deficient cDC1s did not support intracellular growth of *Listeria Monocytogenes* and *Mycl1*-deficient mice were entirely resistant to lethal dose of infection.

7.2.1.7 ZFP366

ZFP366 also named DC-SCRIPT, is a direct target of PU.1 and is expressed only in pre-cDC1s and cDC1s (Chopin et al., 2019). In FLT3L-induced bone marrow cultures, overexpression of DC-SCRIPT promotes cDC1 differentiation at the expense of pDC and cDC2 differentiation. Inversely, DC-SCRIPT loss of function impaired cDC1 development *in vitro* and *in vivo*. Of note, DC-SCRIPT does not regulate ID2, BATF3 and NFIL3 expression, uncovering an additional layer of regulation for cDC1 commitment.

7.2.2 Type 2 DC development

7.2.2.1 IRF4

IRF4 was initially thought to control splenic cDC2 development because of the lack of $CD4^+$ cDCs in spleens of *Irf4*-deficient mice (Suzuki et al., 2004). However, *Irf4*-deficient mice retain a subset of $CD4^-CD11b^+$ cDCs in the spleen (Tamura et al., 2005). IRF4 seems rather to control the terminal differentiation of cDC2s and some functional properties, including their CCR7-dependent migration and CIITA-dependent antigenic presentation by MHC-II molecules (Bajaña et al., 2012; Vander Lugt et al., 2014). Other studies probed the role of IRF4 on the development of non-lymphoid tissue cDC2s (Persson et al., 2013; Schlitzer et al., 2013). Using mixed bone marrow chimeras of WT and *Irf4*-deficient mice they revealed a partial developmental requirement for IRF4 in $CD11b^+CD24^+CD64^-$ cDC2s in the lungs and of $CD11b^+CD103^+$ cDC2s in the gut, and confirmed the absence of requirement of IRF4 for splenic cDC2s (Persson et al., 2013; Schlitzer et al., 2013). Therefore, it seems that IRF4 could have a role in the development of non-lymphoid tissue cDC2s.

7.2.2.2 IRF2

IRF2, another member of the IFN responding factor family, is also involved in the development of splenic $CD4^+$ cDC2s and epidermal Langerhans cells (Ichikawa et al., 2004). Bone marrow chimera experiments demonstrated a cell-intrinsic role for IRF2 in cDC2 development. Interestingly, the defect in cDC2 development is rescued when *Irf2*-deficient mice are crossed to *Ifnar*-deficient mice lacking type I IFN receptor (Ichikawa et al., 2004). This suggests that IRF2 negatively regulates type I IFN signaling to ensure cDC2 development.

7.2.2.3 RELB

RELB, a NF-kB/Rel family TF, was shown to control the development of thymic and splenic cDC2s in a cell-intrinsic manner (Burkly et al., 1995; Wu et al., 1998), as well as the development of T and NKT cells in a cell-extrinsic manner (DeKoning et al., 1997; Elewaut et al., 2003). However, bone marrow chimera experiments in which the bone marrow from *RelB*-deficient animal was adoptively transferred into sub-lethally irradiated WT recipient, showed that RELB expression in radio-resistant non-hematopoietic cells could also contribute to the normal development of cDCs (Burkly et al., 1995). These data suggested a possible cell-extrinsic role for RELB in cDC development. A recent study carefully analyzed the role of RELB in cDC subsets development by using bone marrow chimeras to assess the cell-intrinsic or extrinsic role of RELB (Briseño et al., 2017). This study demonstrated that RELB is required in radio-resistant non-hematopoietic cells since $RelB^{-/-}$ bone marrow into WT recipient chimeras do not present defect in cDC-development. However, using recipient mice reconstituted with a 1:1 ratio of $RelB^{-/-}$ and WT bone marrow, the same study showed a marked reduction of the $ESAM^+CD4^+$NOTCH2-dependent splenic cDC2s, demonstrating a selective cell-intrinsic role for RELB in the development of NOTCH2-dependent cDC2s (Briseño et al., 2017).

7.2.2.4 RBP/J

RBP/J is a DNA-binding TF downstream of NOTCH receptors signaling. RBP/J is involved in various developmental pathways mainly in the lymphoid lineage (Maillard et al., 2005). Caton et al., generated a conditional RBP/J knock-out mouse in CD11c expressing cells and showed that RBP/J controls survival and terminal differentiation of splenic cDC2s (Caton et al., 2007). Later, the same laboratory showed that among the NOTCH receptors, NOTCH2 was controlling cDC2 terminal differentiation in the spleen and the gut (Lewis et al., 2011). Interestingly, NOTCH2 was in fact found to selectively control the development of a subtype of splenic cDC2s expressing CD4 and ESAM. Therefore, this study was the first to provide genetic evidence for the suspected cellular heterogeneity in IRF4-dependent cDC2 subsets. Importantly, NOTCH2-dependent cDC2 subset controls Th17 responses but not Th2 responses. Of note, NOTCH2 also controls the terminal differentiation of cDC1s, for example expression of CD8α by splenic cDC1s (Caton et al., 2007; Kirkling et al., 2018).

7.2.2.5 KLF4

KLF4, a transcriptional regulator involved in monocyte development and macrophage polarization (Feinberg et al., 2007; Kurotaki et al., 2013), was later found to be transiently upregulated in pre-cDCs and to control the development of NOTCH2-independent cDC2s in the spleen responsible for Th2 immunity (Tussiwand et al., 2015).

7.2.2.6 ZEB2

ZEB2, a zinc-finger E homeobox-binding protein, is expressed by the MDP, the CDP, pDCs and cDC2s but not by cDC1s (Miller et al., 2012). Using a conditional *CD11c*-Cre deleter mouse model or *Mx1*-Cre inducible deleter model, two laboratories investigated the role of ZEB2 in cDC development (Scott et al., 2016; Wu et al., 2016). Both studies reported a role for ZEB2 in pDC development. Scott et al. reported that *Zeb2*-deficient mice present a reduction in cDC2s, an increase in cDC1s, and that *Zeb2*-deficient cDC2s aquire features of cDC1s, suggesting a regulatory role for ZEB2 in the commitment of cDC subsets (Scott et al., 2016). In line with this result, the two studies showed that ZEB2 represses the cDC1 TF ID2 in pre-cDC2s and cDC2s. By contrast, Wu et al. reported a partial impact of ZEB2 on cDC2 development and terminal differentiation and propose that ZEB2 mostly acts as a transcriptional repressor of pDC and cDC2 cell fate rather than as a terminal differentiation selector for cDC2s (Wu et al., 2016). Discrepancy between these two studies concerning cDC2 development could be due to the fact that cDC2s are a heterogeneous population and that ZEB2 regulates only a yet unknown subtype of cDC2s.

8. Conclusion and open questions

Since their identification in the early 1970s, our understanding of cDC biology has made tremendous progresses. However, several topics merit further clarification and currently attract research efforts. A list of open questions might include:

(1) What is the underlying molecular basis of cDC2 heterogeneity? In both human and mouse systems, the cDC2 subset is heterogeneous at steady state. This heterogeneity is even more complex in inflammatory settings when the discrimination between inflammatory phagocytes and *bona fide* cDC2s is particularly difficult to establish. Also, it remains to be establish if "inflammatory," cDC-like cells and *bona fide* cDC2s share or not convergent transcriptional programs leading to shared immunological functions.

(2) The molecular pathways driving the acquisition of tissue-specific phenotypes start to be defined for macrophages with a great deal of details. Such knowledge is largely missing for cDC subsets and it is unclear to what extent tissue-specific cues (retinoic acid, GM-CSF, *e.g.*) control the acquisition of specific phenotypes.

(3) Finally, cDC ontogeny has been mostly defined in steady state. It is unclear how this knowledge translates to the very dynamic changes impacting bone marrow progenitors during inflammation. In addition, inflammation is often associated to extra-medullar hematopoiesis, a process of ill-understood relevance for the ontogeny of cDCs.

Acknowledgment

We thank Prudence Pokwai Lui for her carefull reading of the manuscript and her suggestions. The figure was created with Biorender.com. This work was supported by the Institut Curie, Institut National de la Santé et de la Recherche Médicale, Labex DCBIOL (ANR-10-IDEX-0001-02 PSL and ANR-11-LABX0043), SIRIC INCa-DGOS-Inserm_12554, ANR JCJC (ANR-17-CE15-0004), ANR-18-IDEX-0001, ANR-17-CE11-0001-01, CRUK-C57672/A22369, WWCR-18-0422, BBSRC-BB/M029735/1.

References

Adolfsson, J., Borge, O.J., Bryder, D., Theilgaard-Mönch, K., Åstrand-Grundström, I., Sitnicka, E., Sasaki, Y., Jacobsen, S.E.W., 2001. Upregulation of Flt3 expression within the bone marrow Lin − Sca1 + c-kit + stem cell compartment is accompanied by loss of self-renewal capacity. Immunity 15, 659–669. https://doi.org/10.1016/S1074-7613(01)00220-5.

Adolfsson, J., Månsson, R., Buza-Vidas, N., Hultquist, A., Liuba, K., Jensen, C.T., Bryder, D., Yang, L., Borge, O.-J., Thoren, L.A.M., Anderson, K., Sitnicka, E., Sasaki, Y., Sigvardsson, M., Jacobsen, S.E.W., 2005. Identification of Flt3+ lympho-myeloid stem cells lacking erythro-megakaryocytic potential a revised road map for adult blood lineage commitment. Cell 121, 295–306. https://doi.org/10.1016/j.cell.2005.02.013.

Akashi, K., Traver, D., Miyamoto, T., Weissman, I.L., 2000. A clonogenic common myeloid progenitor that gives rise to all myeloid lineages. Nature 404, 193–197. https://doi.org/10.1038/35004599.

Alexandre, Y.O., Ghilas, S., Sanchez, C., Bon, A.L., Crozat, K., Dalod, M., 2016. XCR1+ dendritic cells promote memory CD8+ T cell recall upon secondary infections with Listeria monocytogenes or certain viruses. J. Exp. Med. 213, 75–92. https://doi.org/10.1084/jem.20142350.

Allman, D., Dalod, M., Asselin-Paturel, C., Delale, T., Robbins, S.H., Trinchieri, G., Biron, C.A., Kastner, P., Chan, S., 2006. Ikaros is required for plasmacytoid dendritic cell differentiation. Blood 108, 4025–4034. https://doi.org/10.1182/blood-2006-03-007757.

Anandasabapathy, N., Feder, R., Mollah, S., Tse, S.-W., Longhi, M.P., Mehandru, S., Matos, I., Cheong, C., Ruane, D., Brane, L., Teixeira, A., Dobrin, J., Mizenina, O., Park, C.G., Meredith, M., Clausen, B.E., Nussenzweig, M.C., Steinman, R.M., 2014. Classical Flt3L-dependent dendritic cells control immunity to protein vaccine. J. Exp. Med. 211, 1875–1891. https://doi.org/10.1084/jem.20131397.

Anandasabapathy, N., Breton, G., Hurley, A., Caskey, M., Trumpfheller, C., Sarma, P., Pring, J., Pack, M., Buckley, N., Matei, I., Lyden, D., Green, J., Hawthorne, T., Marsh, H.C., Yellin, M., Davis, T., Keler, T., Schlesinger, S.J., 2015. Efficacy and safety of CDX-301, recombinant human Flt3L, at expanding dendritic cells and hematopoietic stem cells in healthy human volunteers. Bone Marrow Transplant. 50, 924–930. https://doi.org/10.1038/bmt.2015.74.

Anderson, K.L., Perkin, H., Surh, C.D., Venturini, S., Maki, R.A., Torbett, B.E., 2000. Transcription factor PU.1 is necessary for development of thymic and myeloid progenitor-derived dendritic cells. J. Immunol. 164, 1855–1861. https://doi.org/10.4049/jimmunol.164.4.1855.

Ashok, D., Schuster, S., Ronet, C., Rosa, M., Mack, V., Lavanchy, C., Marraco, S.F., Fasel, N., Murphy, K.M., Tacchini-Cottier, F., Acha-Orbea, H., 2014. Cross-presenting dendritic cells are required for control of Leishmania major infection. Eur. J. Immunol. 44, 1422–1432. https://doi.org/10.1002/eji.201344242.

Auffray, C., Fogg, D.K., Narni-Mancinelli, E., Senechal, B., Trouillet, C., Saederup, N., Leemput, J., Bigot, K., Campisi, L., Abitbol, M., Molina, T., Charo, I., Hume, D.A., Cumano, A., Lauvau, G., Geissmann, F., 2009. CX3CR1+ CD115+ CD135+ common macrophage/DC precursors and the role of CX3CR1 in their response to inflammation. J. Exp. Med. 206, 595–606. https://doi.org/10.1084/jem.20081385.

Bachem, A., Güttler, S., Hartung, E., Ebstein, F., Schaefer, M., Tannert, A., Salama, A., Movassaghi, K., Opitz, C., Mages, H.W., Henn, V., Kloetzel, P.-M., Gurka, S., Kroczek, R.A., 2010. Superior antigen cross-presentation and XCR1 expression define human CD11c+CD141+ cells as homologues of mouse CD8+ dendritic cells. J. Exp. Med. 207, 1273–1281. https://doi.org/10.1084/jem.20100348.

Bajaña, S., Roach, K., Turner, S., Paul, J., Kovats, S., 2012. IRF4 promotes cutaneous dendritic cell migration to lymph nodes during homeostasis and inflammation. J. Immunol. 189, 3368–3377. https://doi.org/10.4049/jimmunol.1102613.

Bakri, Y., Sarrazin, S., Mayer, U.P., Tillmanns, S., Nerlov, C., Boned, A., Sieweke, M.H., 2005. Balance of MafB and PU.1 specifies alternative macrophage or dendritic cell fate. Blood 105, 2707–2716. https://doi.org/10.1182/blood-2004-04-1448.

Balan, S., Ollion, V., Colletti, N., Chelbi, R., Montanana-Sanchis, F., Liu, H., Manh, T.-P.V., Sanchez, C., Savoret, J., Perrot, I., Doffin, A.-C., Fossum, E., Bechlian, D., Chabannon, C., Bogen, B., Asselin-Paturel, C., Shaw, M., Soos, T., Caux, C., Valladeau-Guilemond, J., Dalod, M., 2014. Human XCR1+ dendritic cells derived in vitro from CD34+ progenitors closely resemble blood dendritic cells, including their adjuvant responsiveness, contrary to monocyte-derived dendritic cells. J. Immunol. 193, 1622–1635. https://doi.org/10.4049/jimmunol.1401243.

Balan, S., Arnold-Schrauf, C., Abbas, A., Couespel, N., Savoret, J., Imperatore, F., Villani, A.-C., Vu Manh, T.-P., Bhardwaj, N., Dalod, M., 2018. Large-scale human dendritic cell differentiation revealing notch-dependent lineage bifurcation and heterogeneity. Cell Rep. 24, 1902–1915.e6. https://doi.org/10.1016/j.celrep.2018.07.033.

Banchereau, J., Steinman, R.M., 1998. Dendritic cells and the control of immunity. Nature 392, 245–252. https://doi.org/10.1038/32588.

Barry, K.C., Hsu, J., Broz, M.L., Cueto, F.J., Binnewies, M., Combes, A.J., Nelson, A.E., Loo, K., Kumar, R., Rosenblum, M.D., Alvarado, M.D., Wolf, D.M., Bogunovic, D., Bhardwaj, N., Daud, A.I., Ha, P.K., Ryan, W.R., Pollack, J.L., Samad, B., Asthana, S., Chan, V., Krummel, M.F., 2018. A natural killer–dendritic cell axis defines checkpoint therapy–responsive tumor microenvironments. Nat. Med. https://doi.org/10.1038/s41591-018-0085-8.

Beaudin, A.E., Boyer, S.W., Forsberg, E.C., 2014. Flk2/Flt3 promotes both myeloid and lymphoid development by expanding non–self-renewing multipotent hematopoietic progenitor cells. Exp. Hematol. 42, 218–229. https://doi.org/10.1016/j.exphem.2013.11.013. e4.

Becher, B., Tugues, S., Greter, M., 2016. GM-CSF: from growth factor to central mediator of tissue inflammation. Immunity 45, 963–973. https://doi.org/10.1016/j.immuni.2016.10.026.

Becker, A.M., Michael, D.G., Satpathy, A.T., Sciammas, R., Singh, H., Bhattacharya, D., 2012. IRF-8 extinguishes neutrophil production and promotes dendritic cell lineage commitment in both myeloid and lymphoid mouse progenitors. Blood 119, 2003–2012. https://doi.org/10.1182/blood-2011-06-364976.

Becker, M., Güttler, S., Bachem, A., Hartung, E., Mora, A., Jäkel, A., Hutloff, A., Henn, V., Mages, H.W., Gurka, S., Kroczek, R.A., 2014. Ontogenic, phenotypic, and functional characterization of XCR1(+) dendritic cells leads to a consistent classification of intestinal dendritic cells based on the expression of XCR1 and SIRPα. Front. Immunol. 5, 326. https://doi.org/10.3389/fimmu.2014.00326.

Bedoui, S., Whitney, P.G., Waithman, J., Eidsmo, L., Wakim, L., Caminschi, I., Allan, R.S., Wojtasiak, M., Shortman, K., Carbone, F.R., Brooks, A.G., Heath, W.R., 2009. Cross-presentation of viral and self antigens by skin-derived CD103+ dendritic cells. Nat. Immunol. 10, 488–495. https://doi.org/10.1038/ni.1724.

Bigley, V., Haniffa, M., Doulatov, S., Wang, X.-N., Dickinson, R., McGovern, N., Jardine, L., Pagan, S., Dimmick, I., Chua, I., Wallis, J., Lordan, J., Morgan, C., Kumararatne, D.S., Doffinger, R., van der Burg, M., van Dongen, J., Cant, A., Dick, J.E., Hambleton, S., Collin, M., 2011. The human syndrome of dendritic cell, monocyte, B and NK lymphoid deficiency. J. Exp. Med. 208, 227–234. https://doi.org/10.1084/jem.20101459.

Bigley, V., Maisuria, S., Cytlak, U., Jardine, L., Care, M.A., Green, K., Gunawan, M., Milne, P., Dickinson, R., Wiscombe, S., Parry, D., Doffinger, R., Laurence, A., Fonseca, C., Stoevesandt, O., Gennery, A., Cant, A., Tooze, R., Simpson, A.J., Hambleton, S., Savic, S., Doody, G., Collin, M., 2018. Biallelic interferon regulatory factor 8 mutation: a complex immunodeficiency syndrome with dendritic cell deficiency, monocytopenia, and immune dysregulation. J. Allergy Clin. Immunol. 141, 2234–2248. https://doi.org/10.1016/j.jaci.2017.08.044.

Bogunovic, M., Ginhoux, F., Helft, J., Shang, L., Hashimoto, D., Greter, M., Liu, K., Jakubzick, C., Ingersoll, M.A., Leboeuf, M., Stanley, E.R., Nussenzweig, M., Lira, S.A., Randolph, G.J., Merad, M., 2009. Origin of the lamina propria dendritic cell network. Immunity 31, 513–525. https://doi.org/10.1016/j.immuni.2009.08.010.

Bonifaz, L., Bonnyay, D., Mahnke, K., Rivera, M., Nussenzweig, M.C., Steinman, R.M., 2002. Efficient targeting of protein antigen to the dendritic cell receptor DEC-205 in the steady state leads to antigen presentation on major histocompatibility complex class I products and peripheral CD8+ T cell tolerance. J. Exp. Med. 196, 1627–1638. https://doi.org/10.1084/jem.20021598.

Böttcher, J.P., Bonavita, E., Chakravarty, P., Blees, H., Cabeza-Cabrerizo, M., Sammicheli, S., Rogers, N.C., Sahai, E., Zelenay, S., Reis E Sousa, C., 2018. NK cells stimulate recruitment of cDC1 into the tumor microenvironment promoting cancer immune control. Cell 172, 1022–1037.e14. https://doi.org/10.1016/j.cell.2018.01.004.

Bougnères, L., Helft, J., Tiwari, S., Vargas, P., Chang, B.H.-J., Chan, L., Campisi, L., Lauvau, G., Hugues, S., Kumar, P., Kamphorst, A.O., Dumenil, A.-M.L., Nussenzweig, M., MacMicking, J.D., Amigorena, S., Guermonprez, P., 2009. A role for lipid bodies in the cross-presentation of phagocytosed antigens by MHC class I in dendritic cells. Immunity 31, 232–244. https://doi.org/10.1016/j.immuni.2009.06.022.

Breton, G., Lee, J., Zhou, Y.J., Schreiber, J.J., Keler, T., Puhr, S., Anandasabapathy, N., Schlesinger, S., Caskey, M., Liu, K., Nussenzweig, M.C., 2015. Circulating precursors of human CD1c+ and CD141+ dendritic cells. J. Exp. Med. 212, 401–413. https://doi.org/10.1084/jem.20141441.

Breton, G., Zheng, S., Valieris, R., Tojal da Silva, I., Satija, R., Nussenzweig, M.C., 2016. Human dendritic cells (DCs) are derived from distinct circulating precursors that are pre-committed to become CD1c+ or CD141+ DCs. J. Exp. Med. 213, 2861–2870. https://doi.org/10.1084/jem.20161135.

Brewig, N., Kissenpfennig, A., Malissen, B., Veit, A., Bickert, T., Fleischer, B., Mostböck, S., Ritter, U., 2009. Priming of CD8+ and CD4+ T cells in experimental leishmaniasis is initiated by different dendritic cell subtypes. J. Immunol. 182, 774–783. https://doi.org/10.4049/jimmunol.182.2.774.

Brewitz, A., Eickhoff, S., Dähling, S., Quast, T., Bedoui, S., Kroczek, R.A., Kurts, C., Garbi, N., Barchet, W., Iannacone, M., Klauschen, F., Kolanus, W., Kaisho, T., Colonna, M., Germain, R.N., Kastenmüller, W., 2017. CD8 + T cells orchestrate pDC-XCR1 + dendritic cell spatial and functional cooperativity to optimize priming. Immunity 46, 205–219. https://doi.org/10.1016/j.immuni.2017.01.003.

Briseño, C.G., Haldar, M., Kretzer, N.M., Wu, X., Theisen, D.J., Kc, W., Durai, V., Grajales-Reyes, G.E., Iwata, A., Bagadia, P., Murphy, T.L., Murphy, K.M., 2016. Distinct transcriptional programs control cross-priming in classical and monocyte-derived dendritic cells. Cell Rep. 15, 2462–2474. https://doi.org/10.1016/j.celrep.2016.05.025.

Briseño, C.G., Gargaro, M., Durai, V., Davidson, J.T., Theisen, D.J., Anderson, D.A., Novack, D.V., Murphy, T.L., Murphy, K.M., 2017. Deficiency of transcription factor RelB perturbs myeloid and DC development by hematopoietic-extrinsic mechanisms. Proc. Natl. Acad. Sci. U.S.A. 114, 3957–3962. https://doi.org/10.1073/pnas.1619863114.

Briseño, C.G., Satpathy, A.T., Davidson, J.T., Ferris, S.T., Durai, V., Bagadia, P., O'Connor, K.W., Theisen, D.J., Murphy, T.L., Murphy, K.M., 2018. Notch2-dependent DC2s mediate splenic germinal center responses. Proc. Natl. Acad. Sci. U.S.A. 115, 10726–10731. https://doi.org/10.1073/pnas.1809925115.

Burkly, L., Hession, C., Ogata, L., Reilly, C., Marconi, L.A., Olson, D., Tizard, R., Cate, R., Lo, D., 1995. Expression of relB is required for the development of thymic medulla and dendritic cells. Nature 373, 531–536. https://doi.org/10.1038/373531a0.

Cabeza-Cabrerizo, M., van Blijswijk, J., Wienert, S., Heim, D., Jenkins, R.P., Chakravarty, P., Rogers, N., Frederico, B., Acton, S., Beerling, E., van Rheenen, J., Clevers, H., Schraml, B.U., Bajénoff, M., Gerner, M., Germain, R.N., Sahai, E., Klauschen, F., Reis E Sousa, C., 2019. Tissue clonality of dendritic cell subsets and emergency DCpoiesis revealed by multicolor fate mapping of DC progenitors. Sci. Immunol. 4, 1022–1037. https://doi.org/10.1126/sciimmunol.aaw1941.

Carotta, S., Dakic, A., D'Amico, A., Pang, S.H.M., Greig, K.T., Nutt, S.L., Wu, L., 2010. The transcription factor PU.1 controls dendritic cell development and Flt3 cytokine receptor expression in a dose-dependent manner. Immunity 32, 628–641. https://doi.org/10.1016/j.immuni.2010.05.005.

Caton, M.L., Smith-Raska, M.R., Reizis, B., 2007. Notch-RBP-J signaling controls the homeostasis of CD8- dendritic cells in the spleen. J. Exp. Med. 204, 1653–1664. https://doi.org/10.1084/jem.20062648.

Cella, M., Engering, A., Pinet, V., Pieters, J., Lanzavecchia, A., 1997. Inflammatory stimuli induce accumulation of MHC class II complexes on dendritic cells. Nature 388, 782–787. https://doi.org/10.1038/42030.

Chicha, L., Jarrossay, D., Manz, M.G., 2004. Clonal type I interferon-producing and dendritic cell precursors are contained in both human lymphoid and myeloid progenitor populations. J. Exp. Med. 200, 1519–1524. https://doi.org/10.1084/jem.20040809.

Chopin, M., Lun, A.T., Zhan, Y., Schreuder, J., Coughlan, H., D'Amico, A., Mielke, L.A., Almeida, F.F., Kueh, A.J., Dickins, R.A., Belz, G.T., Naik, S.H., Lew, A.M., Bouillet, P., Herold, M.J., Smyth, G.K., Corcoran, L.M., Nutt, S.L., 2019. Transcription factor PU.1 promotes conventional dendritic cell identity and function via induction of transcriptional regulator DC-SCRIPT. Immunity 50, 77–90.e5. https://doi.org/10.1016/j.immuni.2018.11.010.

Christensen, J.L., Weissman, I.L., 2001. Flk-2 is a marker in hematopoietic stem cell differentiation: a simple method to isolate long-term stem cells. Proc. Natl. Acad. Sci. U.S.A. 98, 14541–14546. https://doi.org/10.1073/pnas.261562798.

Colonna, M., Trinchieri, G., Liu, Y.-J., 2004. Plasmacytoid dendritic cells in immunity. Nat. Immunol. 5, 1219–1226. https://doi.org/10.1038/ni1141.

Cook, A.D., Braine, E.L., Hamilton, J.A., 2004. Stimulus-dependent requirement for granulocyte-macrophage colony-stimulating factor in inflammation. J. Immunol. 173, 4643–4651.

Cook, S.J., Lee, Q., Wong, A.C., Spann, B.C., Vincent, J.N., Wong, J.J., Schlitzer, A., Gorrell, M.D., Weninger, W., Roediger, B., 2018. Differential chemokine receptor expression and usage by pre-cDC1 and pre-cDC2. Immunol. Cell Biol. 96, 1131–1139. https://doi.org/10.1111/imcb.12186.

Crozat, K., Guiton, R., Contreras, V., Feuillet, V., Dutertre, C.-A., Ventre, E., Vu Manh, T.-P., Baranek, T., Storset, A.K., Marvel, J., Boudinot, P., Hosmalin, A., Schwartz-Cornil, I., Dalod, M., 2010a. The XC chemokine receptor 1 is a conserved selective marker of mammalian cells homologous to mouse CD8alpha+ dendritic cells. J. Exp. Med. 207, 1283–1292. https://doi.org/10.1084/jem.20100223.

Crozat, K., Guiton, R., Guilliams, M., Henri, S., Baranek, T., Schwartz-Cornil, I., Malissen, B., Dalod, M., 2010b. Comparative genomics as a tool to reveal functional equivalences between human and mouse dendritic cell subsets. Immunol. Rev. 234, 177–198. https://doi.org/10.1111/j.0105-2896.2009.00868.x.

Cytlak, U., Resteu, A., Bogaert, D., Kuehn, H.S., Altmann, T., Gennery, A., Jackson, G., Kumanovics, A., Voelkerding, K.V., Prader, S., Dullaers, M., Reichenbach, J., Hill, H., Haerynck, F., Rosenzweig, S.D., Collin, M., Bigley, V., 2018. Ikaros family zinc finger 1 regulates dendritic cell development and function in humans. Nat. Commun. 9, 1239. https://doi.org/10.1038/s41467-018-02977-8.

D'Amico, A., Wu, L., 2003. The early progenitors of mouse dendritic cells and plasmacytoid predendritic cells are within the bone marrow hemopoietic precursors expressing Flt3. J. Exp. Med. 198, 293–303. https://doi.org/10.1084/jem.20030107.

DeKoning, J., DiMolfetto, L., Reilly, C., Wei, Q., Havran, W.L., Lo, D., 1997. Thymic cortical epithelium is sufficient for the development of mature T cells in relB-deficient mice. J. Immunol. 158, 2558–2566.

DeKoter, R.P., Singh, H., 2000. Regulation of B lymphocyte and macrophage development by graded expression of PU.1. Science 288, 1439–1441. https://doi.org/10.1126/science.288.5470.1439.

Del Fresno, C., Saz-Leal, P., Enamorado, M., Wculek, S.K., Martínez-Cano, S., Blanco-Menéndez, N., Schulz, O., Gallizioli, M., Miró-Mur, F., Cano, E., Planas, A., Sancho, D., 2018. DNGR-1 in dendritic cells limits tissue damage by dampening neutrophil recruitment. Science 362, 351–356. https://doi.org/10.1126/science.aan8423.

del Hoyo, G.M., Martín, P., Vargas, H.H., Ruiz, S., Arias, C.F., Ardavín, C., 2002. Characterization of a common precursor population for dendritic cells. Nature 415, 1043–1047. https://doi.org/10.1038/4151043a.

del Rio, M.-L., Rodriguez-Barbosa, J.-I., Kremmer, E., Forster, R., 2007. CD103- and CD103+ bronchial lymph node dendritic cells are specialized in presenting and cross-presenting innocuous antigen to CD4+ and CD8+ T cells. J. Immunol. 178, 6861–6866. https://doi.org/10.4049/jimmunol.178.11.6861.

Diao, J., Winter, E., Chen, W., Cantin, C., Cattral, M.S., 2004. Characterization of distinct conventional and plasmacytoid dendritic cell-committed precursors in murine bone marrow. J. Immunol. 173, 1826–1833.

Diao, J., Winter, E., Cantin, C., Chen, W., Xu, L., Kelvin, D., Phillips, J., Cattral, M.S., 2006. In situ replication of immediate dendritic cell (DC) precursors contributes to conventional DC homeostasis in lymphoid tissue. J. Immunol. 176, 7196–7206.

Dickinson, R.E., Griffin, H., Bigley, V., Reynard, L.N., Hussain, R., Haniffa, M., Lakey, J.H., Rahman, T., Wang, X.-N., McGovern, N., Pagan, S., Cookson, S., McDonald, D., Chua, I., Wallis, J., Cant, A., Wright, M., Keavney, B., Chinnery, P.F., Loughlin, J., Hambleton, S., Santibanez-Koref, M., Collin, M., 2011. Exome sequencing identifies GATA-2 mutation as the cause of dendritic cell, monocyte, B and NK lymphoid deficiency. Blood 118, 2656–2658. https://doi.org/10.1182/blood-2011-06-360313.

Dickinson, R.E., Milne, P., Jardine, L., Zandi, S., Swierczek, S.I., McGovern, N., Cookson, S., Ferozepurwalla, Z., Langridge, A., Pagan, S., Gennery, A., Heiskanen-Kosma, T., Hämäläinen, S., Seppänen, M., Helbert, M., Tholouli, E., Gambineri, E., Reykdal, S., Gottfreðsson, M., Thaventhiran, J.E., Morris, E., Hirschfield, G., Richter, A.G., Jolles, S., Bacon, C.M., Hambleton, S., Haniffa, M., Bryceson, Y., Allen, C., Prchal, J.T., Dick, J.E., Bigley, V., Collin, M., 2014. The evolution of cellular deficiency in GATA2 mutation. Blood 123, 863–874. https://doi.org/10.1182/blood-2013-07-517151.

Ding, Y., Wilkinson, A., Idris, A., Fancke, B., O'Keeffe, M., Khalil, D., Ju, X., Lahoud, M.H., Caminschi, I., Shortman, K., Rodwell, R., Vuckovic, S., Radford, K.J., 2014. FLT3-ligand treatment of humanized mice results in the generation of large numbers of CD141+ and CD1c+ dendritic cells in vivo. J. Immunol. 192, 1982–1989. https://doi.org/10.4049/jimmunol.1302391.

Dorner, B.G., Dorner, M.B., Zhou, X., Opitz, C., Mora, A., Güttler, S., Hutloff, A., Mages, H.W., Ranke, K., Schaefer, M., Jack, R.S., Henn, V., Kroczek, R.A., 2009. Selective expression of the chemokine receptor XCR1 on cross-presenting dendritic cells determines cooperation with CD8+ T cells. Immunity 31, 823–833. https://doi.org/10.1016/j.immuni.2009.08.027.

Doulatov, S., Notta, F., Eppert, K., Nguyen, L.T., Ohashi, P.S., Dick, J.E., 2010. Revised map of the human progenitor hierarchy shows the origin of macrophages and dendritic cells in early lymphoid development. Nat. Immunol. 11, 585–593. https://doi.org/10.1038/ni.1889.

Dranoff, G., Jaffee, E., Lazenby, A., Golumbek, P., Levitsky, H., Brose, K., Jackson, V., Hamada, H., Pardoll, D., Mulligan, R.C., 1993. Vaccination with irradiated tumor cells engineered to secrete murine granulocyte-macrophage colony-stimulating factor stimulates potent, specific, and long-lasting anti-tumor immunity. Proc. Natl. Acad. Sci. U.S.A. 90, 3539–3543. https://doi.org/10.1073/pnas.90.8.3539.

Dudziak, D., Kamphorst, A.O., Heidkamp, G.F., Buchholz, V.R., Trumpfheller, C., Yamazaki, S., Cheong, C., Liu, K., Lee, H.-W., Park, C.G., Steinman, R.M., Nussenzweig, M.C., 2007. Differential antigen processing by dendritic cell subsets in vivo. Science 315, 107–111. https://doi.org/10.1126/science.1136080.

Durai, V., Murphy, K.M., 2016. Functions of murine dendritic cells. Immunity 45, 719–736. https://doi.org/10.1016/j.immuni.2016.10.010.

Durai, V., Bagadia, P., Briseño, C.G., Theisen, D.J., Iwata, A., Davidson, J.T., Gargaro, M., Fremont, D.H., Murphy, T.L., Murphy, K.M., 2018. Altered compensatory cytokine signaling underlies the discrepancy between Flt3-/- and Flt3l-/- mice. J. Exp. Med. 215, 1417–1435. https://doi.org/10.1084/jem.20171784.

Dzionek, A., Fuchs, A., Schmidt, P., Cremer, S., Zysk, M., Miltenyi, S., Buck, D.W., Schmitz, J., 2000. BDCA-2, BDCA-3, and BDCA-4: three markers for distinct subsets of dendritic cells in human peripheral blood. J. Immunol. 165, 6037–6046.

Edelson, B.T., Kc, W., Juang, R., Kohyama, M., Benoit, L.A., Klekotka, P.A., Moon, C., Albring, J.C., Ise, W., Michael, D.G., Bhattacharya, D., Stappenbeck, T.S., Holtzman, M.J., Sung, S.-S.J., Murphy, T.L., Hildner, K., Murphy, K.M., 2010. Peripheral CD103$^+$ dendritic cells form a unified subset developmentally related to CD8α$^+$ conventional dendritic cells. J. Exp. Med. 207, 823–836. https://doi.org/10.1084/jem.20091627.

Edelson, B.T., Bradstreet, T.R., Hildner, K., Carrero, J.A., Frederick, K.E., Kc, W., Belizaire, R., Aoshi, T., Schreiber, R.D., Miller, M.J., Murphy, T.L., Unanue, E.R., Murphy, K.M., 2011a. CD8α(+) dendritic cells are an obligate cellular entry point for productive infection by Listeria monocytogenes. Immunity 35, 236–248. https://doi.org/10.1016/j.immuni.2011.06.012.

Edelson, B.T., Bradstreet, T.R., Kc, W., Hildner, K., Herzog, J.W., Sim, J., Russell, J.H., Murphy, T.L., Unanue, E.R., Murphy, K.M., 2011b. Batf3-dependent CD11b(low/-) peripheral dendritic cells are GM-CSF-independent and are not required for Th cell priming after subcutaneous immunization. PLoS One 6 e25660 https://doi.org/10.1371/journal.pone.0025660.

Eickhoff, S., Brewitz, A., Gerner, M.Y., Klauschen, F., Komander, K., Hemmi, H., Garbi, N., Kaisho, T., Germain, R.N., Kastenmüller, W., 2015. Robust anti-viral immunity requires multiple distinct T cell-dendritic cell interactions. Cell 162, 1322–1337. https://doi.org/10.1016/j.cell.2015.08.004.

Eisenbarth, S.C., 2019. Dendritic cell subsets in T cell programming: location dictates function. Nat. Rev. Immunol. 19, 89–103. https://doi.org/10.1038/s41577-018-0088-1.

Elewaut, D., Shaikh, R.B., Hammond, K.J.L., De Winter, H., Leishman, A.J., Sidobre, S., Turovskaya, O., Prigozy, T.I., Ma, L., Banks, T.A., Lo, D., Ware, C.F., Cheroutre, H., Kronenberg, M., 2003. NIK-dependent RelB activation defines a unique signaling pathway for the development of V alpha 14i NKT cells. J. Exp. Med. 197, 1623–1633. https://doi.org/10.1084/jem.20030141.

Everts, B., Tussiwand, R., Dreesen, L., Fairfax, K.C., Huang, S.C.-C., Smith, A.M., O'Neill, C.M., Lam, W.Y., Edelson, B.T., Urban, J.F., Murphy, K.M., Pearce, E.J., 2016. Migratory CD103+ dendritic cells suppress helminth-driven type 2 immunity through constitutive expression of IL-12. J. Exp. Med. 213, 35–51. https://doi.org/10.1084/jem.20150235.

Feinberg, M.W., Wara, A.K., Cao, Z., Lebedeva, M.A., Rosenbauer, F., Iwasaki, H., Hirai, H., Katz, J.P., Haspel, R.L., Gray, S., Akashi, K., Segre, J., Kaestner, K.H., Tenen, D.G., Jain, M.K., 2007. The Kruppel-like factor KLF4 is a critical regulator of monocyte differentiation. EMBO J. 26, 4138–4148. https://doi.org/10.1038/sj.emboj.7601824.

Fogg, D.K., Sibon, C., Miled, C., Jung, S., Aucouturier, P., Littman, D.R., Cumano, A., Geissmann, F., 2006. A clonogenic bone marrow progenitor specific for macrophages and dendritic cells. Science 311, 83–87. https://doi.org/10.1126/science.1117729.

Gao, Y., Nish, S.A., Jiang, R., Hou, L., Licona-Limón, P., Weinstein, J.S., Zhao, H., Medzhitov, R., 2013. Control of T helper 2 responses by transcription factor IRF4-dependent dendritic cells. Immunity 39, 722–732. https://doi.org/10.1016/j.immuni.2013.08.028.

Gatto, D., Wood, K., Caminschi, I., Murphy-Durland, D., Schofield, P., Christ, D., Karupiah, G., Brink, R., 2013. The chemotactic receptor EBI2 regulates the homeostasis, localization and immunological function of splenic dendritic cells. Nat. Immunol. 14, 446–453. https://doi.org/10.1038/ni.2555.

Gautier, E.L., Shay, T., Miller, J., Greter, M., Jakubzick, C., Ivanov, S., Helft, J., Chow, A., Elpek, K.G., Gordonov, S., Mazloom, A.R., Ma'ayan, A., Chua, W.-J., Hansen, T.H., Turley, S.J., Merad, M., Randolph, G.J., 2012. Gene-expression profiles and transcriptional regulatory pathways that underlie the identity and diversity of mouse tissue macrophages. Nat. Immunol. 13, 1118–1128. https://doi.org/10.1038/ni.2419.

Georgopoulos, K., Bigby, M., Wang, J.H., Molnar, A., Wu, P., Winandy, S., Sharpe, A., 1994. The Ikaros gene is required for the development of all lymphoid lineages. Cell 79, 143–156.

Giladi, A., Paul, F., Herzog, Y., Lubling, Y., Weiner, A., Yofe, I., Jaitin, D., Cabezas-Wallscheid, N., Dress, R., Ginhoux, F., Trumpp, A., Tanay, A., Amit, I., 2018. Single-cell characterization of haematopoietic progenitors and their trajectories in homeostasis and perturbed haematopoiesis. Nat. Cell Biol. 20, 836–846. https://doi.org/10.1038/s41556-018-0121-4.

Ginhoux, F., Collin, M.P., Bogunovic, M., Abel, M., Leboeuf, M., Helft, J., Ochando, J., Kissenpfennig, A., Malissen, B., Grisotto, M., Snoeck, H., Randolph, G., Merad, M., 2007. Blood-derived dermal langerin + dendritic cells survey the skin in the steady state. J. Exp. Med. 204, 3133–3146. https://doi.org/10.1084/jem.20071733.

Ginhoux, F., Liu, K., Helft, J., Bogunovic, M., Greter, M., Hashimoto, D., Price, J., Yin, N., Bromberg, J., Lira, S.A., Stanley, E.R., Nussenzweig, M., Merad, M., 2009. The origin and development of nonlymphoid tissue CD103 + DCs. J. Exp. Med. 206, 3115–3130. https://doi.org/10.1084/jem.20091756.

Grajales-Reyes, G.E., Iwata, A., Albring, J., Wu, X., Tussiwand, R., Kc, W., Kretzer, N.M., Briseño, C.G., Durai, V., Bagadia, P., Haldar, M., Schönheit, J., Rosenbauer, F., Murphy, T.L., Murphy, K.M., 2015. Batf3 maintains autoactivation of *Irf8* for commitment of a $CD8\alpha^+$ conventional DC clonogenic progenitor. Nat. Immunol. 16, 708–717. https://doi.org/10.1038/ni.3197.

Greter, M., Helft, J., Chow, A., Hashimoto, D., Mortha, A., Agudo-Cantero, J., Bogunovic, M., Gautier, E.L., Miller, J., Leboeuf, M., Lu, G., Aloman, C., Brown, B.D., Pollard, J.W., Xiong, H., Randolph, G.J., Chipuk, J.E., Frenette, P.S., Merad, M., 2012. GM-CSF controls nonlymphoid tissue dendritic cell homeostasis but is dispensable for the differentiation of inflammatory dendritic cells. Immunity 36, 1031–1046. https://doi.org/10.1016/j.immuni.2012.03.027.

Guermonprez, P., Helft, J., Claser, C., Deroubaix, S., Karanje, H., Gazumyan, A., Darasse-Jèze, G., Telerman, S.B., Breton, G., Schreiber, H.A., Frias-Staheli, N., Billerbeck, E., Dorner, M., Rice, C.M., Ploss, A., Klein, F., Swiecki, M., Colonna, M., Kamphorst, A.O., Meredith, M., Niec, R., Takacs, C., Mikhail, F., Hari, A.,

Bosque, D., Eisenreich, T., Merad, M., Shi, Y., Ginhoux, F., Rénia, L., Urban, B.C., Nussenzweig, M.C., 2013. Inflammatory Flt3l is essential to mobilize dendritic cells and for T cell responses during Plasmodium infection. Nat. Med. 19, 730–738. https://doi.org/10.1038/nm.3197.

Guerriero, A., Langmuir, P.B., Spain, L.M., Scott, E.W., 2000. PU.1 is required for myeloid-derived but not lymphoid-derived dendritic cells. Blood 95, 879–885.

Guilliams, M., Ginhoux, F., Jakubzick, C., Naik, S.H., Onai, N., Schraml, B.U., Segura, E., Tussiwand, R., Yona, S., 2014. Dendritic cells, monocytes and macrophages: a unified nomenclature based on ontogeny. Nat. Rev. Immunol. 14, 571–578. https://doi.org/10.1038/nri3712.

Guilliams, M., Dutertre, C.-A., Scott, C.L., McGovern, N., Sichien, D., Chakarov, S., Van Gassen, S., Chen, J., Poidinger, M., De Prijck, S., Tavernier, S.J., Low, I., Irac, S.E., Mattar, C.N., Sumatoh, H.R., Low, G.H.L., Chung, T.J.K., Chan, D.K.H., Tan, K.K., Hon, T.L.K., Fossum, E., Bogen, B., Choolani, M., Chan, J.K.Y., Larbi, A., Luche, H., Henri, S., Saeys, Y., Newell, E.W., Lambrecht, B.N., Malissen, B., Ginhoux, F., 2016. Unsupervised high-dimensional analysis aligns dendritic cells across tissues and species. Immunity 45, 669–684. https://doi.org/10.1016/j.immuni.2016.08.015.

Hacker, C., Kirsch, R.D., Ju, X.-S., Hieronymus, T., Gust, T.C., Kuhl, C., Jorgas, T., Kurz, S.M., Rose-John, S., Yokota, Y., Zenke, M., 2003. Transcriptional profiling identifies Id2 function in dendritic cell development. Nat. Immunol. 4, 380–386. https://doi.org/10.1038/ni903.

Hambleton, S., Salem, S., Bustamante, J., Bigley, V., Boisson-Dupuis, S., Azevedo, J., Fortin, A., Haniffa, M., Ceron-Gutierrez, L., Bacon, C.M., Menon, G., Trouillet, C., McDonald, D., Carey, P., Ginhoux, F., Alsina, L., Zumwalt, T.J., Kong, X.-F., Kumararatne, D., Butler, K., Hubeau, M., Feinberg, J., Al-Muhsen, S., Cant, A., Abel, L., Chaussabel, D., Doffinger, R., Talesnik, E., Grumach, A., Duarte, A., Abarca, K., Moraes-Vasconcelos, D., Burk, D., Berghuis, A., Geissmann, F., Collin, M., Casanova, J.-L., Gros, P., 2011. IRF8 mutations and human dendritic-cell immunodeficiency. N. Engl. J. Med. 365, 127–138. https://doi.org/10.1056/NEJMoa1100066.

Hamilton, J.A., Anderson, G.P., 2004. GM-CSF Biology. Growth Factors 22, 225–231. https://doi.org/10.1080/08977190412331279881.

Haniffa, M., Shin, A., Bigley, V., McGovern, N., Teo, P., See, P., Wasan, P.S., Wang, X.-N., Malinarich, F., Malleret, B., Larbi, A., Tan, P., Zhao, H., Poidinger, M., Pagan, S., Cookson, S., Dickinson, R., Dimmick, I., Jarrett, R.F., Renia, L., Tam, J., Song, C., Connolly, J., Chan, J.K.Y., Gehring, A., Bertoletti, A., Collin, M., Ginhoux, F., 2012. Human tissues contain CD141hi cross-presenting dendritic cells with functional homology to mouse CD103+ nonlymphoid dendritic cells. Immunity 37, 60–73. https://doi.org/10.1016/j.immuni.2012.04.012.

Hawiger, D., Inaba, K., Dorsett, Y., Guo, M., Mahnke, K., Rivera, M., Ravetch, J.V., Steinman, R.M., Nussenzweig, M.C., 2001. Dendritic cells induce peripheral T cell unresponsiveness under steady state conditions in vivo. J. Exp. Med. 194, 769–780. https://doi.org/10.1084/jem.194.6.769.

Helft, J., Manicassamy, B., Guermonprez, P., Hashimoto, D., Silvin, A., Agudo, J., Brown, B.D., Schmolke, M., Miller, J.C., Leboeuf, M., Murphy, K.M., García-Sastre, A., Merad, M., 2012. Cross-presenting $CD103^+$ dendritic cells are protected from influenza virus infection. J. Clin. Invest. 122, 4037–4047. https://doi.org/10.1172/JCI60659.

Helft, J., Böttcher, J., Chakravarty, P., Zelenay, S., Huotari, J., Schraml, B.U., Goubau, D., Reis e Sousa, C., 2015. GM-CSF mouse bone marrow cultures comprise a heterogeneous population of CD11c+MHCII+ macrophages and dendritic cells. Immunity 42, 1197–1211. https://doi.org/10.1016/j.immuni.2015.05.018.

Helft, J., Böttcher, J.P., Chakravarty, P., Zelenay, S., Huotari, J., Schraml, B.U., Goubau, D., Reis e Sousa, C., 2016. Alive but confused: heterogeneity of CD11c(+) MHC class II(+) cells in GM-CSF mouse bone marrow cultures. Immunity 44, 3–4. https://doi.org/10.1016/j.immuni.2015.12.014.

Helft, J., Anjos-Afonso, F., van der Veen, A.G., Chakravarty, P., Bonnet, D., Reis e Sousa, C., 2017. Dendritic cell lineage potential in human early hematopoietic progenitors. Cell Rep. 20, 529–537. https://doi.org/10.1016/j.celrep.2017.06.075.

Henri, S., Poulin, L.F., Tamoutounour, S., Ardouin, L., Guilliams, M., de Bovis, B., Devilard, E., Viret, C., Azukizawa, H., Kissenpfennig, A., Malissen, B., 2010. CD207 $^{+}$ CD103 $^{+}$ dermal dendritic cells cross-present keratinocyte-derived antigens irrespective of the presence of Langerhans cells. J. Exp. Med. 207, 189–206. https://doi.org/10.1084/jem.20091964.

Hettinger, J., Richards, D.M., Hansson, J., Barra, M.M., Joschko, A.-C., Krijgsveld, J., Feuerer, M., 2013. Origin of monocytes and macrophages in a committed progenitor. Nat. Immunol. 14, 821–830. https://doi.org/10.1038/ni.2638.

Hildner, K., Edelson, B.T., Purtha, W.E., Diamond, M., Matsushita, H., Kohyama, M., Calderon, B., Schraml, B.U., Unanue, E.R., Diamond, M.S., Schreiber, R.D., Murphy, T.L., Murphy, K.M., 2008. Batf3 deficiency reveals a critical role for CD8alpha + dendritic cells in cytotoxic T cell immunity. Science 322 (5904), 1097–1100. https://doi.org/10.1126/science.1164206.

Hor, J.L., Whitney, P.G., Zaid, A., Brooks, A.G., Heath, W.R., Mueller, S.N., 2015. Spatiotemporally distinct interactions with dendritic cell subsets facilitates CD4+ and CD8+ T cell activation to localized viral infection. Immunity 43, 615. https://doi.org/10.1016/j.immuni.2015.08.018.

Ichikawa, E., Hida, S., Omatsu, Y., Shimoyama, S., Takahara, K., Miyagawa, S., Inaba, K., Taki, S., 2004. Defective development of splenic and epidermal CD4+ dendritic cells in mice deficient for IFN regulatory factor-2. Proc. Natl. Acad. Sci. U.S.A. 101, 3909–3914. https://doi.org/10.1073/pnas.0400610101.

Idoyaga, J., Suda, N., Suda, K., Park, C.G., Steinman, R.M., 2009. Antibody to Langerin/CD207 localizes large numbers of CD8alpha+ dendritic cells to the marginal zone of mouse spleen. Proc. Natl. Acad. Sci. U.S.A. 106, 1524–1529. https://doi.org/10.1073/pnas.0812247106.

Inaba, K., Inaba, M., Deguchi, M., Hagi, K., Yasumizu, R., Ikehara, S., Muramatsu, S., Steinman, R.M., 1993a. Granulocytes, macrophages, and dendritic cells arise from a common major histocompatibility complex class II-negative progenitor in mouse bone marrow. Proc. Natl. Acad. Sci. U.S.A. 90, 3038–3042. https://doi.org/10.1073/pnas.90.7.3038.

Inaba, K., Inaba, M., Naito, M., Steinman, R.M., 1993b. Dendritic cell progenitors phagocytose particulates, including bacillus Calmette-Guerin organisms, and sensitize mice to mycobacterial antigens in vivo. J. Exp. Med. 178, 479–488. https://doi.org/10.1084/jem.178.2.479.

Inaba, K., Turley, S., Iyoda, T., Yamaide, F., Shimoyama, S., e Sousa, C.R., Germain, R.N., Mellman, I., Steinman, R.M., 2000. The formation of immunogenic major histocompatibility complex class II–peptide ligands in lysosomal compartments of dendritic cells is regulated by inflammatory stimuli. J. Exp. Med. 191, 927–936. https://doi.org/10.1084/jem.191.6.927.

Ivanov, I.I., Frutos, R.d.L., Manel, N., Yoshinaga, K., Rifkin, D.B., Sartor, R.B., Finlay, B.B., Littman, D.R., 2008. Specific microbiota direct the differentiation of IL-17-producing T-helper cells in the mucosa of the small intestine. Cell Host Microbe 4, 337–349. https://doi.org/10.1016/j.chom.2008.09.009.

Jackson, J.T., Hu, Y., Liu, R., Masson, F., D'Amico, A., Carotta, S., Xin, A., Camilleri, M.J., Mount, A.M., Kallies, A., Wu, L., Smyth, G.K., Nutt, S.L., Belz, G.T., 2011. Id2 expression delineates differential checkpoints in the genetic program of CD8α+ and CD103+ dendritic cell lineages. EMBO J. 30, 2690–2704. https://doi.org/10.1038/emboj.2011.163.

Jaiswal, H., Kaushik, M., Sougrat, R., Gupta, M., Dey, A., Verma, R., Ozato, K., Tailor, P., 2013. Batf3 and Id2 have a synergistic effect on Irf8-directed classical CD8α+ dendritic cell development. J. Immunol. 191, 5993–6001. https://doi.org/10.4049/jimmunol.1203541.

Janela, B., Patel, A.A., Lau, M.C., Goh, C.C., Msallam, R., Kong, W.T., Fehlings, M., Hubert, S., Lum, J., Simoni, Y., Malleret, B., Zolezzi, F., Chen, J., Poidinger, M., Satpathy, A.T., Briseno, C., Wohn, C., Malissen, B., Murphy, K.M., Maini, A.A., Vanhoutte, L., Guilliams, M., Vial, E., Hennequin, L., Newell, E., Ng, L.G., Musette, P., Yona, S., Hacini-Rachinel, F., Ginhoux, F., 2019. A subset of type I conventional dendritic cells controls cutaneous bacterial infections through VEGFα-mediated recruitment of neutrophils. Immunity 50 (4), 1069–1083. e8. https://doi.org/10.1016/j.immuni.2019.03.001.

Jongbloed, S.L., Kassianos, A.J., McDonald, K.J., Clark, G.J., Ju, X., Angel, C.E., Chen, C.-J.J., Dunbar, P.R., Wadley, R.B., Jeet, V., Vulink, A.J.E., Hart, D.N.J., Radford, K.J., 2010. Human CD141+ (BDCA-3)+ dendritic cells (DCs) represent a unique myeloid DC subset that cross-presents necrotic cell antigens. J. Exp. Med. 207, 1247–1260. https://doi.org/10.1084/jem.20092140.

Kabashima, K., Banks, T.A., Ansel, K.M., Lu, T.T., Ware, C.F., Cyster, J.G., 2005. Intrinsic lymphotoxin-beta receptor requirement for homeostasis of lymphoid tissue dendritic cells. Immunity 22, 439–450. https://doi.org/10.1016/j.immuni.2005.02.007.

Kamath, A.T., Pooley, J., O'Keeffe, M.A., Vremec, D., Zhan, Y., Lew, A.M., D'Amico, A., Wu, L., Tough, D.F., Shortman, K., 2000. The development, maturation, and turnover rate of mouse spleen dendritic cell populations. J. Immunol. 165, 6762–6770. https://doi.org/10.4049/jimmunol.165.12.6762.

Karamitros, D., Stoilova, B., Aboukhalil, Z., Hamey, F., Reinisch, A., Samitsch, M., Quek, L., Otto, G., Repapi, E., Doondeea, J., Usukhbayar, B., Calvo, J., Taylor, S., Goardon, N., Six, E., Pflumio, F., Porcher, C., Majeti, R., Göttgens, B., Vyas, P., 2018. Single-cell analysis reveals the continuum of human lympho-myeloid progenitor cells. Nat. Immunol. 19, 85–97. https://doi.org/10.1038/s41590-017-0001-2.

Karsunky, H., Merad, M., Cozzio, A., Weissman, I.L., Manz, M.G., 2003. Flt3 ligand regulates dendritic cell development from Flt3+ lymphoid and myeloid-committed progenitors to Flt3+ dendritic cells in vivo. J. Exp. Med. 198, 305–313. https://doi.org/10.1084/jem.20030323.

Kashiwada, M., Pham, N.-L.L., Pewe, L.L., Harty, J.T., Rothman, P.B., 2011. NFIL3/E4BP4 is a key transcription factor for CD8α$^+$ dendritic cell development. Blood 117, 6.

Kastenmüller, W., Brandes, M., Wang, Z., Herz, J., Egen, J.G., Germain, R.N., 2013. Peripheral prepositioning and local CXCL9 chemokine-mediated guidance orchestrate rapid memory CD8+ T cell responses in the lymph node. Immunity 38, 502–513. https://doi.org/10.1016/j.immuni.2012.11.012.

Kawamura, S., Onai, N., Miya, F., Sato, T., Tsunoda, T., Kurabayashi, K., Yotsumoto, S., Kuroda, S., Takenaka, K., Akashi, K., Ohteki, T., 2017. Identification of a human clonogenic progenitor with strict monocyte differentiation potential: a counterpart of mouse cMoPs. Immunity 46, 835–848.e4. https://doi.org/10.1016/j.immuni.2017.04.019.

Kc, W., Satpathy, A.T., Rapaport, A.S., Briseño, C.G., Wu, X., Albring, J.C., Russler-Germain, E.V., Kretzer, N.M., Durai, V., Persaud, S.P., Edelson, B.T., Loschko, J., Cella, M., Allen, P.M., Nussenzweig, M.C., Colonna, M., Sleckman, B.P., Murphy, T.L., Murphy, K.M., 2014. L-Myc expression by dendritic cells is required for optimal T-cell priming. Nature 507, 243–247. https://doi.org/10.1038/nature12967.

King, I.L., Kroenke, M.A., Segal, B.M., 2010. GM-CSF-dependent, CD103+ dermal dendritic cells play a critical role in Th effector cell differentiation after subcutaneous immunization. J. Exp. Med. 207, 953–961. https://doi.org/10.1084/jem.20091844.

Kingston, D., Schmid, M.A., Onai, N., Obata-Onai, A., Baumjohann, D., Manz, M.G., 2009. The concerted action of GM-CSF and Flt3-ligand on in vivo dendritic cell homeostasis. Blood 114, 835–843. https://doi.org/10.1182/blood-2009-02-206318.

Kinnebrew, M.A., Buffie, C.G., Diehl, G.E., Zenewicz, L.A., Leiner, I., Hohl, T.M., Flavell, R.A., Littman, D.R., Pamer, E.G., 2012. Interleukin 23 production by intestinal CD103(+)CD11b(+) dendritic cells in response to bacterial flagellin enhances mucosal innate immune defense. Immunity 36, 276–287. https://doi.org/10.1016/j.immuni.2011.12.011.

Kirkling, M.E., Cytlak, U., Lau, C.M., Lewis, K.L., Resteu, A., Khodadadi-Jamayran, A., Siebel, C.W., Salmon, H., Merad, M., Tsirigos, A., Collin, M., Bigley, V., Reizis, B., 2018. Notch signaling facilitates in vitro generation of cross-presenting classical dendritic cells. Cell Rep. 23, 3658–3672.e6. https://doi.org/10.1016/j.celrep.2018.05.068.

Kondo, M., Weissman, I.L., Akashi, K., 1997. Identification of clonogenic common lymphoid progenitors in mouse bone marrow. Cell 91, 661–672.

Kretzer, N.M., Theisen, D.J., Tussiwand, R., Briseño, C.G., Grajales-Reyes, G.E., Wu, X., Durai, V., Albring, J., Bagadia, P., Murphy, T.L., Murphy, K.M., 2016. RAB43 facilitates cross-presentation of cell-associated antigens by CD8α+ dendritic cells. J. Exp. Med. 213, 2871–2883. https://doi.org/10.1084/jem.20160597.

Kroczek, R.A., Henn, V., 2012. The role of XCR1 and its ligand XCL1 in antigen cross-presentation by murine and human dendritic cells. Front. Immunol. 3, 14. https://doi.org/10.3389/fimmu.2012.00014.

Kurotaki, D., Osato, N., Nishiyama, A., Yamamoto, M., Ban, T., Sato, H., Nakabayashi, J., Umehara, M., Miyake, N., Matsumoto, N., Nakazawa, M., Ozato, K., Tamura, T., 2013. Essential role of the IRF8-KLF4 transcription factor cascade in murine monocyte differentiation. Blood 121, 1839–1849. https://doi.org/10.1182/blood-2012-06-437863.

Kurotaki, D., Yamamoto, M., Nishiyama, A., Uno, K., Ban, T., Ichino, M., Sasaki, H., Matsunaga, S., Yoshinari, M., Ryo, A., Nakazawa, M., Ozato, K., Tamura, T., 2014. IRF8 inhibits C/EBPα activity to restrain mononuclear phagocyte progenitors from differentiating into neutrophils. Nat. Commun. 5, 4978. https://doi.org/10.1038/ncomms5978.

Kurotaki, D., Kawase, W., Sasaki, H., Nakabayashi, J., Nishiyama, A., Morse, H.C., Ozato, K., Suzuki, Y., Tamura, T., 2019. Epigenetic control of early dendritic cell lineage specification by the transcription factor IRF8 in mice. Blood 133, 1803–1813. https://doi.org/10.1182/blood-2018-06-857789.

Laouar, Y., Welte, T., Fu, X.-Y., Flavell, R.A., 2003. STAT3 is required for Flt3L-dependent dendritic cell differentiation. Immunity 19, 903–912.

Lau, C.M., Tiniakou, I., Perez, O.A., Kirkling, M.E., Yap, G.S., Hock, H., Reizis, B., 2018. Transcription factor Etv6 regulates functional differentiation of cross-presenting classical dendritic cells. J. Exp. Med. 215, 2265–2278. https://doi.org/10.1084/jem.20172323.

Lauterbach, H., Bathke, B., Gilles, S., Traidl-Hoffmann, C., Luber, C.A., Fejer, G., Freudenberg, M.A., Davey, G.M., Vremec, D., Kallies, A., Wu, L., Shortman, K., Chaplin, P., Suter, M., O'Keeffe, M., Hochrein, H., 2010. Mouse CD8alpha+ DCs and human BDCA3+ DCs are major producers of IFN-lambda in response to poly IC. J. Exp. Med. 207, 2703–2717. https://doi.org/10.1084/jem.20092720.

Lee, J., Breton, G., Aljoufi, A., Zhou, Y.J., Puhr, S., Nussenzweig, M.C., Liu, K., 2015a. Clonal analysis of human dendritic cell progenitor using a stromal cell culture. J. Immunol. Methods 425, 21–26. https://doi.org/10.1016/j.jim.2015.06.004.

Lee, J., Breton, G., Oliveira, T.Y.K., Zhou, Y.J., Aljoufi, A., Puhr, S., Cameron, M.J., Sékaly, R.-P., Nussenzweig, M.C., Liu, K., 2015b. Restricted dendritic cell and monocyte progenitors in human cord blood and bone marrow. J. Exp. Med. 212, 385–399. https://doi.org/10.1084/jem.20141442.

Lee, J., Zhou, Y.J., Ma, W., Zhang, W., Aljoufi, A., Luh, T., Lucero, K., Liang, D., Thomsen, M., Bhagat, G., Shen, Y., Liu, K., 2017. Lineage specification of human dendritic cells is marked by IRF8 expression in hematopoietic stem cells and multipotent progenitors. Nat. Immunol. 18, 877–888. https://doi.org/10.1038/ni.3789.

Lewis, K.L., Caton, M.L., Bogunovic, M., Greter, M., Grajkowska, L.T., Ng, D., Klinakis, A., Charo, I.F., Jung, S., Gommerman, J.L., Ivanov, I.I., Liu, K., Merad, M., Reizis, B., 2011. Notch2 receptor signaling controls functional differentiation of dendritic cells in the spleen and intestine. Immunity 35, 780–791. https://doi.org/10.1016/j.immuni.2011.08.013.

Lin, D.S., Kan, A., Gao, J., Crampin, E.J., Hodgkin, P.D., Naik, S.H., 2018. DiSNE movie visualization and assessment of clonal kinetics reveal multiple trajectories of dendritic cell development. Cell Rep. 22, 2557–2566. https://doi.org/10.1016/j.celrep.2018.02.046.

Liu, K., Waskow, C., Liu, X., Yao, K., Hoh, J., Nussenzweig, M., 2007. Origin of dendritic cells in peripheral lymphoid organs of mice. Nat. Immunol. 8, 578–583. https://doi.org/10.1038/ni1462.

Liu, K., Victora, G.D., Schwickert, T.A., Guermonprez, P., Meredith, M.M., Yao, K., Chu, F.-F., Randolph, G.J., Rudensky, A.Y., Nussenzweig, M., 2009. In vivo analysis of dendritic cell development and homeostasis. Science 324, 392–397. https://doi.org/10.1126/science.1170540.

Lu, E., Dang, E.V., McDonald, J.G., Cyster, J.G., 2017. Distinct oxysterol requirements for positioning naïve and activated dendritic cells in the spleen. Sci. Immunol. 2, 1–12. https://doi.org/10.1126/sciimmunol.aal5237.

Lundie, R.J., de Koning-Ward, T.F., Davey, G.M., Nie, C.Q., Hansen, D.S., Lau, L.S., Mintern, J.D., Belz, G.T., Schofield, L., Carbone, F.R., Villadangos, J.A., Crabb, B.S., Heath, W.R., 2008. Blood-stage plasmodium infection induces CD8+ T lymphocytes to parasite-expressed antigens, largely regulated by CD8alpha+ dendritic cells. Proc. Natl. Acad. Sci. U.S.A. 105, 14509–14514. https://doi.org/10.1073/pnas.0806727105.

Mach, N., Gillessen, S., Wilson, S.B., Sheehan, C., Mihm, M., Dranoff, G., 2000. Differences in dendritic cells stimulated in vivo by tumors engineered to secrete granulocyte-macrophage colony-stimulating factor or Flt3-ligand. Cancer Res. 60, 3239–3246.

Maillard, I., Fang, T., Pear, W.S., 2005. Regulation of lymphoid development, differentiation, and function by the Notch pathway. Annu. Rev. Immunol. 23, 945–974. https://doi.org/10.1146/annurev.immunol.23.021704.115747.

Martínez-López, M., Iborra, S., Conde-Garrosa, R., Sancho, D., 2015. Batf3-dependent CD103+ dendritic cells are major producers of IL-12 that drive local Th1 immunity against Leishmania major infection in mice. Eur. J. Immunol. 45, 119–129. https://doi.org/10.1002/eji.201444651.

Martinez-Moczygemba, M., Huston, D.P., 2003. Biology of common beta receptor-signaling cytokines: IL-3, IL-5, and GM-CSF. J. Allergy Clin. Immunol. 112, 653–665. quiz 666. https://doi.org/10.1016/S0091.

Mashayekhi, M., Sandau, M.M., Dunay, I.R., Frickel, E.M., Khan, A., Goldszmid, R.S., Sher, A., Ploegh, H.L., Murphy, T.L., Sibley, L.D., Murphy, K.M., 2011. CD8α(+) dendritic cells are the critical source of interleukin-12 that controls acute infection by *Toxoplasma gondii* tachyzoites. Immunity 35, 249–259. https://doi.org/10.1016/j.immuni.2011.08.008.

Mayer, C.T., Ghorbani, P., Nandan, A., Dudek, M., Arnold-Schrauf, C., Hesse, C., Berod, L., Stuve, P., Puttur, F., Merad, M., Sparwasser, T., 2014. Selective and efficient

generation of functional Batf3-dependent CD103+ dendritic cells from mouse bone marrow. Blood 124, 3081–3091. https://doi.org/10.1182/blood-2013-12-545772.
McKenna, H.J., Stocking, K.L., Miller, R.E., Brasel, K., De Smedt, T., Maraskovsky, E., Maliszewski, C.R., Lynch, D.H., Smith, J., Pulendran, B., Roux, E.R., Teepe, M., Lyman, S.D., Peschon, J.J., 2000. Mice lacking flt3 ligand have deficient hematopoiesis affecting hematopoietic progenitor cells, dendritic cells, and natural killer cells. Blood 95, 3489–3497.
Melillo, J.A., Song, L., Bhagat, G., Blazquez, A.B., Plumlee, C.R., Lee, C., Berin, C., Reizis, B., Schindler, C., 2010. Dendritic cell (DC)-specific targeting reveals Stat3 as a negative regulator of DC function. J. Immunol. 184, 2638–2645. https://doi.org/10.4049/jimmunol.0902960.
Menezes, S., Melandri, D., Anselmi, G., Perchet, T., Loschko, J., Dubrot, J., Patel, R., Gautier, E.L., Hugues, S., Longhi, M.P., Henry, J.Y., Quezada, S.A., Lauvau, G., Lennon-Duménil, A.-M., Gutiérrez-Martínez, E., Bessis, A., Gomez-Perdiguero, E., Jacome-Galarza, C.E., Garner, H., Geissmann, F., Golub, R., Nussenzweig, M.C., Guermonprez, P., 2016. The heterogeneity of Ly6Chi monocytes controls their differentiation into iNOS+ macrophages or monocyte-derived dendritic cells. Immunity 45, 1205–1218. https://doi.org/10.1016/j.immuni.2016.12.001.
Meredith, M.M., Liu, K., Darrasse-Jeze, G., Kamphorst, A.O., Schreiber, H.A., Guermonprez, P., Idoyaga, J., Cheong, C., Yao, K.-H., Niec, R.E., Nussenzweig, M.C., 2012a. Expression of the zinc finger transcription factor zDC (Zbtb46, Btbd4) defines the classical dendritic cell lineage. J. Exp. Med. 209, 1153–1165. https://doi.org/10.1084/jem.20112675.
Meredith, M.M., Liu, K., Kamphorst, A.O., Idoyaga, J., Yamane, A., Guermonprez, P., Rihn, S., Yao, K.-H., Silva, I.T., Oliveira, T.Y., Skokos, D., Casellas, R., Nussenzweig, M.C., 2012b. Zinc finger transcription factor zDC is a negative regulator required to prevent activation of classical dendritic cells in the steady state. J. Exp. Med. 209, 1583–1593. https://doi.org/10.1084/jem.20121003.
Miller, J.C., Brown, B.D., Shay, T., Gautier, E.L., Jojic, V., Cohain, A., Pandey, G., Leboeuf, M., Elpek, K.G., Helft, J., Hashimoto, D., Chow, A., Price, J., Greter, M., Bogunovic, M., Bellemare-Pelletier, A., Frenette, P.S., Randolph, G.J., Turley, S.J., Merad, M., The Immunological Genome Consortium, Gautier, E.L., Jakubzick, C., Randolph, G.J., Best, A.J., Knell, J., Goldrath, A., Miller, J., Brown, B., Merad, M., Jojic, V., Koller, D., Cohen, N., Brennan, P., Brenner, M., Shay, T., Regev, A., Fletcher, A., Elpek, K., Bellemare-Pelletier, A., Malhotra, D., Turley, S., Jianu, R., Laidlaw, D., Collins, J., Narayan, K., Sylvia, K., Kang, J., Gazit, R., Rossi, D.J., Kim, F., Rao, T.N., Wagers, A., Shinton, S.A., Hardy, R.R., Monach, P., Bezman, N.A., Sun, J.C., Kim, C.C., Lanier, L.L., Heng, T., Kreslavsky, T., Painter, M., Ericson, J., Davis, S., Mathis, D., Benoist, C., 2012. Deciphering the transcriptional network of the dendritic cell lineage. Nat. Immunol. 13, 888–899. https://doi.org/10.1038/ni.2370.
Mortha, A., Chudnovskiy, A., Hashimoto, D., Bogunovic, M., Spencer, S.P., Belkaid, Y., Merad, M., 2014. Microbiota-dependent crosstalk between macrophages and ILC3 promotes intestinal homeostasis. Science 343, 1249288. https://doi.org/10.1126/science.1249288.
Murphy, T.L., Grajales-Reyes, G.E., Wu, X., Tussiwand, R., Briseño, C.G., Iwata, A., Kretzer, N.M., Durai, V., Murphy, K.M., 2016. Transcriptional control of dendritic cell development. Annu. Rev. Immunol. 34, 93–119. https://doi.org/10.1146/annurev-immunol-032713-120204.
Naik, S.H., Proietto, A.I., Wilson, N.S., Dakic, A., Schnorrer, P., Fuchsberger, M., Lahoud, M.H., O'Keeffe, M., Shao, Q., Chen, W., Villadangos, J.A., Shortman, K., Wu, L., 2005. Cutting edge: generation of splenic $CD8^+$ and $CD8^-$ dendritic cell

equivalents in Fms-like tyrosine kinase 3 ligand bone marrow cultures. J. Immunol. 174, 6592–6597. https://doi.org/10.4049/jimmunol.174.11.6592.
Naik, S.H., Metcalf, D., van Nieuwenhuijze, A., Wicks, I., Wu, L., O'Keeffe, M., Shortman, K., 2006. Intrasplenic steady-state dendritic cell precursors that are distinct from monocytes. Nat. Immunol. 7, 663–671. https://doi.org/10.1038/ni1340.
Naik, S.H., Sathe, P., Park, H.-Y., Metcalf, D., Proietto, A.I., Dakic, A., Carotta, S., O'Keeffe, M., Bahlo, M., Papenfuss, A., Kwak, J.-Y., Wu, L., Shortman, K., 2007. Development of plasmacytoid and conventional dendritic cell subtypes from single precursor cells derived in vitro and in vivo. Nat. Immunol. 8, 1217–1226. https://doi.org/10.1038/ni1522.
Naik, S.H., Perié, L., Swart, E., Gerlach, C., van Rooij, N., de Boer, R.J., Schumacher, T.N., 2013. Diverse and heritable lineage imprinting of early haematopoietic progenitors. Nature 496, 229–232. https://doi.org/10.1038/nature12013.
Nestorowa, S., Hamey, F.K., Pijuan Sala, B., Diamanti, E., Shepherd, M., Laurenti, E., Wilson, N.K., Kent, D.G., Göttgens, B., 2016. A single-cell resolution map of mouse hematopoietic stem and progenitor cell differentiation. Blood 128, e20–e31. https://doi.org/10.1182/blood-2016-05-716480.
Neuenhahn, M., Kerksiek, K.M., Nauerth, M., Suhre, M.H., Schiemann, M., Gebhardt, F.E., Stemberger, C., Panthel, K., Schröder, S., Chakraborty, T., Jung, S., Hochrein, H., Rüssmann, H., Brocker, T., Busch, D.H., 2006. CD8alpha+ dendritic cells are required for efficient entry of Listeria monocytogenes into the spleen. Immunity 25, 619–630. https://doi.org/10.1016/j.immuni.2006.07.017.
Nurieva, R.I., Chung, Y., Martinez, G.J., Yang, X.O., Tanaka, S., Matskevitch, T.D., Wang, Y.-H., Dong, C., 2009. Bcl6 mediates the development of T follicular helper cells. Science 325, 1001–1005. https://doi.org/10.1126/science.1176676.
O'Keeffe, M., Hochrein, H., Vremec, D., Scott, B., Hertzog, P., Tatarczuch, L., Shortman, K., 2003. Dendritic cell precursor populations of mouse blood: identification of the murine homologues of human blood plasmacytoid pre-DC2 and CD11c+ DC1 precursors. Blood 101, 1453–1459. https://doi.org/10.1182/blood-2002-03-0974.
Ohtsuka, H., Sakamoto, A., Pan, J., Inage, S., Horigome, S., Ichii, H., Arima, M., Hatano, M., Okada, S., Tokuhisa, T., 2011. Bcl6 Is required for the development of mouse CD4+ and CD8α+ dendritic cells. J. Immunol. 186, 255–263. https://doi.org/10.4049/jimmunol.0903714.
Onai, N., Obata-Onai, A., Schmid, M.A., Ohteki, T., Jarrossay, D., Manz, M.G., 2007. Identification of clonogenic common Flt3+M-CSFR+ plasmacytoid and conventional dendritic cell progenitors in mouse bone marrow. Nat. Immunol. 8, 1207–1216. https://doi.org/10.1038/ni1518.
Onai, N., Kurabayashi, K., Hosoi-Amaike, M., Toyama-Sorimachi, N., Matsushima, K., Inaba, K., Ohteki, T., 2013. A clonogenic progenitor with prominent plasmacytoid dendritic cell developmental potential. Immunity 38, 943–957. https://doi.org/10.1016/j.immuni.2013.04.006.
Onodera, K., Fujiwara, T., Onishi, Y., Itoh-Nakadai, A., Okitsu, Y., Fukuhara, N., Ishizawa, K., Shimizu, R., Yamamoto, M., Harigae, H., 2016. GATA2 regulates dendritic cell differentiation. Blood 128, 508–518. https://doi.org/10.1182/blood-2016-02-698118.
Ozato, K., Tailor, P., Kubota, T., 2007. The interferon regulatory factor family in host defense: mechanism of action. J. Biol. Chem. 282, 20065–20069. https://doi.org/10.1074/jbc.R700003200.
Pack, M., Trumpfheller, C., Thomas, D., Park, C.G., Granelli-Piperno, A., Münz, C., Steinman, R.M., 2008. DEC-205/CD205+ dendritic cells are abundant in the white pulp of the human spleen, including the border region between the red and white pulp. Immunology 123, 438–446. https://doi.org/10.1111/j.1365-2567.2007.02710.x.

Paul, F., Arkin, Y., Giladi, A., Jaitin, D.A., Kenigsberg, E., Keren-Shaul, H., Winter, D., Lara-Astiaso, D., Gury, M., Weiner, A., David, E., Cohen, N., Lauridsen, F.K.B., Haas, S., Schlitzer, A., Mildner, A., Ginhoux, F., Jung, S., Trumpp, A., Porse, B.T., Tanay, A., Amit, I., 2016. Transcriptional heterogeneity and lineage commitment in myeloid progenitors. Cell 164, 325. https://doi.org/10.1016/j.cell.2015.12.046.

Perié, L., Duffy, K.R., Kok, L., de Boer, R.J., Schumacher, T.N., 2015. The branching point in erythro-myeloid differentiation. Cell 163, 1655–1662. https://doi.org/10.1016/j.cell.2015.11.059.

Persson, E.K., Uronen-Hansson, H., Semmrich, M., Rivollier, A., Hägerbrand, K., Marsal, J., Gudjonsson, S., Håkansson, U., Reizis, B., Kotarsky, K., Agace, W.W., 2013. IRF4 transcription-factor-dependent CD103+CD11b+ dendritic cells drive mucosal T helper 17 cell differentiation. Immunity 38, 958–969. https://doi.org/10.1016/j.immuni.2013.03.009.

Poulin, L.F., Henri, S., de Bovis, B., Devilard, E., Kissenpfennig, A., Malissen, B., 2007. The dermis contains langerin+ dendritic cells that develop and function independently of epidermal Langerhans cells. J. Exp. Med. 204, 3119–3131. https://doi.org/10.1084/jem.20071724.

Poulin, L.F., Salio, M., Griessinger, E., Anjos-Afonso, F., Craciun, L., Chen, J.-L., Keller, A.M., Joffre, O., Zelenay, S., Nye, E., Le Moine, A., Faure, F., Donckier, V., Sancho, D., Cerundolo, V., Bonnet, D., Reis e Sousa, C., 2010. Characterization of human DNGR-1 $^{+}$ BDCA3 $^{+}$ leukocytes as putative equivalents of mouse CD8α $^{+}$ dendritic cells. J. Exp. Med. 207, 1261–1271. https://doi.org/10.1084/jem.20092618.

Poulin, L.F., Reyal, Y., Uronen-Hansson, H., Schraml, B.U., Sancho, D., Murphy, K.M., Håkansson, U.K., Moita, L.F., Agace, W.W., Bonnet, D., Sousa, C.R.e., 2012. DNGR-1 is a specific and universal marker of mouse and human Batf3-dependent dendritic cells in lymphoid and nonlymphoid tissues. Blood 119, 6052–6062. https://doi.org/10.1182/blood-2012-01-406967.

Proietto, A.I., Mittag, D., Roberts, A.W., Sprigg, N., Wu, L., 2012. The equivalents of human blood and spleen dendritic cell subtypes can be generated in vitro from human CD34(+) stem cells in the presence of fms-like tyrosine kinase 3 ligand and thrombopoietin. Cell. Mol. Immunol. 9, 446–454. https://doi.org/10.1038/cmi.2012.48.

Pulendran, B., Lingappa, J., Kennedy, M.K., Smith, J., Teepe, M., Rudensky, A., Maliszewski, C.R., Maraskovsky, E., 1997. Developmental pathways of dendritic cells in vivo: distinct function, phenotype, and localization of dendritic cell subsets in FLT3 ligand-treated mice. J. Immunol. 159, 2222–2231.

Pulendran, B., Smith, J.L., Caspary, G., Brasel, K., Pettit, D., Maraskovsky, E., Maliszewski, C.R., 1999. Distinct dendritic cell subsets differentially regulate the class of immune response in vivo. Proc. Natl. Acad. Sci. U.S.A. 96, 1036–1041. https://doi.org/10.1073/pnas.96.3.1036.

Pulendran, B., Dillon, S., Joseph, C., Curiel, T., Banchereau, J., Mohamadzadeh, M., 2004. Dendritic cells generated in the presence of GM-CSF plus IL-15 prime potent CD8+ Tc1 responses in vivo. Eur. J. Immunol. 34, 66–73. https://doi.org/10.1002/eji.200324567.

Qiu, C.-H., Miyake, Y., Kaise, H., Kitamura, H., Ohara, O., Tanaka, M., 2009. Novel subset of CD8{alpha}+ dendritic cells localized in the marginal zone is responsible for tolerance to cell-associated antigens. J. Immunol. 182, 4127–4136. https://doi.org/10.4049/jimmunol.0803364.

Radtke, F., MacDonald, H.R., Tacchini-Cottier, F., 2013. Regulation of innate and adaptive immunity by Notch. Nat. Rev. Immunol. 13, 427–437. https://doi.org/10.1038/nri3445.

Rathinam, C., Geffers, R., Yücel, R., Buer, J., Welte, K., Möröy, T., Klein, C., 2005. The transcriptional repressor Gfi1 controls STAT3-dependent dendritic cell development and function. Immunity 22, 717–728. https://doi.org/10.1016/j.immuni.2005.04.007.

Reizis, B., 2019. Plasmacytoid dendritic cells: development, regulation, and function. Immunity 50, 37–50. https://doi.org/10.1016/j.immuni.2018.12.027.

Robbins, S.H., Walzer, T., Dembélé, D., Thibault, C., Defays, A., Bessou, G., Xu, H., Vivier, E., Sellars, M., Pierre, P., Sharp, F.R., Chan, S., Kastner, P., Dalod, M., 2008. Novel insights into the relationships between dendritic cell subsets in human and mouse revealed by genome-wide expression profiling. Genome Biol. 9, R17. https://doi.org/10.1186/gb-2008-9-1-r17.

Rodrigues, N.P., Janzen, V., Forkert, R., Dombkowski, D.M., Boyd, A.S., Orkin, S.H., Enver, T., Vyas, P., Scadden, D.T., 2005. Haploinsufficiency of GATA-2 perturbs adult hematopoietic stem-cell homeostasis. Blood 106, 477–484. https://doi.org/10.1182/blood-2004-08-2989.

Rodrigues, P.F., Alberti-Servera, L., Eremin, A., Grajales-Reyes, G.E., Ivanek, R., Tussiwand, R., 2018. Distinct progenitor lineages contribute to the heterogeneity of plasmacytoid dendritic cells. Nat. Immunol. 19, 711–722. https://doi.org/10.1038/s41590-018-0136-9.

Salem, S., Langlais, D., Lefebvre, F., Bourque, G., Bigley, V., Haniffa, M., Casanova, J.-L., Burk, D., Berghuis, A., Butler, K.M., Leahy, T.R., Hambleton, S., Gros, P., 2014. Functional characterization of the human dendritic cell immunodeficiency associated with the IRF8K108E mutation. Blood 124, 1894–1904. https://doi.org/10.1182/blood-2014-04-570879.

Sallusto, F., Lanzavecchia, A., 1994. Efficient presentation of soluble antigen by cultured human dendritic cells is maintained by granulocyte/macrophage colony-stimulating factor plus interleukin 4 and downregulated by tumor necrosis factor alpha. J. Exp. Med. 179, 1109–1118. https://doi.org/10.1084/jem.179.4.1109.

Salmon, H., Idoyaga, J., Rahman, A., Leboeuf, M., Remark, R., Jordan, S., Casanova-Acebes, M., Khudoynazarova, M., Agudo, J., Tung, N., Chakarov, S., Rivera, C., Hogstad, B., Bosenberg, M., Hashimoto, D., Gnjatic, S., Bhardwaj, N., Palucka, A.K., Brown, B.D., Brody, J., Ginhoux, F., Merad, M., 2016. Expansion and activation of CD103 + dendritic cell progenitors at the tumor site enhances tumor responses to therapeutic PD-L1 and BRAF inhibition. Immunity 44, 924–938. https://doi.org/10.1016/j.immuni.2016.03.012.

Samie, M., Cresswell, P., 2015. The transcription factor TFEB acts as a molecular switch that regulates exogenous antigen-presentation pathways. Nat. Immunol. 16, 729–736. https://doi.org/10.1038/ni.3196

Sancho, D., Joffre, O.P., Keller, A.M., Rogers, N.C., Martínez, D., Hernanz-Falcón, P., Rosewell, I., Reis e Sousa, C., 2009. Identification of a dendritic cell receptor that couples sensing of necrosis to immunity. Nature 458, 899–903. https://doi.org/10.1038/nature07750.

Sander, J., Schmidt, S.V., Cirovic, B., McGovern, N., Papantonopoulou, O., Hardt, A.-L., Aschenbrenner, A.C., Kreer, C., Quast, T., Xu, A.M., Schmidleithner, L.M., Theis, H., Thi Huong, L.D., Sumatoh, H.R.B., Lauterbach, M.A.R., Schulte-Schrepping, J., Günther, P., Xue, J., Baßler, K., Ulas, T., Klee, K., Katzmarski, N., Herresthal, S., Krebs, W., Martin, B., Latz, E., Händler, K., Kraut, M., Kolanus, W., Beyer, M., Falk, C.S., Wiegmann, B., Burgdorf, S., Melosh, N.A., Newell, E.W., Ginhoux, F., Schlitzer, A., Schultze, J.L., 2017. Cellular differentiation of human monocytes is regulated by time-dependent interleukin-4 signaling and the transcriptional regulator NCOR2. Immunity 47, 1051–1066.e12. https://doi.org/10.1016/j.immuni.2017.11.024.

Sathaliyawala, T., O'Gorman, W.E., Greter, M., Bogunovic, M., Konjufca, V., Hou, Z.E., Nolan, G.P., Miller, M.J., Merad, M., Reizis, B., 2010. Mammalian target of rapamycin controls dendritic cell development downstream of Flt3 ligand signaling. Immunity 33, 597–606. https://doi.org/10.1016/j.immuni.2010.09.012.

Sathe, P., Pooley, J., Vremec, D., Mintern, J., Jin, J.-O., Wu, L., Kwak, J.-Y., Villadangos, J.A., Shortman, K., 2011. The acquisition of antigen cross-presentation function by newly formed dendritic cells. J. Immunol. 186, 5184–5192. https://doi.org/10.4049/jimmunol.1002683.

Sathe, P., Metcalf, D., Vremec, D., Naik, S.H., Langdon, W.Y., Huntington, N.D., Wu, L., Shortman, K., 2014. Lymphoid tissue and plasmacytoid dendritic cells and macrophages do not share a common macrophage-dendritic cell-restricted progenitor. Immunity 41, 104–115. https://doi.org/10.1016/j.immuni.2014.05.020.

Satpathy, A.T., Kc, W., Albring, J.C., Edelson, B.T., Kretzer, N.M., Bhattacharya, D., Murphy, T.L., Murphy, K.M., 2012. Zbtb46 expression distinguishes classical dendritic cells and their committed progenitors from other immune lineages. J. Exp. Med. 209, 1135–1152. https://doi.org/10.1084/jem.20120030.

Satpathy, A.T., Briseño, C.G., Lee, J.S., Ng, D., Manieri, N.A., Kc, W., Wu, X., Thomas, S.R., Lee, W.-L., Turkoz, M., McDonald, K.G., Meredith, M.M., Song, C., Guidos, C.J., Newberry, R.D., Ouyang, W., Murphy, T.L., Stappenbeck, T.S., Gommerman, J.L., Nussenzweig, M.C., Colonna, M., Kopan, R., Murphy, K.M., 2013. Notch2-dependent classical dendritic cells orchestrate intestinal immunity to attaching-and-effacing bacterial pathogens. Nat. Immunol. 14, 937–948. https://doi.org/10.1038/ni.2679.

Satpathy, A.T., Bhattacharya, D., Speck, N.A., Egawa, T., 2014. Runx1 and Cbfβ regulate the development of Flt3+ dendritic cell progenitors and restrict myeloproliferative disorder. Blood 123, 11.

Savina, A., Peres, A., Cebrian, I., Carmo, N., Moita, C., Hacohen, N., Moita, L.F., Amigorena, S., 2009. The small GTPase Rac2 controls phagosomal alkalinization and antigen crosspresentation selectively in CD8(+) dendritic cells. Immunity 30, 544–555. https://doi.org/10.1016/j.immuni.2009.01.013.

Scheffler, J.M., Sparber, F., Tripp, C.H., Herrmann, C., Humenberger, A., Blitz, J., Romani, N., Stoitzner, P., Huber, L.A., 2014. LAMTOR2 regulates dendritic cell homeostasis through FLT3-dependent mTOR signalling. Nat. Commun. 5, 5138. https://doi.org/10.1038/ncomms6138.

Schiavoni, G., Mattei, F., Sestili, P., Borghi, P., Venditti, M., Morse, H.C., Belardelli, F., Gabriele, L., 2002. ICSBP is essential for the development of mouse type I interferon-producing cells and for the generation and activation of CD8α+ dendritic cells. J. Exp. Med. 196, 1415–1425. https://doi.org/10.1084/jem.20021263.

Schlenner, S.M., Madan, V., Busch, K., Tietz, A., Läufle, C., Costa, C., Blum, C., Fehling, H.J., Rodewald, H.-R., 2010. Fate mapping reveals separate origins of T cells and myeloid lineages in the thymus. Immunity 32, 426–436. https://doi.org/10.1016/j.immuni.2010.03.005.

Schlitzer, A., McGovern, N., Teo, P., Zelante, T., Atarashi, K., Low, D., Ho, A.W.S., See, P., Shin, A., Wasan, P.S., Hoeffel, G., Malleret, B., Heiseke, A., Chew, S., Jardine, L., Purvis, H.A., Hilkens, C.M.U., Tam, J., Poidinger, M., Stanley, E.R., Krug, A.B., Renia, L., Sivasankar, B., Ng, L.G., Collin, M., Ricciardi-Castagnoli, P., Honda, K., Haniffa, M., Ginhoux, F., 2013. IRF4 transcription factor-dependent CD11b+ dendritic cells in human and mouse control mucosal IL-17 cytokine responses. Immunity 38, 970–983. https://doi.org/10.1016/j.immuni.2013.04.011.

Schlitzer, A., Sivakamasundari, V., Chen, J., Sumatoh, H.R.B., Schreuder, J., Lum, J., Malleret, B., Zhang, S., Larbi, A., Zolezzi, F., Renia, L., Poidinger, M., Naik, S., Newell, E.W., Robson, P., Ginhoux, F., 2015. Identification of cDC1- and

cDC2-committed DC progenitors reveals early lineage priming at the common DC progenitor stage in the bone marrow. Nat. Immunol. 16, 718–728. https://doi.org/10.1038/ni.3200.
Schönheit, J., Kuhl, C., Gebhardt, M.L., Klett, F.F., Riemke, P., Scheller, M., Huang, G., Naumann, R., Leutz, A., Stocking, C., Priller, J., Andrade-Navarro, M.A., Rosenbauer, F., 2013. PU.1 level-directed chromatin structure remodeling at the Irf8 gene drives dendritic cell commitment. Cell Rep. 3, 1617–1628. https://doi.org/10.1016/j.celrep.2013.04.007.
Schotte, R., Schmidlin, H., Nagasawa, M., Dontje, W., Karrich, J.J., Uittenbogaart, C., Spits, H., Blom, B., 2010. Isolation and in vitro generation of gene-manipulated human plasmacytoid and conventional dendritic cells. Methods Mol. Biol. 595, 67–85. https://doi.org/10.1007/978-1-60761-421-0_5.
Schraml, B.U., van Blijswijk, J., Zelenay, S., Whitney, P.G., Filby, A., Acton, S.E., Rogers, N.C., Moncaut, N., Carvajal, J.J., Reis e Sousa, C., 2013. Genetic tracing via DNGR-1 expression history defines dendritic cells as a hematopoietic lineage. Cell 154, 843–858. https://doi.org/10.1016/j.cell.2013.07.014.
Scott, C.L., Soen, B., Martens, L., Skrypek, N., Saelens, W., Taminau, J., Blancke, G., Van Isterdael, G., Huylebroeck, D., Haigh, J., Saeys, Y., Guilliams, M., Lambrecht, B.N., Berx, G., 2016. The transcription factor Zeb2 regulates development of conventional and plasmacytoid DCs by repressing Id2. J. Exp. Med. 213, 897–911. https://doi.org/10.1084/jem.20151715.
See, P., Dutertre, C.-A., Chen, J., Günther, P., McGovern, N., Irac, S.E., Gunawan, M., Beyer, M., Händler, K., Duan, K., Sumatoh, H.R.B., Ruffin, N., Jouve, M., Gea-Mallorquí, E., Hennekam, R.C.M., Lim, T., Yip, C.C., Wen, M., Malleret, B., Low, I., Shadan, N.B., Fen, C.F.S., Tay, A., Lum, J., Zolezzi, F., Larbi, A., Poidinger, M., Chan, J.K.Y., Chen, Q., Rénia, L., Haniffa, M., Benaroch, P., Schlitzer, A., Schultze, J.L., Newell, E.W., Ginhoux, F., 2017. Mapping the human DC lineage through the integration of high-dimensional techniques. Science 356, eaag3009. https://doi.org/10.1126/science.aag3009.
Segura, E., Touzot, M., Bohineust, A., Cappuccio, A., Chiocchia, G., Hosmalin, A., Dalod, M., Soumelis, V., Amigorena, S., 2013. Human inflammatory dendritic cells induce Th17 cell differentiation. Immunity 38, 336–348. https://doi.org/10.1016/j.immuni.2012.10.018.
Seillet, C., Jackson, J.T., Markey, K.A., Brady, H.J.M., Hill, G.R., MacDonald, K.P.A., Nutt, S.L., Belz, G.T., 2013. CD8 + DCs can be induced in the absence of transcription factors Id2, Nfil3, and Batf3. Blood 121, 1574–1583. https://doi.org/10.1182/blood-2012-07-445650.
Serbina, N.V., Salazar-Mather, T.P., Biron, C.A., Kuziel, W.A., Pamer, E.G., 2003. TNF/iNOS-producing dendritic cells mediate innate immune defense against bacterial infection. Immunity 19, 59–70.
Shibata, Y., Berclaz, P.Y., Chroneos, Z.C., Yoshida, M., Whitsett, J.A., Trapnell, B.C., 2001. GM-CSF regulates alveolar macrophage differentiation and innate immunity in the lung through PU.1. Immunity 15, 557–567.
Sichien, D., Scott, C.L., Martens, L., Vanderkerken, M., Van Gassen, S., Plantinga, M., Joeris, T., De Prijck, S., Vanhoutte, L., Vanheerswynghels, M., Van Isterdael, G., Toussaint, W., Madeira, F.B., Vergote, K., Agace, W.W., Clausen, B.E., Hammad, H., Dalod, M., Saeys, Y., Lambrecht, B.N., Guilliams, M., 2016. IRF8 transcription factor controls survival and function of terminally differentiated conventional and plasmacytoid dendritic cells, respectively. Immunity 45, 626–640. https://doi.org/10.1016/j.immuni.2016.08.013.
Silvin, A., Yu, C.I., Lahaye, X., Imperatore, F., Brault, J.-B., Cardinaud, S., Becker, C., Kwan, W.-H., Conrad, C., Maurin, M., Goudot, C., Marques-Ladeira, S.,

Wang, Y., Pascual, V., Anguiano, E., Albrecht, R.A., Iannacone, M., García-Sastre, A., Goud, B., Dalod, M., Moris, A., Merad, M., Palucka, A.K., Manel, N., 2017. Constitutive resistance to viral infection in human CD141+ dendritic cells. Sci. Immunol. 2, 1–12. https://doi.org/10.1126/sciimmunol.aai8071.

Spranger, S., Bao, R., Gajewski, T.F., 2015. Melanoma-intrinsic β-catenin signalling prevents anti-tumour immunity. Nature 523, 231–235. https://doi.org/10.1038/nature14404.

Spranger, S., Dai, D., Horton, B., Gajewski, T.F., 2017. Tumor-residing Batf3 dendritic cells are required for effector T cell trafficking and adoptive T cell therapy. Cancer Cell 31, 711–723.e4. https://doi.org/10.1016/j.ccell.2017.04.003.

Steinman, R.M., 1973. Identification of a novel cell type in peripheral lymphoid organs of mice: I. morphology, quantitation, tissue distribution. J. Exp. Med. 137, 1142–1162. https://doi.org/10.1084/jem.137.5.1142.

Steinman, R.M., Cohn, Z.A., 1974. Identification of a novel cell type in peripheral lymphoid organs of mice: II. Functional properties in vitro. J. Exp. Med. 139, 380–397. https://doi.org/10.1084/jem.139.2.380.

Steinman, R.M., Witmer, M.D., 1978. Lymphoid dendritic cells are potent stimulators of the primary mixed leukocyte reaction in mice. Proc. Natl. Acad. Sci. U.S.A. 75, 5132–5136.

Steinman, R.M., Lustig, D.S., Cohn, Z.A., 1974. Identification of a novel cell type in peripheral lymphoid organs of mice: III. Functional properties in vivo. J. Exp. Med. 139, 1431–1445. https://doi.org/10.1084/jem.139.6.1431.

Steinman, R.M., Adams, J.C., Cohn, Z.A., 1975. Identification of a novel cell type in peripheral lymphoid organs of mice. IV. Identification and distribution in mouse spleen. J. Exp. Med. 141, 804–820.

Suzuki, S., Honma, K., Matsuyama, T., Suzuki, K., Toriyama, K., Akitoyo, I., Yamamoto, K., Suematsu, T., Nakamura, M., Yui, K., Kumatori, A., 2004. Critical roles of interferon regulatory factor 4 in CD11bhighCD8alpha- dendritic cell development. Proc. Natl. Acad. Sci. U.S.A. 101, 8981–8986. https://doi.org/10.1073/pnas.0402139101.

Tailor, P., Tamura, T., Morse, H.C., Ozato, K., 2008. The BXH2 mutation in IRF8 differentially impairs dendritic cell subset development in the mouse. Blood 111, 1942–1945. https://doi.org/10.1182/blood-2007-07-100750.

Tamura, T., Nagamura-Inoue, T., Shmeltzer, Z., Kuwata, T., Ozato, K., 2000. ICSBP directs bipotential myeloid progenitor cells to differentiate into mature macrophages. Immunity 13, 155–165. https://doi.org/10.1016/S1074-7613(00)00016-9.

Tamura, T., Tailor, P., Yamaoka, K., Kong, H.J., Tsujimura, H., O'Shea, J.J., Singh, H., Ozato, K., 2005. IFN regulatory factor-4 and -8 govern dendritic cell subset development and their functional diversity. J. Immunol. 174, 2573–2581.

Theisen, D.J., Davidson, J.T., Briseño, C.G., Gargaro, M., Lauron, E.J., Wang, Q., Desai, P., Durai, V., Bagadia, P., Brickner, J.R., Beatty, W.L., Virgin, H.W., Gillanders, W.E., Mosammaparast, N., Diamond, M.S., Sibley, L.D., Yokoyama, W., Schreiber, R.D., Murphy, T.L., Murphy, K.M., 2018. WDFY4 is required for cross-presentation in response to viral and tumor antigens. Science 362, 694–699. https://doi.org/10.1126/science.aat5030.

Traggiai, E., Chicha, L., Mazzucchelli, L., Bronz, L., Piffaretti, J.-C., Lanzavecchia, A., Manz, M.G., 2004. Development of a human adaptive immune system in cord blood cell-transplanted mice. Science 304, 104–107. https://doi.org/10.1126/science.1093933.

Traver, D., Akashi, K., Manz, M., Merad, M., Miyamoto, T., Engleman, E.G., Weissman, I.L., 2000. Development of CD8alpha-positive dendritic cells from a common myeloid progenitor. Science 290, 2152–2154.

Tussiwand, R., Lee, W.-L., Murphy, T.L., Mashayekhi, M., Kc, W., Albring, J.C., Satpathy, A.T., Rotondo, J.A., Edelson, B.T., Kretzer, N.M., Wu, X., Weiss, L.A., Glasmacher, E., Li, P., Liao, W., Behnke, M., Lam, S.S.K., Aurthur, C.T.,

Leonard, W.J., Singh, H., Stallings, C.L., Sibley, L.D., Schreiber, R.D., Murphy, K.M., 2012. Compensatory dendritic cell development mediated by BATF-IRF interactions. Nature 490, 502–507. https://doi.org/10.1038/nature11531.
Tussiwand, R., Everts, B., Grajales-Reyes, G.E., Kretzer, N.M., Iwata, A., Bagaitkar, J., Wu, X., Wong, R., Anderson, D.A., Murphy, T.L., Pearce, E.J., Murphy, K.M., 2015. Klf4 expression in conventional dendritic cells is required for T helper 2 cell responses. Immunity 42, 916–928. https://doi.org/10.1016/j.immuni.2015.04.017.
Vander Lugt, B., Khan, A.A., Hackney, J.A., Agrawal, S., Lesch, J., Zhou, M., Lee, W.P., Park, S., Xu, M., DeVoss, J., Spooner, C.J., Chalouni, C., Delamarre, L., Mellman, I., Singh, H., 2014. Transcriptional programming of dendritic cells for enhanced MHC class II antigen presentation. Nat. Immunol. 15, 161–167. https://doi.org/10.1038/ni.2795.
Velten, L., Haas, S.F., Raffel, S., Blaszkiewicz, S., Islam, S., Hennig, B.P., Hirche, C., Lutz, C., Buss, E.C., Nowak, D., Boch, T., Hofmann, W.-K., Ho, A.D., Huber, W., Trumpp, A., Essers, M.A.G., Steinmetz, L.M., 2017. Human haematopoietic stem cell lineage commitment is a continuous process. Nat. Cell Biol. 19, 271–281. https://doi.org/10.1038/ncb3493.
Villani, A.-C., Satija, R., Reynolds, G., Sarkizova, S., Shekhar, K., Fletcher, J., Griesbeck, M., Butler, A., Zheng, S., Lazo, S., Jardine, L., Dixon, D., Stephenson, E., Nilsson, E., Grundberg, I., McDonald, D., Filby, A., Li, W., De Jager, P.L., Rozenblatt-Rosen, O., Lane, A.A., Haniffa, M., Regev, A., Hacohen, N., 2017. Single-cell RNA-seq reveals new types of human blood dendritic cells, monocytes, and progenitors. Science 356, eaah4573. https://doi.org/10.1126/science.aah4573.
Vremec, D., Lieschke, G.J., Dunn, A.R., Robb, L., Metcalf, D., Shortman, K., 1997. The influence of granulocyte/macrophage colony-stimulating factor on dendritic cell levels in mouse lymphoid organs. Eur. J. Immunol. 27, 40–44. https://doi.org/10.1002/eji.1830270107.
Vremec, D., Pooley, J., Hochrein, H., Wu, L., Shortman, K., 2000. CD4 and CD8 expression by dendritic cell subtypes in mouse thymus and spleen. J. Immunol. 164, 2978–2986.
Wang, H., Yan, M., Sun, J., Jain, S., Yoshimi, R., Abolfath, S.M., Ozato, K., Coleman, W.G., Ng, A.P., Metcalf, D., DiRago, L., Nutt, S.L., Morse, H.C., 2014a. A reporter mouse reveals lineage-specific and heterogeneous expression of IRF8 during lymphoid and myeloid cell differentiation. J. Immunol. 193, 1766–1777. https://doi.org/10.4049/jimmunol.1301939.
Wang, P., Xue, Y., Han, Y., Lin, L., Wu, C., Xu, S., Jiang, Z., Xu, J., Liu, Q., Cao, X., 2014b. The STAT3-binding long noncoding RNA lnc-DC controls human dendritic cell differentiation. Science 344, 310–313. https://doi.org/10.1126/science.1251456.
Waskow, C., Liu, K., Darrasse-Jèze, G., Guermonprez, P., Ginhoux, F., Merad, M., Shengelia, T., Yao, K., Nussenzweig, M., 2008. The receptor tyrosine kinase Flt3 is required for dendritic cell development in peripheral lymphoid tissues. Nat. Immunol. 9, 676–683. https://doi.org/10.1038/ni.1615.
Watchmaker, P.B., Lahl, K., Lee, M., Baumjohann, D., Morton, J., Kim, S.J., Zeng, R., Dent, A., Ansel, K.M., Diamond, B., Hadeiba, H., Butcher, E.C., 2014. Comparative transcriptional and functional profiling defines conserved programs of intestinal DC differentiation in humans and mice. Nat. Immunol. 15, 98–108. https://doi.org/10.1038/ni.2768.
Wu, L., Nichogiannopoulou, A., Shortman, K., Georgopoulos, K., 1997. Cell-autonomous defects in dendritic cell populations of ikaros mutant mice point to a developmental relationship with the lymphoid lineage. Immunity 7, 483–492. https://doi.org/10.1016/S1074-7613(00)80370-2.

Wu, L., D'Amico, A., Winkel, K.D., Suter, M., Lo, D., Shortman, K., 1998. RelB is essential for the development of myeloid-related CD8α− dendritic cells but not of lymphoid-related CD8α+ dendritic cells. Immunity 9, 839–847. https://doi.org/10.1016/S1074-7613(00)80649-4.

Wu, X., Briseño, C.G., Grajales-Reyes, G.E., Haldar, M., Iwata, A., Kretzer, N.M., Kc, W., Tussiwand, R., Higashi, Y., Murphy, T.L., Murphy, K.M., 2016. Transcription factor Zeb2 regulates commitment to plasmacytoid dendritic cell and monocyte fate. Proc. Natl. Acad. Sci. U.S.A. 113, 14775–14780. https://doi.org/10.1073/pnas.1611408114.

Xu, W., Domingues, R.G., Fonseca-Pereira, D., Ferreira, M., Ribeiro, H., Lopez-Lastra, S., Motomura, Y., Moreira-Santos, L., Bihl, F., Braud, V., Kee, B., Brady, H., Coles, M.C., Vosshenrich, C., Kubo, M., Di Santo, J.P., Veiga-Fernandes, H., 2015. NFIL3 orchestrates the emergence of common helper innate lymphoid cell precursors. Cell Rep. 10, 2043–2054. https://doi.org/10.1016/j.celrep.2015.02.057.

Yamazaki, C., Sugiyama, M., Ohta, T., Hemmi, H., Hamada, E., Sasaki, I., Fukuda, Y., Yano, T., Nobuoka, M., Hirashima, T., Iizuka, A., Sato, K., Tanaka, T., Hoshino, K., Kaisho, T., 2013. Critical roles of a dendritic cell subset expressing a chemokine receptor, XCR1. J. Immunol. 190, 6071–6082. https://doi.org/10.4049/jimmunol.1202798.

Yarovinsky, F., Zhang, D., Andersen, J.F., Bannenberg, G.L., Serhan, C.N., Hayden, M.S., Hieny, S., Sutterwala, F.S., Flavell, R.A., Ghosh, S., Sher, A., 2005. TLR11 activation of dendritic cells by a protozoan profilin-like protein. Science 308, 1626–1629. https://doi.org/10.1126/science.1109893.

Yoshida, T., Ng, S.Y.-M., Zuniga-Pflucker, J.C., Georgopoulos, K., 2006. Early hematopoietic lineage restrictions directed by Ikaros. Nat. Immunol. 7, 382–391. https://doi.org/10.1038/ni1314.

Yoshio, S., Kanto, T., Kuroda, S., Matsubara, T., Higashitani, K., Kakita, N., Ishida, H., Hiramatsu, N., Nagano, H., Sugiyama, M., Murata, K., Fukuhara, T., Matsuura, Y., Hayashi, N., Mizokami, M., Takehara, T., 2013. Human blood dendritic cell antigen 3 (BDCA3)(+) dendritic cells are a potent producer of interferon-λ in response to hepatitis C virus. Hepatology 57, 1705–1715. https://doi.org/10.1002/hep.26182.

Zhang, T., Liu, D., Calabro, S., Eisenbarth, S.C., Cattoretti, G., Haberman, A.M., 2014. Dynamic expression of BCL6 in murine conventional dendritic cells during in vivo development and activation. PLoS One 9, e101208. https://doi.org/10.1371/journal.pone.0101208.

Further reading

Igyártó, B.Z., Haley, K., Ortner, D., Bobr, A., Gerami-Nejad, M., Edelson, B.T., Zurawski, S.M., Malissen, B., Zurawski, G., Berman, J., Kaplan, D.H., 2011. Skin-resident murine dendritic cell subsets promote distinct and opposing antigen-specific T helper cell responses. Immunity 35, 260–272. https://doi.org/10.1016/j.immuni.2011.06.005.

Theisen, D.J., Ferris, S.T., Briseño, C.G., Kretzer, N., Iwata, A., Murphy, K.M., Murphy, T.L., 2019. BATF3-dependent genes control tumor rejection induced by dendritic cells independently of cross-presentation. Cancer Immunol. Res. 7, 29–39. canimm.0138.2018. https://doi.org/10.1158/2326-6066.CIR-18-0138.

CHAPTER TWO

Transcriptional control of dendritic cell development and functions

Lukas Amon, Christian H.K. Lehmann, Anna Baranska†, Janina Schoen, Lukas Heger, Diana Dudziak*
Laboratory of Dendritic Cell Biology, Department of Dermatology, University Hospital Erlangen, Friedrich-Alexander University of Erlangen-Nürnberg, Erlangen, Germany
*Corresponding author: e-mail address: diana.dudziak@uk-erlangen.de

Contents

1. Introduction 56
2. The unique functions of DCs 58
3. Overview about antigen uptake, processing and presentation by DCs 59
   3.1 Classical MHC I presentation in comparison to cross-presentation 60
   3.2 Classical MHC II presentation in comparison to autophagy 61
4. Phenotypical characterization of cDCs and pDCs 62
5. Identification and functional specialization of the cDC1 subset 63
6. Identification and functional specialization of the cDC2 subset 65
7. cDC1 and cDC2 and the induction and maintenance of tolerance 68
8. Identification and functional specialization of pDCs 69
9. Origin, identification and functional specialization of Langerhans cells 77
10. Identification and functional specialization of MoDCs 79
11. Phenotypical characterization, functional specialization, and presence of DC subsets in human tissues 80
12. The development of early DC progenitors within the bone marrow 82
13. Crucial factors for DC development: Flt3 and Flt3L, GM-CSF and GM-CSFR 86
14. Transcriptional regulation of DC differentiation and function 87
   14.1 Crucial regulatory transcription factors of multiple DC subsets 89
   14.2 Specific factors regulating transcription in cDC1 94
   14.3 Specific factors for the transcriptional regulation of cDC2 98
   14.4 Specific factors in transcriptional regulation of pDCs 103
15. Impact of post-transcriptional regulation by microRNAs 104
   15.1 Systems to investigate miRNA function in specific cell types 105
   15.2 Important functions of miRNAs in murine BMDCs and human MoDCs 106

† Current address: Miltenyi Biotec B.V. & Co. KG, Friedrich-Ebert-Straße 68, 51429 Bergisch Gladbach, Germany.

International Review of Cell and Molecular Biology, Volume 349
ISSN 1937-6448
https://doi.org/10.1016/bs.ircmb.2019.10.001

15.3 miRNAs involved in murine and human primary DC development 109
15.4 miRNAs regulating DC function 110
16. Epigenetic regulation of DCs 113
16.1 Organization of the epigenome 113
16.2 The epigenetic code of DCs 114
17. Conclusions 115
Acknowledgments 115
Statement 115
References 116

Abstract

Dendritic cells (DCs) are major regulators of adaptive immunity, as they are not only capable to induce efficient immune responses, but are also crucial to maintain peripheral tolerance and thereby inhibit autoimmune reactions. DCs bridge the innate and the adaptive immune system by presenting peptides of self and foreign antigens as peptide MHC complexes to T cells. These properties render DCs as interesting target cells for immunomodulatory therapies in cancer, but also autoimmune diseases. Several subsets of DCs with special properties and functions have been described. Recent achievements in understanding transcriptional programs on single cell level, together with the generation of new murine models targeting specific DC subsets, advanced our current understanding of DC development and function. Thus, DCs arise from precursor cells in the bone marrow with distinct progenitor cell populations splitting the monocyte populations and macrophage populations from the DC lineage, which upon lineage commitment can be separated into conventional cDC1, cDC2, and plasmacytoid DCs (pDCs). The DC populations harbor intrinsic programs enabling them to react for specific pathogens in dependency on the DC subset, and thereby orchestrate T cell immune responses. Similarities, but also varieties, between human and murine DC subpopulations are challenging, and will require further investigation of human specimens under consideration of the influence of the tissue micromilieu and DC subset localization in the future.

1. Introduction

Dendritic cells are key regulators of the immune system. This is due to their ability to efficiently prime and activate T cell immune responses in the defense against invading pathogens, but also due to their ability to maintain peripheral tolerance in the steady state (Mellman, 2013; Steinman and Banchereau, 2007). The dual role of DCs in this delicate balance within the immune system renders them an interesting target cell type for the development of new immunotherapeutic approaches (Bol et al., 2016; Lehmann et al., 2016; Palucka et al., 2010; Steinman and Banchereau, 2007; Tacken et al., 2007). DCs can be separated into different DC subsets using phenotypical and transcriptional markers. In order to harness the potential of DCs, many efforts have been undertaken to identify distinct DC subpopulations in lymphoid and non-lymphoid tissues in mouse and man and their respective functions in immune responses.

Today, we distinguish conventional DCs (cDCs) and plasmacytoid DCs (pDCs), but also account monocyte-derived DCs (MoDCs) and Langerhans cells (LCs) as professional antigen presenting cells (Guilliams et al., 2014; Kaplan, 2017; Kashem et al., 2017; Merad et al., 2008, 2013; Murphy et al., 2016; Segura and Amigorena, 2013). All of these cell types display a distinct expression profile of pattern recognition receptors (PRRs), such as Toll like receptors (TLRs) and C-type lectin receptors (CLRs) (Akira et al., 2006; Figdor et al., 2002; Hoving et al., 2014; Kaufmann, 2007; Medzhitov, 2001; Mildner and Jung, 2014; Monteiro and Lepenies, 2017; Van Dinther et al., 2017; Vijay, 2018). Additionally, they demonstrate unique functions in the initiation, modulation, and dampening of T cell immune responses and other immunological defense mechanisms in the control of pathogens, cancer and autoimmunity (Banchereau and Steinman, 1998; Coutant and Miossec, 2016; Mellman, 2013; Mildner and Jung, 2014; Ohnmacht et al., 2009; Palucka and Banchereau, 2012; Saxena and Bhardwaj, 2018; Serra and Santamaria, 2019). Beside others, especially DC researchers advanced the development of single-cell RNA sequencing and computational based methods for the understanding of transcriptional profiles on single cell level of these rare cell populations (Alcantara-Hernandez et al., 2017; Guilliams et al., 2016; Jaitin et al., 2014; Schaum et al., 2018; See et al., 2017; Villani et al., 2017). Moreover, the characterization of DCs on a cellular level by flow cytometry, immunofluorescence, and mass-spectrometry based technologies on isolated cells or cells in the tissues further improved the investigation of scarce cell populations (Alcantara-Hernandez et al., 2017; Becher et al., 2014; Cabeza-Cabrerizo et al., 2019; Gerner et al., 2012, 2017; Granot et al., 2017; Heidkamp et al., 2016a; See et al., 2017; Villani et al., 2017). Together with the generation of new mouse models targeting transcriptional regulators of DCs and their progenitors (see this review, below), these new methodologies led to a strong increase in understanding of DC development from progenitor cells to mature antigen-presenting DCs and their respective functions.

In this review, we will first focus on the current state-of-the-art literature about the rather non-redundant functional role of the different DC subpopulations in the immune system. Second, we will summarize recent findings on transcriptional regulators, including miRNA-based and epigenetic regulators, responsible for the development of pluripotent hematopoietic progenitor cells into a DC progenitor population and finally into the differentiated and committed DC subsets in mice and men. Our focus will be on cDCs and pDCs, and we will only briefly discuss identification, function and transcriptional control of LCs and MoDCs.

2. The unique functions of DCs

Ralph Steinman and Zanvil Cohn were the first researchers to describe a phagocytic and dendrite-like pseudopod-possessing cell as dendritic cell (DC) (Steinman and Cohn, 1973). Further pioneer work of Ralph Steinman elaborated the functional role of DCs as potent stimulators of T cell immune responses (Banchereau and Steinman, 1998; Steinman and Witmer, 1978). Since then, groundbreaking research on DC development, DC migration, and immune cell activation has been published.

DCs play a pivotal role in innate and adaptive immune responses. The contact with immunostimulatory or pathogenic agents (i.e., viruses, bacteria) induces a variety of changes within the DCs, a process called DC maturation. These changes include: (i) the upregulation of costimulatory and activating molecules (e.g., CD40, CD80, CD86), (ii) the upregulation of chemokine receptors (e.g., CCR7), (iii) the secretion of cytokines (e.g., IL6, IL12, TNFα), (iv) the processing of engulfed antigens, and (v) finally the presentation of antigens as peptide MHC complexes (pMHC) on the DC surface (Banchereau and Steinman, 1998; Gottschalk et al., 2015; Merad et al., 2013). During these processes, the DCs are losing their cell-cell contacts in the periphery, where they have encountered the pathogen (e.g., in the skin), allowing for their faster migration through afferent lymphatics into the T cell zones of secondary lymphatic tissues, primarily regional lymph nodes (Heidkamp et al., 2016a; Jiang et al., 2007). Here, the DC is staying within the T cell zone, where it meets antigen-specific T cells. This interaction is leading to T cell activation, proliferation and differentiation before the effector T cells finally leave the lymph node to fight the pathogen in the periphery (Delamarre et al., 2003; Heidkamp et al., 2016a; Lindquist et al., 2004; Mellman, 2013; Shakhar et al., 2005).

Besides their function in inducing and regulating potent immune responses against invading pathogens, DCs are playing an important role in the maintenance of tolerance. Pioneer work of Nussenzweig and Steinman demonstrated that antigen-loaded DCs are indeed capable to prime

T cells in the periphery even if strong activation signals by the DCs are missing. T cells reacting to antigens that are presented in a non-inflammatory situation either become anergic or are deleted. The antigen presentation by DCs in the steady-state is therefore an important checkpoint in the periphery to circumvent autoimmune responses against self-antigens (e.g., from acquired material of dying cells) (Dudziak et al., 2007; Hawiger et al., 2001; Kretschmer et al., 2005; Steinman and Nussenzweig, 2002). The exact mechanisms of maintaining peripheral self-tolerance are not fully understood. There are hints that DCs in the periphery can not only expand antigen-specific regulatory T cells (T_{reg} cells), but can also trigger the differentiation of T cells from the naïve $CD4^+$ T cell pool into T_{reg} cells (Coombes et al., 2007; Sun et al., 2007; Yamazaki et al., 2008). In addition to TGF-β and/or retinoic acid secreted by DCs, various other factors contribute to the induction of peripheral T cell tolerance. These other factors are influencing the activation state of DCs or T cells either by (i) the expression of negative co-receptors, such as PD-1 and CTLA4 on the T cell side and their counterparts PD-L1/PD-L2 or CD80/CD86 on the DC side; by (ii) specific metabolic activities, such as indoleamine 2,3-dioxygenase (IDO); or by (iii) the tissue microenvironment, such as mesenchymal cells in the placenta (Coutant and Miossec, 2016; De Sousa Linhares et al., 2018; Fallarino et al., 2003; Guram et al., 2019; Heidkamp et al., 2016a; McGovern et al., 2017; Mellor and Munn, 2004; Xing and Hogquist, 2012).

3. Overview about antigen uptake, processing and presentation by DCs

The process of antigen uptake, degradation and loading of peptides onto MHC class I and II molecules is called antigen presentation and necessary for T cell stimulation. Here, we will only touch on the most prominent regulatory processes of antigen processing and presentation. A more detailed description can be found in the cited articles. In general, antigens that are synthesized intracellularly (e.g., viral or tumor peptides) are presented on MHC class I molecules to trigger cytotoxic $CD8^+$ T cell responses, whereas peptides of engulfed and processed extracellular antigens are presented on MHC class II molecules to activate $CD4^+$ helper T cells (Guermonprez et al., 2002; Heidkamp et al., 2016a; Villadangos et al., 2001; Villadangos and Schnorrer, 2007). Beside these functions, DCs are capable of presenting external antigens on MHC class I via a process called cross-presentation (Den Haan et al., 2000; Desch et al., 2014;

Dudziak et al., 2007; Gutierrez-Martinez et al., 2015; Heath et al., 2004; Joffre et al., 2012; Kurts et al., 2010; Montealegre and Van Endert, 2019; Segura and Villadangos, 2011; Van Endert, 2016) or cytoplasmic antigens on MHC class II via autophagy (Nimmerjahn et al., 2003; Schmid and Munz, 2007).

Antigens are acquired by DCs via phagocytosis, micropinocytosis or receptor-mediated endocytosis (Guermonprez et al., 2002; Roche and Furuta, 2015). Following antigen recognition, DCs stop sampling their surrounding and undergo maturation. This process includes upregulation of co-stimulatory molecules and the MHC machinery as well as the migration toward secondary lymphoid organs directed by a new chemokine receptor profile (Banchereau and Steinman, 1998; Benvenuti et al., 2004; Eisenbarth, 2019; Guermonprez et al., 2002). There, antigens are presented to naïve T cells.

3.1 Classical MHC I presentation in comparison to cross-presentation

The classical MHC I pathway is designed to present foreign (mostly viral) and self-antigens (mostly defective ribosomal products (DRiPs)) located in the cytoplasm (Kloetzel, 2001; Yewdell et al., 1996, 2003). Following proteasomal degradation of ubiquitinated proteins to peptides with a length of 3 to 22 amino acids (AA) (Kisselev et al., 1999), peptides of 8 to 16 AA are then translocated into the endoplasmatic reticulum (ER). This process mediated by the ATP-dependent transporter associated with antigen processing (TAP) is leading to peptide loading on MHC I molecules associated to TAP via different chaperone proteins (Blum et al., 2013; Neefjes et al., 2011). The chaperones facilitate MHC I assembly and stabilize MHC I after its ribosomal translation within the ER membrane (Blum et al., 2013; Kloetzel, 2001; Neefjes et al., 2011). Peptides too long for the MHC I binding cleft are further processed by the endoplasmatic reticulum aminopeptidase associated with antigen processing, short ERAAP, to a length of 8–10 AA (Saric et al., 2002; Serwold et al., 2002; Van Endert, 2011). Subsequently, proper peptide binding enables the complete assembly and transport of the MHC I:peptide (pMHCI) complex to the cell surface allowing for $CD8^+$ T cell priming (Blum et al., 2013; Neefjes et al., 2011). MHC I expression is found on nearly all nucleated cells, thereby ensuring that these cells are able to signal viral infection (Riedhammer and Weissert, 2015).

In addition to the classical MHC I presentation pathway, DCs can cross-present extracellular antigens on MHC I molecules (Bevan, 1976; Den Haan et al., 2000; Dudziak et al., 2007; Heidkamp et al., 2016a).

During evolution, this mechanism may have developed to ensure cytotoxic CD8$^+$ T cell (CTL) responses against viruses even if they do not infect DCs themselves (Rock et al., 2016). Furthermore, the cross-priming function of cDC1 appears particularly important for the initiation of productive anti-tumor CD8$^+$ T cell responses (Fuertes et al., 2011; Hildner et al., 2008; Neubert et al., 2014; Roberts et al., 2016; Salmon et al., 2016; Spranger et al., 2017). How acquired extracellular antigens are directed from the endosomal compartments into the cytoplasm for proteasomal degradation is still a matter of debate (Gros and Amigorena, 2019; Vyas et al., 2008). However, it is suggested that the ER translocation protein Sec61 is recruited to endosomes enabling the export of proteins into the cytosol (Embgenbroich and Burgdorf, 2018; Zehner et al., 2015). Furthermore, the SNARE protein Sec22b is thought to be important for the recruitment of ER proteins including Sec61 to endosomes (Alloatti et al., 2017; Cebrian et al., 2011; Embgenbroich and Burgdorf, 2018). Notably, the indispensability of Sec22b for cross-presentation is still a matter of debate (Alloatti et al., 2017; Cebrian et al., 2011; Embgenbroich and Burgdorf, 2018; Wu et al., 2017). Additionally, it was observed that phagosomes carrying microbial products, which induce TLR signaling, demonstrate a specific accumulation of MHC I molecules (Nair-Gupta et al., 2014).

3.2 Classical MHC II presentation in comparison to autophagy

In contrast to MHC I presentation, MHC II presentation is an intrinsic feature of antigen presenting cells (APCs) designed to present engulfed antigens to CD4$^+$ T cells (Blum et al., 2013; Neefjes et al., 2011; Roche and Furuta, 2015). Uptaken antigens are transported in phagosomes via the endosomal compartment to fuse with lysosomes. Thereby, phagolysosomes are generated, in which a pH and enzyme dependent degradation of proteins occurs (Blum et al., 2013; Neefjes et al., 2011; Roche and Furuta, 2015). While the closed ends of the MHC I binding cleft selectively enables binding of peptide fragments with 8–10 AA in length, the open ended MHC II binding cleft allows for more flexibility concerning the peptide length (Mohan and Unanue, 2012; Neefjes et al., 2011; Roche and Furuta, 2015). To prevent self-antigen presentation, the MHC II binding cleft is blocked by the invariant chain li (Neefjes et al., 2011; Roche and Furuta, 2015; Villadangos et al., 1997). Following MHC II synthesis into the ER membrane, MHC II:invariant chain complexes are transported via the

Golgi network into MHC class II compartments (MIIC) (Neefjes et al., 2011). In the MIIC, processed peptides meet nascent MHC II molecules and following several cleaving events only the class II-associated invariant chain peptide (CLIP) is left blocking the MHC II binding groove (Garstka and Neefjes, 2013; Neefjes et al., 2011; Roche and Furuta, 2015). This CLIP fragment is then exchanged by the human (h) HLA-DM or murine (m) H2-M chaperone with an antigenic peptide (Denzin, 2013; Denzin et al., 1996; Garstka and Neefjes, 2013; Neefjes et al., 2011; Roche and Furuta, 2015). To prevent overloading with non-pathogenic peptides during steady-state conditions, hHLA-DM/mH2-M activity is controlled by its negative regulator hHLA-DO/mH2-O (Roche and Furuta, 2015; Walter et al., 2000) by a yet unknown mechanism. Evolutionary, this multistep regulation may be important to avoid systemic autoantibody responses and autoimmune disorders since a tight regulation of the MHC II presentation is critical for maintaining $CD4^+$ T cell tolerance (Riedhammer and Weissert, 2015). Additionally, these enzymes conduct peptide editing, thereby ensuring that only high affinity peptide MHC II (pMHCII) complexes reach the cell surface (Roche and Furuta, 2015). Beneath the MIIC compartment, peptide loading can also occur on recycling MHC II molecules without CLIP involvement (Roche and Furuta, 2015). Cell surface pMHCII complexes finally prime $CD4^+$ T cell responses. While cross-presentation allows for presentation of extracellular peptides in the context of MHC I, autophagy enables the presentation of intracellular antigens by MHC II and the subsequent priming of $CD4^+$ T cells (Crotzer and Blum, 2010; Neefjes et al., 2011; Nimmerjahn et al., 2003; Paludan et al., 2005). In contrast to most cells, where autophagy is observed following metabolic stress, this process is of particular importance in DCs involved in central tolerance in the thymus (Crotzer and Blum, 2010; Nedjic et al., 2008).

4. Phenotypical characterization of cDCs and pDCs

Most of the acquired knowledge about DC biology is based on their phenotypical characterization by cell surface markers and various profiling techniques. DCs are a family of different cell subpopulations and originate from DC precursors of the bone marrow. In recent years, efforts to reduce the complexity in the known DC subsets (Guilliams et al., 2014) and to provide a guideline for the comparison of the various DC populations found in different species were undertaken (Guilliams et al., 2016). Phenotypically, mature primary DCs can be separated into three major populations consisting of two conventional (also named classical) DC types (cDC1 and cDC2) as well as plasmacytoid DCs (pDCs) (Chopin et al., 2012; Guilliams et al., 2014, 2016; Merad et al., 2013; Murphy et al., 2016). In this review, we will apply the nomenclature of cDC1 and cDC2. Depending on their tissue origin, both of the two conventional DC subpopulations can be further distinguished into tissue-resident and migratory cDCs (Merad et al., 2013).

In steady state, all lineage-committed cDCs express the major histocompatibility antigen II molecule (MHC II; in humans: human leucocyte antigen II, HLA II) and most of them also the integrin CD11c (Murphy et al., 2016). All DC subsets do usually not express markers that define B cells, T cells, NK cells, as well as granulocytes (such as CD3, CD19, NKp46, Ly-6G (mouse), Siglec-F (mouse), CD66b (human)). Therefore, DCs are described as lineage negative (lin_{neg}) (Alcantara-Hernandez et al., 2017; Granot et al., 2017; Heidkamp et al., 2016b; Liu and Nussenzweig, 2010; Merad and Manz, 2009; See et al., 2017; Shortman and Naik, 2007; Villani et al., 2017; Waskow et al., 2008). Additional markers are used to further stratify the various DC subsets phenotypically. Here we account CD8α, CD103, CD11b, CD4, CD317 (PDCA-1; Bst2), CD207 (Langerin), CD326 (Epcam), CD64 (FcγRI), Xcr1, Clec9A, CD172a (Sirpα) and specifically in humans CD1c (BDCA-1), CD303 (BDCA-2), CD141 (BDCA-3), CD304 (BDCA-4) as well as CD123 (Guilliams et al., 2014; Heidkamp et al., 2016a; Merad et al., 2013; Schraml and Reis e Sousa, 2015). A comprehensive summary is given in Fig. 1.

5. Identification and functional specialization of the cDC1 subset

Depending on their localization and tissue origin, tissue-resident cDC1 cells are positive for the marker CD8α (e.g., in the spleen and lymph nodes), while migratory cDC1 cells are positive for CD103 (Edelson et al., 2010). The latter cells can be found in the non-lymphoid tissues (e.g., in the intestine or the skin), but also in the lymph nodes. Of note, other immune cell populations express these receptors as well (e.g., T cells). Both populations, $CD8\alpha^+$ and $CD103^+$ DCs, share the marker Xcr1 in a maturation and organ-location independent manner (Gurka et al., 2015). Recent studies demonstrate that all of the cDC1 across tissues express the marker CD26, which is not changing upon stimulation (Guilliams et al., 2016),

	pDCs		cDC1		cDC2		moDCs		Langerhans	
	TFs	Marker	TFs	Marker	TFs	Marker	TFs	Marker	TFs	Marker
Mouse	Zeb2*	Siglec-H$^+$ Ly6C$^+$ B220$^+$ (CD45R) CD317$^+$ (Bst2; PDCA-1)	Bcl6* PU.1* Nfil3* DC-SCRIPT* β-Catenin*	Langerin$^+$ (Skin) CD8α$^+$ (LT) CD103$^+$ (NLT) CD205$^+$ CD24$^+$ CD11b$^-$	Bcl6* RelB & Traf6* Irf2*	CD4$^+$ (LT) DCIR2$^+$ (LT) Esam* CD8α$^-$		Ly6C$^+$ CD107b$^+$ CD64$^+$ (lung) F4/80$^+$ CD24$^+$ (lung)	Irf8* Irf2*	F4/80$^+$ CD205$^+$ CD103$^-$
Shared	Irf8 Irf4 E2-2 Ikaros SpiB Bcl11a	CD45RA$^+$ CD14$^+$ XCR1$^-$ Sirpα$^+$ MHC II$^+$	Zbtb46* Irf8 BatF3 Id2 Etv6*	CD11c$^+$ XCR1$^+$ Clec9A$^+$ BTLA$^+$ Necl2$^+$ Sirpα$^-$ MHC II^{high} CD14$^-$	Zbtb46* Irf4 Id2 Notch/Kfl4 Zeb2*	CD11c$^+$ Sirpα$^+$ CD11b$^+$ XCR1$^-$ MHC II^{high}	PU.1 Zbtb46*	CD11c$^+$ CD206$^+$ CD11b$^+$ FcεRI$^+$ M-CSFR$^+$ GM-CSFR$^+$ MHC II$^+$ Sirpα$^+$	PU.1 Id2 Runx3 β-Catenin	Langerin$^+$ Epcam$^+$ CD11b$^+$ E-Cadherin$^+$ Sirpα$^+$ XCR1$^-$ MHC II^{high}
Human		CD303$^+$ (BDCA-2) CD304$^+$ (BDCA-4) CD123$^+$		CD141$^+$ (BDCA-3)		CD1c$^+$ (BDCA-1) CD1a$^+$ (Skin) CD1b$^+$ Clec10A$^+$	VentX* Irf4*	CD1c$^+$ (BDCA-1) CD1a$^+$ CD14$^+$ CD209$^+$	Stat5* C/EBP*	CD1a$^+$ CD1c$^+$

Fig. 1 Phenotypic characterization of murine and human antigen presenting cells by transcription factor and marker expression. This figure shows a direct comparison of transcription factor and marker expression in mice and men. pDC, plasmacytoid dendritic cells; cDC, conventional dendritic cells; MoDC, monocyte-derived dendritic cell; TF, transcription factor; LT, lymphoid tissue; NLT, non-lymphoid tissue. * indicate transcription factors where reports for human or murine equivalents are lacking or only a similar expression pattern but not the function is described in mice or man, respectively. Please find references substantiating Fig. 1 in Table 1. *Cellular images are provided and adapted from Servier Medical Art (smart.servier.com) and this images are licensed under the Creative Commons Attribution 3.0 Unported License (creativecommons.org/licenses/by/3.0/).*

as well as Clec9A and Xcr1 (see Fig. 1) (Caminschi et al., 2008; Contreras et al., 2010; Crozat et al., 2011; Kroczek and Henn, 2012; Sancho et al., 2008). The genetic knock out of transcriptional regulators demonstrated that cDC1 cells are dependent on the expression of the transcription factors Batf3 (Basic Leucine Zipper ATF-Like Transcription Factor 3) (Hildner et al., 2008), Irf8 (Interferon regulating factor-8) (Sichien et al., 2016), but also Bcl6 (B-cell lymphoma 6) (Watchmaker et al., 2013; Zeng et al., 2016; Zhang et al., 2014), Id2 (Inhibitor of DNA binding 2) (Hacker et al., 2003; Kusunoki et al., 2003), Nfil3 (Nuclear factor interleukin-3-regulated protein) (Kashiwada et al., 2011), and Etv6 (ETS variant 6) (Lau et al., 2018). Due to their dependency on the transcription factors Irf8 and Batf3 as well as the expression of surface markers, the literature names cDC1 also as Irf8 dependent DCs, Batf3 dependent DCs, CD8$^+$ DCs (most common until RNA sequencing era), as well as CD103$^+$ DCs.

One of the most important functions of cDC1 is their capability to cross-present antigens, a process leading to the cross-priming of $CD8^+$ T cell responses (Bevan, 1976; Bozzacco et al., 2007; den Haan et al., 2000; Dudziak et al., 2007; Hildner et al., 2008; Kurts et al., 2010; Lehmann et al., 2017). This mechanism is of critical importance since it allows for the generation of cytotoxic T cell (CTL) responses to tumor cells and viruses (even when the DCs are not infected directly).

Besides its supreme role in antigen cross-presentation, the cDC1 subset ensures the induction of $CD4^+$ helper T cell responses as well (Boscardin et al., 2006; Lehmann et al., 2017; Neubert et al., 2014). Recent data demonstrate that naïve $CD4^+$ T cells are predominantly differentiating into Th1 helper T cells (Draheim et al., 2017; Harpur et al., 2019; Igyarto et al., 2011; Martinez-Lopez et al., 2015). This is mainly due to the secretion of IL-12 upon stimulation (Farrand et al., 2009; Maldonado-LÓpez et al., 1999; Martinez-Lopez et al., 2015). Furthermore, cDC1 cells were demonstrated to secrete Interferon-λ after poly(I:C) stimulation (Lauterbach et al., 2010). All described functions underline the importance of cDC1 cells to fight intracellular pathogens. The functions of cDC1 cells are summarized in Fig. 2 and Table 2.

6. Identification and functional specialization of the cDC2 subset

The other major subset of conventional DCs are cDC2. This subset is representing a more functionally diverse subpopulation of cDCs. Tissue resident and migratory cDC2 are positive for the marker Sirpα (CD172a) (Guilliams et al., 2016; Gurka et al., 2015). Because this marker is not unique to DCs and can also be found on monocytes and macrophages (Barclay and Brown, 2006; De Vries et al., 2002), it is important to combine this cell surface marker with other hallmark DC markers, such as MHC II and CD11c. Importantly, cDC2 are lacking CD8, Xcr1, and Clec9A, but are positive for CD11b. Resident splenic cDC2 can be formally split into $CD4^+$ and $CD4^-$ DCs (also known as $CD4^+CD8^-$, $CD4^-CD8^-$) (Cebrian et al., 2011; Merad et al., 2013; Murphy et al., 2016). These two potential cDC2 subsets identified in the murine spleen may also be characterized by their differential expression of ESAM, Csf1R, Csf-3R, and CCR2 (see Fig. 1) (Merad et al., 2013; Murphy et al., 2016). However, not all of their functional capacities are formally established.

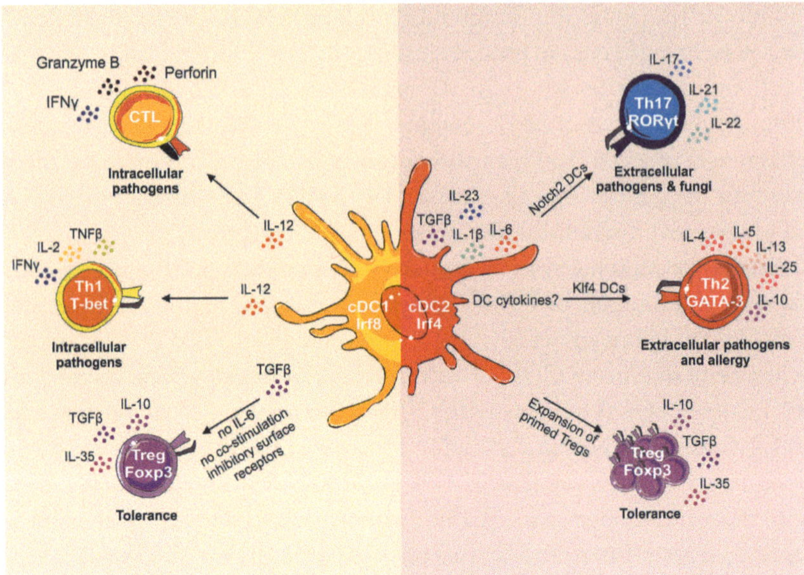

Fig. 2 Conventional DCs orchestrate the development of naïve T cells to defined effector populations. Murine conventional dendritic cell (cDC) subpopulations drive defined effector pathways during immune responses. Irf8 dependent cDC1 cells excel in priming immune responses directed against intracellular pathogens. This is either achieved by priming cytotoxic T cell responses via the cross-presentation machinery or Th1 cells following MHC II presentation and IL-12 secretion. In an anti-inflammatory milieu, cDC1 cells are able to convert T cells to T_{reg} cells. Irf4 cDC2 cells govern effector T cell responses directed against extracellular pathogens including priming of Th2 or Th17 cells following IL-23 secretion. Without inflammatory stimuli, cDC2 cells are able to expand already primed T_{reg} cells. CTL, cytotoxic T lymphocyte; Th1, Type 1 T helper cells; Th2, Type 2 T helper cells; Th17, Type 17 T helper cells; T_{reg} cells, regulatory T cells. The information in this figure is substantiated by the references in Table 2. *Cellular images are provided and adapted from Servier Medical Art (smart.servier.com) and this images are licensed under the Creative Commons Attribution 3.0 Unported License (creativecommons.org/licenses/by/3.0/).*

In the different non-lymphatic tissues, it is often difficult to separate monocyte/macrophage as well as MoDC populations from cDC2 as they share several markers (Merad et al., 2013). Whereas MoDCs and cDC2 in muscle tissue can be distinguished by the high expression of CD64 (FcγRI) on MoDCs (Langlet et al., 2012), in lamina propria, cDC2 cells are positive for CD103 expression and negative for CX_3CR_1 (Bogunovic et al., 2009; Schulz et al., 2009).

Although, cDC2 have been described to prime cytotoxic $CD8^+$ T cell responses by cross-presentation in an FcγR dependent manner (Den Haan and Bevan, 2002; Platzer et al., 2014), cDC1 clearly excel in cross-presentation

(Bonifaz et al., 2004; Den Haan et al., 2000; Dudziak et al., 2007; Idoyaga et al., 2008; Lehmann et al., 2017; Lin et al., 2008; Neubert et al., 2014). However, upon direct virus infection of cDC2, they are able to present the respective antigens on their surface in a MHC I context, thus leading to cytotoxic $CD8^+$ T cell responses (Ballesteros-Tato et al., 2010; Kastenmuller et al., 2011; Kim and Braciale, 2009; Merad et al., 2013).

cDC2 cells have been suggested to play an important role in the protection from extracellular pathogens. They are specialized for the presentation of antigens in MHC II context. In general, the preferential presentation of antigens as pMHCII complexes by cDC2 cells can be partly explained by their higher expression level of proteins involved in the MHCII presentation machinery (Dudziak et al., 2007). This is pivotal for the activation, proliferation and differentiation of naïve $CD4^+$ T cells into $CD4^+$ helper T cell subsets (Merad et al., 2013). Here, cDC2 are described to excel in the differentiation of Th2 as well as Th17 helper T cells (Gao et al., 2013; Idoyaga et al., 2008; Kumamoto et al., 2013; Lewis et al., 2011; Maldonado-LÓpez et al., 1999; Merad et al., 2013; Satpathy et al., 2013; Schlitzer et al., 2013, 2015a; Schlitzer and Ginhoux, 2014). Currently, the mode of Th2 response orchestration by cDC2 is under debate. Even though various studies have demonstrated the importance of cDC2 for Th2 responses (Dillon et al., 2004; Kitajima and Ziegler, 2013; Plantinga et al., 2013; Pulendran et al., 1999), it is unclear, if these DCs contribute by lack of Th1/Th17 cytokines to Th2 polarization or by an yet unknown signal. Since primary cDC2 are thought to not produce the Th2 master cytokine IL-4, one can hypothesize that cDC2 either stimulate a positive IL-4 feedback loop leading to autocrine IL-4 signaling in naïve T cells and/or IL-4 is provided by other immune cells (Murphy and Reiner, 2002; Na et al., 2016; Paul and Zhu, 2010). Potentially, the absence of the Th1 master cytokine IL-12 allows Gata3 expression and therefore Th2 differentiation (Murphy and Reiner, 2002). However, within the cDC2 cells a Klf4 dependent subpopulation was described to be essential for Th2 responses (Tussiwand et al., 2015), whereas a Notch2 dependent subset was described to be important for the promotion and formation of follicular helper T cells (Tfh) and germinal center B cells important for the establishment of humoral immune responses (BriseÑO et al., 2018; Calabro et al., 2016; Shin et al., 2015). An overview of the functional T cell modules induced by cDC2 cells is given in Fig. 2 and Table 2.

7. cDC1 and cDC2 and the induction and maintenance of tolerance

As described in the introduction, cDCs do not only prime new T cell responses, they are also able to dampen and modulate immune responses (Audiger et al., 2017; Manicassamy and Pulendran, 2011). Interestingly, in these processes, cDC1 and cDC2 divide their work by functional specialization. In this respect, cDC1 have been attributed to be able to convert naïve $CD4^+$ T cells into $CD4^+$ $FoxP3^+$ Treg cells in the periphery (see Fig. 2 and Table 2). Within cDC1 cells, a $BTLA^+CD205^+CD8\alpha^+CD11c^+$ subpopulation was found to induce extrathymic regulatory T cells, whereas $BTLA^-$ DCs failed to promote Treg generation (Jones et al., 2016). This process is pivotal for the induction of peripheral tolerance to antigens not included in the negative selection process within the thymus. Therefore, upon uptake of self-antigens in a non-inflammatory context (meaning without any danger signals), these antigens are presented on MHC molecules without increased co-stimulatory signals (Blander, 2007; Bonifaz et al., 2002; Coombes et al., 2007; Hawiger et al., 2001; Kretschmer et al., 2005; Yamazaki et al., 2008). Interestingly, cDC1 have been described to efficiently take up dead cells (Liu et al., 2002, 2005; Qiu et al., 2009; Zelenay et al., 2012). Furthermore, cDC1 regulate $CD8^+$ T cell tolerance by inducing clonal deletion or anergy following cross-presentation (Blander and Nair-Gupta, 2013; Joffre et al., 2012; Luckashenak et al., 2008). This and the induction of peripheral T cell tolerance might help to prevent our body from detrimental autoimmune responses (Steinman and Banchereau, 2007; Steinman and Nussenzweig, 2002). However, this might also have a negative influence on the induction of immune responses to cancer cells rich in tolerated self-antigens (Fu and Jiang, 2018).

cDC2 cells, especially migratory $CD103^+CD11b^+$ cDC2 from the lamina propria, but also dermal cDC2, are potent inducers of peripheral tolerance, which might be due to their high expression of ALDH (aldehyde dehydrogenase). This enzyme is pivotal for the conversion of vitamin A into tolerogenic retinoic acid (Coombes et al., 2007; Guilliams et al., 2010; Sun et al., 2007). Accordingly, these subsets were shown to induce antigen-specific $CD4^+FoxP3^+$ Treg cells in vitro (Coombes et al., 2007; Guilliams et al., 2010; Sun et al., 2007; Vitali et al., 2012). In contrast to cDC1, which have been attributed to differentiate $CD4^+FoxP3^+$ Treg cells from naïve $CD4^+$ T cells, cDC2 are thought to be rather responsible for the expansion of already differentiated $CD4^+$ $FoxP3^+$ Treg cells in the periphery in vivo (Yamazaki et al., 2008) (see Fig. 2 and Table 2). Besides their role in peripheral tolerance mechanisms, cDC2 can re-circulate to the thymus (Hadeiba and Butcher, 2013; Li et al., 2009), the organ involved in central T cell tolerance. There, these cells might not only teach $CD4^+$ Treg cells, but they could also ensure the deletion of autoreactive T cells from the naïve T cell pool (Hadeiba and Butcher, 2013).

8. Identification and functional specialization of pDCs

Plasmacytoid dendritic cells (pDCs) act as sensors for viral infection (mostly via TLR7 and TLR9) and are major producers of type I interferons (IFN α and β) (Cella et al., 1999; Chistiakov et al., 2014; Murphy et al., 2016; Reizis, 2019; Reizis et al., 2011; Sawai et al., 2013; Siegal et al., 1999). Murine and human pDCs share the marker CD45RA. While murine pDCs express Siglec-H, CD317 (PDCA-1 or Bst2), and Ly6C, human pDCs are characterized by CD123 (IL-3R), CD303 and CD304 (see Fig. 1 and Table 1) (Murphy et al., 2016).

After their development, pDCs exit the bone marrow and home to peripheral tissues via the blood stream (Sawai et al., 2013; Segura et al., 2012). Similar to cDCs, pDCs migrate from the periphery into the lymph nodes. This migration process is driven by a high expression of the chemokine receptors CCR7, CCR2, and CCR5 (Sawai et al., 2013). CCR7 is sensing the chemokine ligands CCL19 and CCL21, which are produced at high levels in the peripheral LNs (Seth et al., 2011). Even if not directly infected, pDCs can respond to viruses with type I IFN production due to TLR recognition. This is achieved by a positive type I IFN feedback loop, which is thereby maximizing TLR sensitivity that enables type I IFN responses without cytoplasmic viral infections of the pDCs themselves. This is in turn responsible for a general reduction of the viral titers (Kim et al., 2014; Kumagai et al., 2009).

In the literature, the ability and efficiency of pDCs to present antigens to T cells is intensively discussed. The lack of a bona fide pDC knock out mouse model might be one explanation, why the role of pDCs in antigen presentation is not fully clear, yet (Ali et al., 2019; Musumeci et al., 2019; Van Blijswijk et al., 2013). Owing to the fact that pDCs express MHC II molecules, they have been described to induce $CD4^+$ T cell responses when TLRs and other sensors are triggered (Loschko et al., 2011; Reizis, 2019; Reizis et al., 2011; Sage et al., 2014; Villadangos and Young, 2008).

Table 1 Literature substantiating Fig. 1. Phenotypic characterization of murine and human antigen presenting cells by transcription factor and marker expression.

Cell type	Species	Molecule	Reference
Transcription Factors			
pDCs	Mouse	Zeb2	Scott et al. (2016)
	Mouse/ Human	Irf8	Sichien et al. (2016) and Hambleton et al. (2011)
		Irf4	Collin and Bigley (2018) and Nagasawa et al. (2008)
		E2-2	Cisse et al. (2008), Kee (2009), Nagasawa et al. (2008), and Ghosh et al. (2010)
		Ikaros	Allman et al. (2006) and Cytlak et al. (2018)
		SpiB	Nagasawa et al. (2008), Sasaki et al. (2012), Schotte et al. (2004), and Karrich et al. (2012)
		Bcl11a	Chistiakov et al. (2014), Marafioti et al. (2008), Pulford et al. (2006), and Ippolito et al. (2014)
cDC1	Mouse	Bcl6	Ohtsuka et al. (2011)
		PU.1	Chopin et al. (2019)
		DC-SCRIPT	Chopin et al. (2019)
		β Catenin	Cohen et al. (2015)
		Nfil3	Kashiwada et al. (2011)
	Mouse/ Human	Zbtb46	Meredith et al. (2012a,b), and Satpathy et al. (2012)
		Irf8	Becker et al. (2012), Sichien et al. (2016), Hambleton et al. (2011), Grajales-Reyes et al. (2015), Sontag et al. (2017), and Schiavoni et al. (2004)
		BatF3	Jaiswal et al. (2013), Hildner et al. (2008), Grajales-Reyes et al. (2015), Poulin et al. (2012), and Edelson et al. (2010)
		Etv6	Lau et al. (2018) and Mabbott et al. (2013)
		Id2	Hacker et al. (2003) and Jaiswal et al. (2013)
cDC2	Mouse	Bcl6	Ohtsuka et al. (2011)
		Irf2	Ichikawa et al. (2004)
		RelB and Traf6	Briseno et al. (2017), Kobayashi et al. (2003), and Wu et al. (1998)

Table 1 Literature substantiating Fig. 1. Phenotypic characterization of murine and human antigen presenting cells by transcription factor and marker expression.—cont'd

Cell type	Species	Molecule	Reference
	Mouse/Human	Zbtb46	Meredith et al. (2012a,b), and Satpathy et al. (2012)
		Irf4	Ma et al. (2019), Schlitzer et al. (2013), Breton et al. (2016), Williams et al. (2013), Vander Lugt et al. (2014), and Suzuki et al. (2004)
		Id2	Hacker et al. (2003), Breton et al. (2016), and Merad et al. (2013)
		Zeb2	Scott et al. (2016) and Breton et al. (2016)
		Notch/Klf4	Caton et al. (2007), Lewis et al. (2011), Satpathy et al. (2013), Tussiwand et al. (2015), Martin-Gayo et al. (2017), and Czakai et al. (2016)
moDC	Mouse/Human	PU.1	Bakri et al. (2005)
		Zbtb46	Satpathy et al. (2012), Segura et al. (2013b) and Zigmond et al. (2012)
	Human	Irf4	Lehtonen et al. (2005)
		VentX	Wu et al. (2014)
Langerhans	Mouse	Irf8	Merad et al. (2013), Zhang et al. (2016), and Schiavoni et al. (2004)
		Irf2	Merad et al. (2013), Zhang et al. (2016), and Ichikawa et al. (2004)
	Mouse/Human	PU.1	Chopin et al. (2013), Iwama et al. (2002), and Heinz et al., (2006)
		β Catenin	Becker et al. (2011) and Yasmin et al. (2013)
		Id2	Hacker et al. (2003), Chopin et al., (2013), Zhang et al. (2016), Heinz et al. (2006), and Merad et al. (2013)
		Runx3	Chopin et al. (2013), Merad et al. (2013), and Hashimoto-Hill et al. (2018)
	Human	Stat5	Van de Laar et al. (2011)
		C/EBP	Iwama et al. (2002) and Hashimoto-Hill et al. (2018)

Continued

Table 1 Literature substantiating Fig. 1. Phenotypic characterization of murine and human antigen presenting cells by transcription factor and marker expression.—cont'd

Cell type	Species	Molecule	Reference
Markers			
pDCs	Mouse	Siglec-H+	Zhang et al. (2006), Blasius et al. (2006), and Schlitzer et al. (2015b)
		CD317+ (PDCA-1, Bst2)	Blasius et al. (2006), Tavano et al. (2013), and Guilliams et al. (2016)
		B220+ (CD45R)	Musumeci et al. (2019) and Nakano et al. (2001)
		Ly6C+	Asselin-Paturel et al. (2001) and Guilliams et al. (2016)
	Mouse/ Human	CD45RA+	Asselin-Paturel et al. (2001), Guilliams et al. (2016), and Schlitzer et al. (2015b)
		MHC II+	Sadaka et al. (2009), Sapoznikov et al. (2007), Irla et al. (2010), and LeibundGut-Landmann et al. (2004)
		CD14+	Murphy et al. (2016)
		XCR1−	Murphy et al. (2016)
		CD172a (Sirpα)	Murphy et al. (2016)
	Human	CD303+ (BDCA-2)	Dzionek et al. (2000), Dzionek et al. (2001), and Boiocchi et al. (2013)
		CD123+	Rissoan et al. (2002), Garnache-Ottou et al. (2005), Zhang et al. (2016), and Colonna et al. (2004)
		CD304+ (BDCA-4)	Dzionek et al. (2000) and Zhang et al. (2016)
		CD2	Matsui et al. (2009) and Zhang et al. (2016)
cDC1	Mouse	Langerin+ (Skin)	Romani et al. (2010), Henri et al. (2010), Ginhoux et al. (2007), and Sathe et al. (2011)
		CD8α+ (LT)	Dudziak et al. (2007), Sathe et al. (2011), Edelson et al. (2010), Kirkling et al. (2018), Schlitzer et al. (2015a,b), Jones et al. (2016), and Hildner et al. (2008)

Table 1 Literature substantiating Fig. 1. Phenotypic characterization of murine and human antigen presenting cells by transcription factor and marker expression.—cont'd

Cell type	Species	Molecule	Reference
		CD103+ (NLT)	Henri et al. (2010), de Mingo Pulido et al. (2018), Semmrich et al. (2012), Sathe et al. (2011), Edelson et al. (2010), and Kirkling et al. (2018)
		CD205+	Dudziak et al. (2007), Edelson et al. (2010), Kirkling et al. (2018), and Jones et al. (2016)
		CD24+	Guilliams et al. (2016) and Schlitzer et al. (2015b)
	Mouse/ Human	CD11c+	Guilliams et al. (2016), Wu et al. (2018), Jones et al. (2016), Jongbloed et al. (2010), and Watchmaker et al. (2013)
		XCR1+	Bachem et al. (2012), Crozat et al. (2010), and Guilliams et al. (2016)
		Clec9A+	Poulin et al. (2012), Salvermoser et al. (2018), Chu et al. (2012), and Watchmaker et al. (2013)
		BTLA+	Jones et al. (2016)
		Necl2+	Chu et al. (2012)
		CD172a (Sirpα)	Greter et al. (2012), Guilliams et al. (2016), and Watchmaker et al. (2013)
		CD14−	Watchmaker et al. (2013),
		MHC II high	Guilliams et al. (2016), Granot et al. (2017), MacDonald et al. (2002), and Watchmaker et al. (2013)
	Human	CD141+ (BDCA-3)	Haniffa et al. (2012), Kirkling et al. (2018), Chu et al. (2012), Granot et al. (2017), Jongbloed et al. (2010), Watchmaker et al. (2013), and Dzionek et al. (2000)
cDC2	Mouse	CD4+ (LT)	Schlitzer et al. (2015b)
		DCIR2+ (LT)	Dudziak et al. (2007)
	Mouse/ Human	CD11c+	Guilliams et al. (2016), Wu et al. (2018), Jongbloed et al. (2010), and Watchmaker et al. (2013)
		CD172a (Sirpα)	Guilliams et al. (2016), Granot et al. (2017), and Watchmaker et al. (2013)

Continued

Table 1 Literature substantiating Fig. 1. Phenotypic characterization of murine and human antigen presenting cells by transcription factor and marker expression.—cont'd

Cell type	Species	Molecule	Reference
		CD11b+	Guilliams et al. (2016), Schlitzer et al. (2015b), Schlitzer et al. (2013), and Watchmaker et al. (2013)
		XCR1−	Guilliams et al. (2016)
		MHC II high	Guilliams et al. (2016), Granot et al. (2017), MacDonald et al. (2002), and Watchmaker et al. (2013)
	Human	CD1c+ (BDCA-1)	Granot et al. (2017), Jongbloed et al. (2010), MacDonald et al. (2002), and Watchmaker et al. (2013)
		CD1a+ (skin)	Bigley et al. (2015)
		CD1b+	MacDonald et al. (2002)
		Clec10A+	Heger et al. (2018)
moDCs	Mouse	Ly6C+	Leon et al. (2007), Greter et al. (2012) and Naik et al. (2006)
		CD24+ (Lung)	Greter et al. (2012)
		MAR1+ (Lung)	Greter et al. (2012) and Plantinga et al. (2013)
		F4/80+	Greter et al. (2012) and Guilliams et al. (2016)
		CD107b+ (Mac-3)	Greter et al. (2012) and Naik et al. (2006)
		CD64+ (Muscle, Lung)	Plantinga et al. (2013), Langlet et al. (2012), Min et al. (2018), and Guilliams et al. (2016)
	Mouse/Human	CD11c+	Leon et al. (2007), Greter et al. (2012), Guttman-Yassky et al. (2007), and Guilliams et al. (2016)
		CD11b+	Greter et al. (2012), Plantinga et al. (2013), Naik et al. (2006), and Guttman-Yassky et al. (2007)
		CD115+ (M-CSFR)	Greter et al. (2012)
		CD172a (Sirpα)	Greter et al. (2012), Naik et al. (2006), and Guilliams et al. (2016)
		FcεRI+	Hammad et al. (2010) and Guttman-Yassky et al. (2007)

Table 1 Literature substantiating Fig. 1. Phenotypic characterization of murine and human antigen presenting cells by transcription factor and marker expression.—cont'd

Cell type	Species	Molecule	Reference
		GM-CSFR+	Greter et al. (2012), Segura et al. (2013b) and Zigmond et al. (2012)
		CD206+	Wollenberg et al. (2002), Guttman-Yassky et al. (2007), Segura et al. (2013b), Zigmond et al. (2012), and MacDonald et al. (2002)
		MHC II+	Leon et al. (2007), Greter et al. (2012), Guttman-Yassky et al. (2007), and Guilliams et al. (2016)
	Human	CD1c+	Guttman-Yassky et al. (2007)
		CD209 (DC-SIGN)	Guttman-Yassky et al. (2007) and MacDonald et al. (2002)
		CD14+	Segura et al. (2013b) and Zigmond et al. (2012)
		CD1a+	Guttman-Yassky et al. (2007)
Langerhans	Mouse	F4/80+	Kamata et al. (2017), Hume et al. (1983), and Guilliams et al. (2016)
		CD205+	Henri et al. (2001) and Sere et al. (2012)
		CD103−	Kamata et al. (2017)
	Mouse/Human	Langerin+ (Skin)	Guttman-Yassky et al. (2007), Valladeau et al., (2000), Stoitzner et al., (2003), Gaiser et al., (2012), and Guilliams et al. (2016)
		CD326+ (Epcam)	Ouchi et al. (2016) and Gaiser et al. (2012)
		CD11b+	Murphy et al. (2016)
		E-Cadherin+	Ruedl et al. (2000) and Gaiser et al. (2012)
		Birbeck granules	Gaiser et al. (2012)
		CD172a (Sirpα)	Murphy et al. (2016)
		XCR1−	Murphy et al. (2016) and Guilliams et al. (2016)
		MHC II high	Ruedl et al. (2000) and Sere et al. (2012)
	Human	CD1a+	Guttman-Yassky et al. (2007) and Hunger et al. (2004)
		CD1c+	Murphy et al. (2016) and Peiser et al. (2004)

Table 2 Literature substantiating Fig. 2. Conventional DCs orchestrate the development of naïve T cells to defined effector populations.

Cell type	Cytokine type	Molecule	Reference
CTL	Produces	IFNγ	Henry et al. (2008) and Wilson et al. (2008)
		Granzyme B	Wilson et al. (2008), Janas et al. (2005), and Johnson et al. (2003)
		Perforin	Janas et al. (2005) and Johnson et al. (2003)
	Requires	IL-12	Henry et al. (2008) and Wilson et al. (2008)
Th1	Produces	TNFβ	Darrah et al. (2007)
		IL-2	Darrah et al. (2007)
		IFN-γ	Suzuki et al. (1988), Darrah et al. (2007), and Kano et al. (2008)
	Requires	T-bet	Szabo et al. (2000)
		IL-12	Kano et al. (2008) and Henry et al. (2008)
Th2	Produces	IL-4	Swain et al. (1990), Urban et al. (1998), Garlisi et al. (1995), and Kopf et al. (1996)
		IL-5	Garlisi et al. (1995), Kopf et al. (1996), and Ikutani et al. (2012)
		IL-13	Urban et al. (1998)
		IL-25	Angkasekwinai et al. (2007) and Fort et al. (2001)
		IL-10	Fiorentino et al. (1989) and Kopf et al. (1996)
	Requires	GATA-3	Kurata et al. (1999) and Zhu et al. (2004, 2006)
Th17	Produces	IL-17	Harrington et al. (2005), Park et al. (2005), Veldhoen et al. (2006), Bettelli et al. (2006), Veldhoen et al. (2006), and Zhou et al. (2007)
		IL-21	Zhou et al. (2007) and Korn et al. (2007)
		IL-22	Liang et al. (2006) and Zheng et al. (2007)
	Requires	RORγt	Ivanov et al. (2006)
		IL-6	Bettelli et al. (2006), Veldhoen et al. (2006), and Zheng et al. (2007)
		IL-23	Harrington et al. (2005), Park et al. (2005), and Zheng et al. (2007)
		TGFβ	Bettelli et al. (2006), Veldhoen et al. (2006), Zhou et al. (2007), and Zheng et al. (2007)
		IL-1β	Veldhoen et al. (2006)

Table 2 Literature substantiating Fig. 2. Conventional DCs orchestrate the development of naïve T cells to defined effector populations.—cont'd

Cell type	Cytokine type	Molecule	Reference
Treg	Produces	TGFβ	Li et al. (2007)
		IL-10	Li et al. (2007), Belkaid et al. (2002), Rubtsov et al. (2008), and Asseman et al. (1999)
		IL-35	Collison et al. (2007)
	Requires	Foxp3	Chen et al. (2003) and Fontenot et al. (2003)
		TGFβ	Chen et al. (2003), Li et al. (2007), and Konkel et al. (2017)
		No IL-6	Bettelli et al. (2006) and Zhou et al. (2007)

Studies in humans indicated that pDCs may be able to efficiently cross-present antigens to $CD8^+$ T cells (Tel et al., 2013). Furthermore, pDCs were described to produce IL-12 and induce proliferation of naïve allogeneic T cells (Matsui et al., 2009). However only recently, the group of Ginhoux identified a human pre-cDC population expressing the typical pDC markers CD123, CD303 and CD304 (See et al., 2017). Previously, these cells might eventually be mistaken as pDCs. In the study of See et al., this pre-cDC population was identified as the IL-12 producing population able to induce proliferation and polarization of naïve $CD4^+$ T cells (See et al., 2017). In this light, results of previous studies might need a more careful interpretation.

9. Origin, identification and functional specialization of Langerhans cells

Langerhans cells (LCs) are antigen presenting cells (APCs), which reside within the epidermis (Kaplan, 2017). In both, mice and humans, LCs share the expression of MHC II (human HLA II), Langerin (CD207), Epcam (CD326), CD11b, Sirpα (CD172a) (see Fig. 1) and contain Birbeck granules (Kashem et al., 2017; Murphy et al., 2016). Further, murine LCs express low levels of F4/80 and DEC205 (CD205), whereas human LCs express CD1a and CD1c (Kashem et al., 2017; Murphy et al., 2016). Of note, Langerin is also expressed by dermal DC subpopulations. Therefore, besides their localization, the expression of CD103 is used to dissect most dermal $Langerin^+$ DCs from $Epcam^+$ LCs (as well as on LCs migrating trough the dermis) (Malissen et al., 2014; Merad et al., 2008). A summary of LC surface markers and characteristic transcription factors is provided in Fig. 1. Historically, due to their sentinel function including antigen uptake, migration, antigen processing and presentation as well as T cell priming, LCs, residing at the borderline of the body to the environment, were recognized as prime example for DCs in the skin (Kaplan, 2017). For a long time, the ontogeny of LCs was a matter of debate, but recent evidence assigns them to the macrophage lineage owing to their yolk sac origin (Ginhoux and Jung, 2014; Gomez Perdiguero et al., 2014; Guilliams et al., 2014; Hoeffel et al., 2012, 2015; Schulz et al., 2012). Data indicate that primitive erythro-myeloid progenitors (EMPs) in the yolk sac, which do not produce HSCs (Collin and Milne, 2016), develop to LCs via two main pathways. While early EMPs give rise to yolk sac macrophages and further to LCs, late EMPs develop to c-myb dependent fetal liver monocytes and subsequently to LCs (Ginhoux et al., 2006; Gomez Perdiguero et al., 2014; Hoeffel et al., 2012, 2015; Schulz et al., 2012). LCs from both developmental pathways continue to exist in parallel in the adult organism (Ginhoux and Jung, 2014). Once settled in the epidermis, LCs have self-renewing potential, are partially resistant to irradiation, and can remain of host origin in bone marrow chimeras (Merad et al., 2002; Merad et al., 2008). This was also the first indication that LC ontogeny diverges from DCs.

In general, LC differentiation is dependent on macrophage colony stimulating factor 1 receptor (M-CSFR, CD115) and TGFβ (Borkowski et al., 1996; Ginhoux et al., 2006; Greter et al., 2012; Hacker et al., 2003; Wang et al., 2012), but is independent of Flt3-L (Hemmi et al., 2001). Moreover, LC differentiation is dependent on the transcription factor PU.1 under both steady-state and inflammatory conditions, while Id2 was found to be important during steady-state LC homeostasis (Chopin et al., 2013; Hacker et al., 2003). Since PU.1 controls the expression of the TGFβ responsive transcription factor Runx3 in LCs, PU.1 is a central mediator of LC development (Chopin et al., 2013). IL-34 produced by epidermal keratinocytes also contributes to LC development via binding to the M-CSFR (Wang et al., 2012). Under inflammatory conditions, LCs can be replaced by circulating Gr-1$^+$ (Ly6C/G) monocytes (Ginhoux et al., 2006; Kaplan, 2017).

Functionally, LCs were thought to cross-prime CD8$^+$ T cells, even though being less efficient than dermal cDC1 cells (Kaplan, 2017; Stoitzner et al., 2006; Zaric et al., 2015), and to induce cross-tolerance (Flacher et al., 2014; Strandt et al., 2017). LCs were also found to induce

Th17 responses (Igyarto et al., 2011; Mathers et al., 2007). Furthermore, LCs contribute to the generation of antigen-specific, class-switched antibodies by inducing follicular T helper cells (T_{FH}) (Levin et al., 2017; Ouchi et al., 2011; Yao et al., 2015). Besides that, there is evidence, that LCs promote tolerance via T_{reg} cells, but only in specific situations, where the antigen is self-derived (Idoyaga et al., 2013; Yao et al., 2015) or restricted to the epidermis (Gomez De Agüero et al., 2012). LC ontogeny, homeostasis and function are reviewed nicely elsewhere (Clayton et al., 2017; Ginhoux and Jung, 2014; Kaplan, 2017; Kashem et al., 2017; Merad et al., 2008; Romani et al., 2010).

10. Identification and functional specialization of MoDCs

Under inflammatory conditions, the development of monocyte-derived DCs (MoDCs, also called inflammatory DCs) from monocytes was observed in mice and men (Leon et al., 2007; Segura and Amigorena, 2013). In the steady-state, MoDCs are not present in peripheral tissues or in lymphoid organs (Segura and Amigorena, 2013). Murine and human MoDCs share the expression of MHC II (HLA DR), CD11c, CD11b, MMR (Macrophage mannose receptor, CD206), FcεRI, Sirpα (CD172a), M-CSFR (Macrophage colony stimulating factor receptor, CD115), GM-CSFR the (Granulocyte-Macrophage colony stimulating factor receptor, CD116), as well as the transcription factor Zbtb46 (Greter et al., 2012; Guttman-Yassky et al., 2007; Hammad et al., 2010; Leon et al., 2007; Naik et al., 2006; Satpathy et al., 2012; Segura et al., 2009, 2013b; Segura and Amigorena, 2013; Van Den Berg and Miossec, 2009; Wollenberg et al., 2002; Zigmond et al., 2012). In mice, MoDCs additionally express F4/80, Ly6C, Mac-3 (CD107b) and CD64 (FcγRI) (De Trez et al., 2009; Leon et al., 2007; Plantinga et al., 2013), whereas human MoDCs display surface CD1c (BDCA-1), CD1a and CD14 expression (Guttman-Yassky et al., 2007; Segura et al., 2013b; Wollenberg et al., 2002). Since many of these markers are found on primary DCs and macrophages, FcεRI is supposed to be a superior marker for MoDCs, especially in combination with CD64 (Plantinga et al., 2013). In case of CD64, it was previously reported that CD64 nicely separates MoDCs from cDCs (Langlet et al., 2012), but recently two $CD64^+$ DC populations that developed following *Listeria* infection were identified in mice (Min et al., 2018). These included $CD64^+$ MoDCs and a $CD64^+$ cDC population, which was characterized by its Flt3 dependence and transcriptome analysis showing hallmark cDC transcription factor expression (RelB, Id2, Irf4 and BatF3) (Min et al., 2018). MoDC surface markers and transcription factors are summarized in Fig. 1.

In general, MoDC differentiation is not affected in Flt3L deficient mice separating them from the cDC lineage (Plantinga et al., 2013). Even though, GM-CSF is used to differentiate MoDCs in vitro from murine bone marrow cultures or human blood (Schaft et al., 2005; Schuler, 2003; Xu et al., 2007), one study suggested that MoDCs develop in vivo from GM-CSFR deficient bone marrow chimeras following pathogen-dependent inflammation, whereas lack of the M-CSFR negatively influenced MoDC numbers (Greter et al., 2012). Furthermore, several studies displayed that the migratory capacity of MoDCs is driven by CCR2 and CCR7 (Bosschaerts et al., 2010; Hohl et al., 2009; Leon et al., 2007; Plantinga et al., 2013).

Dependent on the inflammatory milieu in different infections or autoimmune disease models, murine MoDCs were reported to present antigens and polarize Th1 cells (Hohl et al., 2009; Iijima et al., 2011; Leon et al., 2007), Th2 cells (Hammad et al., 2010; Kool et al., 2008; Plantinga et al., 2013), and to cross-present antigens to $CD8^+$ T cells (Segura et al., 2009; Wakim et al., 2008). Furthermore, it was reported that MoDCs produce IL-12, IL-23, IL-1α and IL-1β, as well as IFNβ (Mayer-Barber et al., 2011; Nakano et al., 2009; Plantinga et al., 2013). MoDC appearance and their functional properties are reviewed thoroughly elsewhere (Collin and Bigley, 2018; Hespel and Moser, 2012; Merad et al., 2013; Segura and Amigorena, 2013).

11. Phenotypical characterization, functional specialization, and presence of DC subsets in human tissues

In human blood, three main DC subpopulations, namely cDC1, cDC2, and pDCs, have been identified, displaying a distinct expression pattern of blood dendritic cell antigens (BDCA) (Dzionek et al., 2000). Beside their distinct expression pattern of BDCA molecules, all human DC subpopulations express HLA-DR on their surface. cDC2 are characterized by the highest expression level of CD11c, whereas cDC1 display an intermediate expression of CD11c, while pDCs are rather negative for this integrin. Additionally, cDC1 express BDCA-3 (CD141), whereas cDC2 are positive for BDCA-1 (CD1c). pDCs carry high amounts of BDCA-2

(CD303), BDCA-4 (CD304), and the IL-3Rα chain (CD123) on their surface (Dzionek et al., 2000; Heidkamp et al., 2016b). In humans, cDC1 are known as $CD141^{+}$ DCs, which are homologues to the murine cDC1 cells (Bachem et al., 2010; Crozat et al., 2010; Dorner et al., 2009; Robbins et al., 2008; Yamazaki et al., 2010). Murine and human cDC1 share the surface markers Xcr1 and Clec9A (Bachem et al., 2010; Caminschi et al., 2008; Contreras et al., 2010; Crozat et al., 2010, 2011; Sancho et al., 2008). Human cDC2 cells are also known as $CD1c^{+}$ DCs. Similar to cDC1, also cDC2 share markers in mice and man, e.g., Sirpα (CD172a). pDCs in the murine and human system display homologies in the surface markers IL-3Rα (CD123) and CD45RA (see Fig. 1 and Table 1) (Guilliams et al., 2016; Robbins et al., 2008).

The three main DC subpopulations, cDC1, cDC2 and pDCs, were found to be present in all human lymphoid tissues, such as spleen, tonsils, lymph nodes, thymus, cord blood, and bone marrow (Alcantara-Hernandez et al., 2017; Heidkamp et al., 2016a; Mittag et al., 2011; Poulin et al., 2010; Segura et al., 2012; VelÁsquez-Lopera et al., 2008). Additionally, cDC1 and cDC2 were identified in liver and lung (Bamboat et al., 2009; Demedts et al., 2005; Haniffa et al., 2012; Schlitzer et al., 2013). Besides cDC1 and cDC2 in the dermis, the skin was suggested to contain additional DC subpopulations, such as LCs in the epidermis, and $CD14^{+}$ DCs in the dermis (Romani et al., 2010; Stoitzner et al., 2014; Zaba et al., 2007, 2009).

Although the transcription factors Irf8 and Irf4 have been identified to distinguish human cDC1 and cDC2 (Haniffa et al., 2012; Heidkamp et al., 2016a; Schlitzer et al., 2013), comparative RNA analyses between human and murine DCs subsets revealed that especially conventional human DC subsets display various differences in their expression profiles compared to their murine counterparts (Guilliams et al., 2016; Robbins et al., 2008; Watchmaker et al., 2013). This divergent transcriptional profile might be the reason for the discordant literature regarding the function of human conventional DCs in orchestrating T cell immune responses. This is especially the case when the DCs' capacity to cross-present antigens is analyzed. Various studies suggest that cross-presentation might not be restricted to one human cDC subpopulation (Cohn et al., 2013; Segura et al., 2013a). Moreover, human and murine cDC2 were proposed to be superior in the induction of Th17 $CD4^{+}$ T cell responses (Schlitzer et al., 2013). However, others suggested that human and murine $CD103^{+}$ DC subsets induced Th17 $CD4^{+}$ T cell responses, while human $CD103^{-}Sirp\alpha^{+}$ DCs (equivalent of cDC2) were inducing Th1 $CD4^+$ T cell responses. Also in the human system, pDCs were the main producers of type I IFNs (Asselin-Paturel and Trinchieri, 2005; Laustsen et al., 2018). However, the influence of tissue microenvironment, age of the samples, processing of the tissues, and the culturing of human DCs on our current functional assays has still not been elucidated completely.

12. The development of early DC progenitors within the bone marrow

Nearly all immune cells, including DCs, develop from a hematopoietic stem cell (HSC) residing in the bone marrow (Weissman and Shizuru, 2008). Exceptions are specialized tissue-resident macrophages, including LCs, which develop from the yolk sac (Davies et al., 2013). The differentiation pathways of HSCs and cells from yolk sac origin are depicted in Fig. 3. Differentiation, functional specialization, as well as migration to the tissues is tightly regulated by transcription factor networks. Although modern technologies and huge research efforts helped to shed light on these regulatory networks, not all active processes, regulatory mechanisms and their hierarchy in DC development are fully understood, which is especially true for the human system.

So far, HSCs can be distinguished into long-term, intermediate-term, as well as short-term HSCs. They give rise to multipotent progenitors (MPPs) and subsequently to lymphoid-primed multipotent progenitors (so-called LMPPs). Both populations display intrinsic heterogeneity and are partially biased to their offspring cell populations (Adolfsson et al., 2005; Pietras et al., 2015; Wilson et al., 2015). This means that a given LMPP cannot give rise to all lymphoid populations, but is rather committed to certain offspring populations (Naik et al., 2013). LMPPs are the progenitors of the common myeloid progenitors (CMPs) as well as the common lymphoid progenitors (CLPs) (Liu and Nussenzweig, 2010; Shortman and Naik, 2007). It has been proposed that the decision between the myeloid and B cell lineage is based on the interplay between the expression of the transcription factor PU.1 and the cell cycle. More precisely, the proliferation rate of B cell precursors is high and PU.1 might be incapable to reach a certain threshold. In contrast, the division rate of myeloid precursors is lower, allowing for PU.1 accumulation over time, which leads to a cell cycle shut down and the differentiation into the corresponding subsets (Kueh et al., 2013).

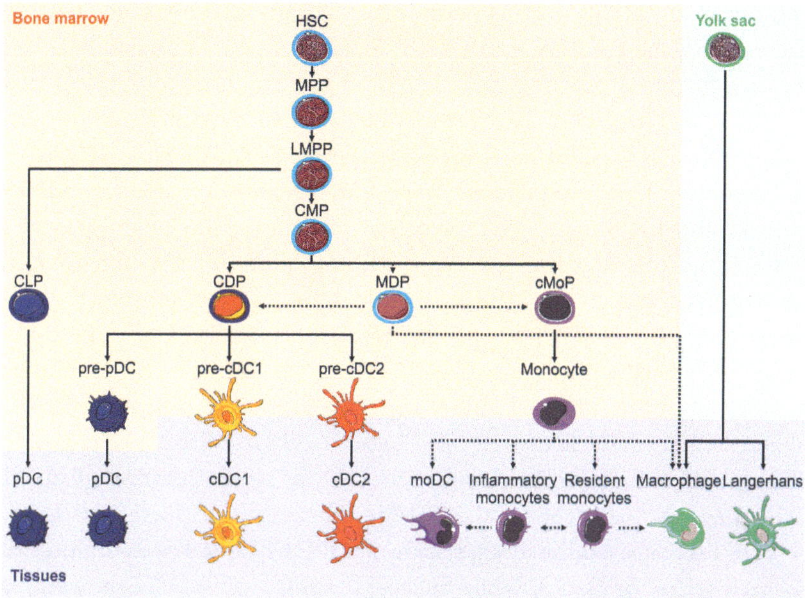

Fig. 3 Summary of the differentiation pathway of important antigen presenting cells. In the bone marrow, hematopoietic stem cells (HSCs) give rise to a set of downstream progenitors including the multipotent progenitor (MPP), and subsequently the lymphoid-primed multipotent progenitor (LMPP). This population differentiates to either the common lymphocyte progenitor (CLP), which is one source of plasmacytoid dendritic cells (pDCs), and the common myeloid progenitor (CMP). The CMP further develops to the common dendritic cell progenitor (CDP), which gives rise to pre-conventional dendritic cells (pre-cDC) 1, pre-cDC2 as well as pDCs. While pDCs terminally develop within the bone marrow, pre-cDCs leave the bone marrow and settle in the tissues, where the final differentiation to cDCs takes place. Additionally, CMPs develop to the common monocyte progenitor (cMoP), which is the source of all subsequent monocyte populations including inflammatory and resident monocytes, monocyte derived dendritic cells (MoDCs) and monocyte-derived macrophages. In the past, another differentiation step from CMPs to macrophage dendritic cell precursors (MDPs) potentially giving rise to both DCs and macrophages was proposed, but its existence has been challenged. In contrast to a HSC developmental pathway, resident macrophage populations and Langerhans cells are of yolk sac origin and perform self-maintenance within tissues. *Cellular images are provided and adapted from Servier Medical Art (smart.servier.com) and this images are licensed under the Creative Commons Attribution 3.0 Unported License (creativecommons.org/licenses/by/3.0/).*

CMPs are the progenitors of most DC subpopulations. In the past, a common macrophage-DC progenitor (MDP) has been proposed, with the potential to give rise to monocytes/macrophages as well as DCs (Auffray et al., 2009; Fogg et al., 2006; Liu and Nussenzweig, 2010;

Waskow et al., 2008), but its definitive ontogeny and existence has been challenged (Merad and Ginhoux, 2007; Sathe et al., 2014). In contrast, common dendritic cell precursors (CDPs) have the potential to develop either into the cDC1 and cDC2 subsets or into pDCs (Naik et al., 2007; Onai et al., 2007, 2013). CDPs are characterized by Fms-like-tyrosine kinase 3 (FLT3 or CD135) expression and the absence of lineage markers (Swiecki and Colonna, 2015). A recent study revealed heterogeneity in the transcriptional signatures at the CDP stage fostering DC lineage commitment (Schlitzer et al., 2015b). Interestingly, one progeny of the CDP population, the pre-cDCs populating all peripheral organs, already displays a bias to either cDC1 or cDC2 (Geissmann et al., 2010; Liu and Nussenzweig, 2010; Merad and Manz, 2009). Within murine pre-cDCs, Schlitzer et al. identified Siglec-H^{-}Ly6C^{-} cells as direct progenitors of cDC1, whereas Siglec-H^{-}Ly6C^{+} pre-cDCs were committed to the cDC2 lineage (Schlitzer et al., 2015b). Furthermore, Siglec-H^{-}Ly6C^{-} pre-cDC1 already exhibited expression of cDC1 hallmark transcription factors such as Id2 or BatF3, while Siglec-H^{-}Ly6C^{+} pre-cDC2 showed high Irf4 expression. Interestingly, Siglec-H^{+}Ly6C^{-} and Siglec-H^{+}Ly6C^{+} cells displayed high expression of E2-2 and Spi-B, indicating that they might maintain their pDC potential (Schlitzer et al., 2015b). These results imply that the core identity of cDCs is established in the bone marrow and not following homing to peripheral tissues. Similar results obtained by Grajales-Reyes et al. support this conclusion. Utilizing Zbtb46$^{GFP/+}$ mice, CD135^{+} CD11c^{+}CD115^{-}CD117^{int} MHC II^{int} Zbtb46-GFP^{+} and CD135^{+} CD11c^{+} CD115^{+}CD117^{-}MHC II^{-}Zbtb46-GFP^{+} pre-cDC populations were identified in the bone marrow, exclusively giving rise to either CD24^{+} DCs (cDC1) or Sirpα (CD172a^{+}) DCs (cDC2), respectively (Grajales-Reyes et al., 2015). Neither of the two pre-cDC populations developed into pDCs (Grajales-Reyes et al., 2015).

In the past, CDPs were described as the major source for pDCs. pDCs developing from CDPs fully differentiate in the bone marrow. Subsequently to the CDP, a lineage negative, Kit^{+}FLT3^{+}IL-7Rα^{-}E2-2^{high} pDC progenitor was identified (Cisse et al., 2008; Onai et al., 2013). This circulating progenitor further develops into a precursor population exhibiting expression of CD11c, B220 and Siglec-H while lacking expression of CCR9. This precursor cell remains plastic and therefore allows for the differentiation into either pDCs or cDCs depending on the tissue microenvironment (Dursun et al., 2016; Musumeci et al., 2019; Schlitzer et al., 2011, 2012). Hence, in some cases, it may happen that the decision between pDC and cDC is not made at the CDP stage within the bone marrow, but rather after circulation and arrival in the homing tissue. Moreover, recent data suggest the contribution of CLPs to the pDC lineage (Chen et al., 2013b; Rodrigues et al., 2018; Sathe et al., 2013). These cells were excluded in previous analysis since only $Kit^{+}FLT3^{+}IL\text{-}7R\alpha^{-}E2\text{-}2^{high}$ were thought to give rise to pDCs, while $IL\text{-}7R\alpha^{+}$ were thought to be bona fide B cell progenitors (Onai et al., 2013; Rodrigues et al., 2018). Only recently, this $IL\text{-}7R\alpha^{+}$ population was described to be the major source of pDCs (Rodrigues et al., 2018). Within the $IL\text{-}7R\alpha^{+}$ expressing lymphoid progenitor pool, $SiglecH^{+}Ly6D^{+}$ exhibited exclusive pDC potential. This finding was further reinforced by the group of Ginhoux, who identified a $Ly6D^{+}IL\text{-}7R\alpha^{+}Flt3^{+}CD2^{int}$ $CD81^{int}$ $CD115^{-}$ pre-pDC population from the CLP pool developing independently of myeloid $CD115^{+}$ cDC precursors (Dress et al., 2019). In their study, they identified CLPs as the sole source for pDCs. With respect to the striking resemblance of the identified $CD115^{+}$ CDPs and $CD115^{-}$ CLPs, which could be identified by single cell transcriptomic profiling approaches, divergent results of previous reports may be explainable by used surface markers and gating strategies during flow cytometry. Therefore, pre-pDCs potentially contaminated the myeloid precursor cell pool in earlier studies. Furthermore, the found close resemblance of B and pDC progenitors was strengthened by a FateID study of Herman et al. (Dress et al., 2019; Herman et al., 2018) and may explain previous reports claiming Rag1 activity and IgH rearrangement to happen in pDCs. This study attributes these features exclusively to the B cell lineage. Taken together, these results indicate that pDCs may arise solely from CLPs in the bone marrow (Dress et al., 2019).

Besides the development into DCs, CMPs or MDPs give rise to monocytes via a committed monocyte progenitor (cMoP) (Hettinger et al., 2013). Beneath the described model based on the progressive loss of differentiation potential, new single-cell transcriptomic analyses of bone marrow cells challenged the concept of mixed-lineage states and revealed new epigenetic landscapes and a tight transcription factor regulation during hematopoiesis (Paul et al., 2015).

Similar to mice, a gradual development of human $CD34^{+}$ HSCs to increasingly committed progenitors was observed in sophisticated cell culture systems (Lee et al., 2015b). This stepwise differentiation includes the development of a monocyte-DC progenitor from a granulocyte-monocyte-DC subsequently giving rise to either monocytes or CDPs (Lee et al., 2015b). While CDPs were only found in the bone marrow and cord blood, but not in lymphoid tissues or the blood (Lee et al., 2015b), Breton et al., identified the progeny of these CDPs in the bone marrow, cord blood, blood and peripheral lymphoid organs (Breton et al., 2015). These cells were direct precursors of CD141$^+$ DCs (cDC1) and CD1c$^+$ DCs (cDC2) with migratory capacity and lacked pDC and monocyte potential (Breton et al., 2015). These results were strengthened by single cell RNA-Seq and CyTOF data by See et al., further defining one uncommitted and two subset committed pre-cDC populations in the bone marrow, spleen and blood (See et al., 2017). Notably, the percentage of uncommitted cells was lower in the blood than in the bone marrow. Taken together, these studies support a conserved model of gradual DC differentiation from early progenitor stages in mice and men. Comparable to mice, DC core programs are established in the bone marrow, while the tissue microenvironment further refines DC identity and function (Heidkamp et al., 2016b).

13. Crucial factors for DC development: Flt3 and Flt3L, GM-CSF and GM-CSFR

Not only cDCs, but also pDCs crucially depend on the Fms-like tyrosine kinase 3 (Flt3) and its ligand Flt3L (Ginhoux et al., 2009; Haniffa et al., 2013; Schraml and Reis e Sousa, 2015). Flt3 is expressed by various precursor cells in the bone marrow as well as pre-cDCs, pre-pDCs, and differentiated DC subpopulations (Mackarehtschian et al., 1995; Matthews et al., 1991). Thus, the administration of Flt3L leads to an expansion of a mixture of cDCs and pDCs in vivo (Brasel et al., 1996; Maraskovsky et al., 1996; Waskow et al., 2008). In vitro cultures of murine bone marrow precursors in the presence of Flt3L generated cDC and pDC equivalents, with limited numbers of CD103$^+$ cDC1 (Mayer et al., 2014; Naik et al., 2005). The incubation of isolated bone marrow precursors in the presence of GM-CSF and/or IL-4 mostly results in the differentiation of monocyte-derived DCs (Helft et al., 2015; Mayer et al., 2014). Recent refinements of the culture protocols, combining Flt3L and GM-CSF application in a timely concerted way was recently demonstrated to be efficient for the generation of CD103$^+$ DCs (termed iCD103$^+$ DCs) with a CD8α^+ cDC1-like phenotype. These cells develop from the heterogeneity of isolated cells of the bone marrow (Menezes et al., 2016) and not only from monocytes in the culture dish and display CD8α/CD103 DC signature genes. These DC-like cells do not express Sirpα or CD8α itself, but can be identified by CD24 and Clec9A expression (Mayer et al., 2014). It was further noted that the iCD103$^+$

CD8α like cDC1 cells depend on Irf8 and Batf3, whereas GM-CSF derived DCs could be generated in normal numbers. (Mayer et al., 2014). The culture-derived data suggest that DCs rely on a fine-tuned mixture of growth factors for their development. Thus, such a delicate balance might be also needed to further improve cultures to generate human primary-like DC subsets from their precursors (Haniffa et al., 2013; Lee et al., 2015a; Maraskovsky et al., 2000).

The dependency of DCs on the Flt3L-Flt3-axis becomes evident in vivo, when mice lacking Flt3L or the receptor Flt3 (CD135) have been analyzed. In such mice, the DC numbers were reduced, which seemed to be more pronounced in Flt3L mice, then in mice, in which Flt3 was deleted or its function was blocked (McKenna et al., 2000; Tussiwand et al., 2005; Whartenby et al., 2005). Beside reduced DC numbers, in Flt3L deficient mice also other hematopoietic lineages, such as B, T and NK cells, were influenced (McKenna et al., 2000). In line with tissue culture analyses, it was observed that the absence of Flt3 (CD135) was not influencing the amount of precursor cells, such as MDPs in the bone marrow and pre-DCs in spleen (Waskow et al., 2008).

The culture protocols using GM-CSF to expand monocyte derived DCs are not resembling the in vivo situation as mice lacking GM-CSF or the GM-CSF receptor display normal development of MoDCs (after induction) and display no altered development of DCs within the bone marrow in vivo (Greter et al., 2012). However, there were effects on several DC subpopulations in some tissues suggesting a role of the GM-CSF/GM-CSFR in the maintenance of peripheral and lymphoid tissue DCs. This data indicate a concerted role of GM-CSF with other factors, most probably Flt3L, for the development and homeostasis of DCs (Greter et al., 2012; Vremec et al., 1997; Waskow et al., 2008). Interestingly, GM-CSF might be rather crucial for the development of non-lymphoid tissue $CD103^+$ DCs (Edelson et al., 2011; Greter et al., 2012; Kingston et al., 2009), which would be in accordance to the in vitro culture protocol of Mayer et al. (Mayer et al., 2014).

14. Transcriptional regulation of DC differentiation and function

The differentiation from CDPs to cDC1, cDC2 or pDCs is regulated by diverse sets of transcription factors. Before we give an overview of transcription factors important for the development and function of one specific subset, transcription factors involved in lineage commitment and regulation of multiple subsets will be discussed. A comprehensive summary of the transcriptional networks driving DC subset differentiation from CDPs and CLPs is shown in Fig. 4.

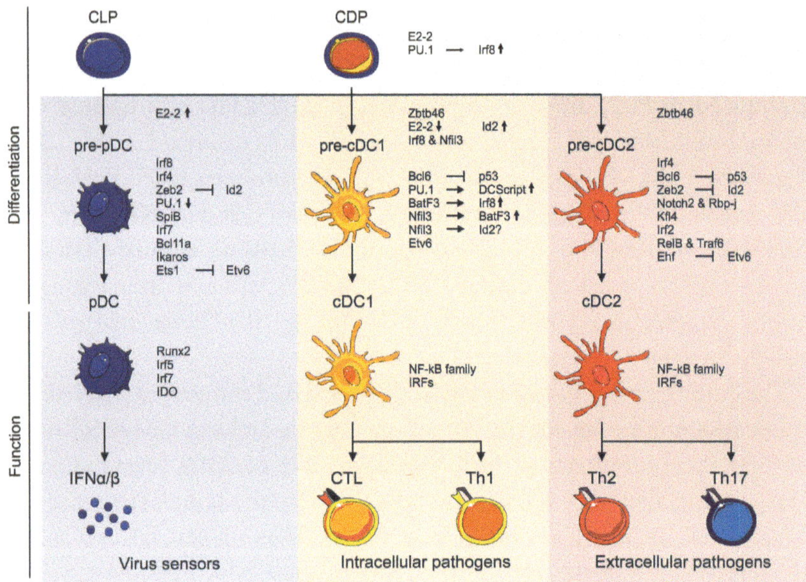

Fig. 4 Transcription factor networks during dendritic cell differentiation pathways. cDC1 and cDC2 cells as well as pDCs develop from CDPs via a pre-cDC1, pre-cDC2 and pre-pDC state. In addition, pDCs originate from CLPs. Key transcription factors within these differentiation processes and regulatory mechanisms are indicated for different developmental stages. Furthermore, transcription factors involved in DC function are shown. pDCs are the major producer of IFNα/β during viral infections acting as virus sensors. cDC1 cells excel in the priming of CTL responses, but also prime Th1 cells, thereby acting in the defense against intracellular pathogens. cDC2 cells are major inducers of Th2 and Th17 cells in response to extracellular pathogens. CDP, common dendritic cell precursor; CLP, common lymphocyte progenitor; pDC, plasmacytoid dendritic cell; cDC, conventional dendritic cell; CTL, cytotoxic T lymphocytes; Th1, Type 1 T helper cell; Th2, Type 2 T helper cell; Th17, Type 17 T helper cell. Horizontal arrows between transcription factors indicate directional regulatory functions. Vertical arrows between transcription factors indicate up- or downregulation. Please find references for transcription factor literature in Table 1. *Cellular images are provided and adapted from Servier Medical Art (smart.servier.com) and this images are licensed under the Creative Commons Attribution 3.0 Unported License (creativecommons.org/licenses/by/3.0/).*

14.1 Crucial regulatory transcription factors of multiple DC subsets

14.1.1 Zbtb46 in cDCs

Broad complex tramtrack bric-a-brac zinc finger (BTB-ZF) transcription factors comprise more than 45 members in mice and men and several of these transcription factors are specifically involved in the transcriptional regulation of DC subsets. In general, BTB-ZF proteins possess an N-terminal BTB domain allowing for protein interactions, while the C-terminal zinc finger domain is responsible for DNA binding. In collaboration with chromatin remodeling complexes recruited by the BTB domain, transcription of target genes can be initiated or repressed (Beaulieu and Sant'angelo, 2011).

Zbtb46 is the best characterized member of the BTB-ZF family in cDCs. This transcription factor was identified as marker for both cDC subsets (cDC1 and cDC2) in mice and men (Meredith et al., 2012a,b; Satpathy et al., 2012). Beside cDCs and pre-DCs in the bone marrow, Zbtb46 is also expressed in non-hematopoietic cells, such as endothelial cells and erythroid precursors (Satpathy et al., 2012; Wang et al., 2019). The Zbtb46 promotor was suggested to be superior compared to CD11c for the generation of conditional knockout mice since other hematopoietic cell lineages, such as pDCs, NK cells, activated monocytes, macrophages and other lymphoid cells, can express CD11c to some extent (Loschko et al., 2016). Due to the fact that Zbtb46 is also expressed in non-hematopoietic cells, the application of diphtheria toxin (DT) was found to be lethal in $Zbtb46^{DTR}$ mice (Meredith et al., 2012a). Further analyses revealed that lethality can be circumvented by the generation of bone marrow chimeras, in which $Zbtb46^{DTR}$ bone marrow is transferred into irradiated wild-type mice (Meredith et al., 2012a). This enables the specific depletion of cDCs, while cells with no hematopoietic origin were unaffected. When utilizing $Zbtb46^{GFP/+}$ mice, it could be shown that GFP expression was restricted to cDCs and pre-cDCs (Satpathy et al., 2012). Overexpression of Zbtb46 impeded granulocyte differentiation from bone marrow-located progenitor cells, but fostered cDC development (Satpathy et al., 2012). Conceivably, mode of action of Zbtb46 in controlling cDC lineage fate may be the limitation of the expression of growth factor receptors usually required for the development of other lineages (Satpathy et al., 2012). Nevertheless, replacement of Zbtb46 with GFP on both alleles ($Zbtb46^{GFP/GFP}$ mice) still allowed for cDC development. Therefore, Zbtb46 may not be critical for DC development but its promotor represents a strong alternative to the CD11c promotor for cDC specific gene expression or knock down, since

$CD11c^{DTR}$ or CD11c-Cre mice usually require controls ($LysM^{DTR}$ or LysM-Cre) for the contribution of monocytes and macrophages (Clausen et al., 1999; Satpathy et al., 2012; Schreiber et al., 2013). Functionally, Chromatin Immunoprecipitation (ChIP) DNA sequencing identified Zbtb46 as transcriptional repressor of cDC genes associated with DC activation, such as the MHC II genes and the vascular endothelial growth factor (VEGF) (Meredith et al., 2012b). Zbtb46 expression rapidly declined following TLR stimuli (Meredith et al., 2012b). In combination, Zbtb46 may serve to repress cDC activation during steady-state conditions.

14.1.2 Bcl6 in cDCs

Another prominent member of the BTB-ZF transcription factor family is Bcl6: a potent repressor of protein and microRNA transcription. Bcl6 exerts important functions in many other cell types, such as macrophages, B and T cells (Johnston et al., 2009; Niu et al., 2003), but also acts in pre-cDCs as repressor of apoptosis. This is potentially mediated by a repression of the cell cycle arrest protein p53, thereby contributing to the survival of cDC1 and cDC2 in general (Ohtsuka et al., 2011). Complete ablation of Bcl6 in mice leads to a severe reduction of both, cDC1 and cDC2 populations (Ohtsuka et al., 2011). Nevertheless, in these studies, the Bcl6 knock out was not specific to DCs and defects in other cell populations might affect the cDC development.

14.1.3 The E2-2 to Id2 axis determines pDC vs. cDC1 lineage decision

While Zbtb46 is important for cDCs, the E2-2 and Id2 axis determines pDC and cDC1 lineage commitment. Together with several other proteins, E2-2 is a class I basic helix loop helix transcription factor family member. These transcription factors exert their regulatory influence by *E*-box DNA motif binding after protein dimerization (Satpathy et al., 2011). In combination with inhibitor of DNA binding (Id) proteins, E-proteins regulate many differentiation processes within the hematopoietic system (Kee, 2009). If an E-protein dimerizes with an Id protein instead of another member of the E-protein-transcription factor family, the DNA binding capability is lost, thereby negatively regulating E2-2 target genes (Satpathy et al., 2011). The importance of E- and Id proteins in the development of DC lineage members was first demonstrated in an in vitro study by utilizing ectopic expression of Id2 in precursor cells. This led to a block of pDC differentiation, most likely by E2-2 inhibition (Spits et al., 2000). High protein levels of E2-2 can be found within pDCs, where this transcription factor directly activates genes associated with pDC differentiation and activity, such as Irf8, Spi-B, Bcl11a, Irf7, CC-chemokine receptor 9 (Ccr9), BDCA2, TLR7 and TLR9 (Cisse et al., 2008; Kee, 2009; Murphy et al., 2016).

While E2-2 is considered as a driver of pDC development, its antagonist Id2 was found to be important in LC and cDC development. Furthermore, reduction of E2-2 levels in mice carrying only one functional E2-2 allele already reduced pDC numbers and led to functional aberrations (Cisse et al., 2008). This might be due to the loss of Id2 repression by E2-2 (Satpathy et al., 2011). Interestingly, complete deletion of E2-2 in pDCs results in cDC-like reprogramming of pDCs, which is involving phenotype, morphology and gene expression signature (Ghosh et al., 2010). Mice deficient for Id2 specifically lacked LCs and cDC1, while pDC numbers were increased (Hacker et al., 2003). Interestingly, Id2 was also found to suppress many B cell related genes shared with pDCs (Becker-Herman et al., 2002; Hacker et al., 2003; Robbins et al., 2008). Altogether, this may imply that E2-2 dependent pDC differentiation is a default pathway, which has to be actively repressed during cDC development. Nevertheless, the E2-2:Id2 interaction cannot completely account for the cDC-pDC lineage choice, since the generation of cDC2 remains unaffected in Id2 deficient mice (Hacker et al., 2003). Furthermore, Id2 expression in pre-DCs is only observed at low levels, while in fully differentiated DCs it is expressed at high levels. This is in contrast to its partner E2-2, which is present as early as the CDP state (Hacker et al., 2003; Murphy et al., 2016). However, the E2-2: Id2 axis depicts an important antagonistic checkpoint feedback-loop during pDC:cDC1 lineage decision. Interestingly, an early study discovered a novel potential pDC progenitor cell indicating a different mechanism of pDC differentiation (Onai et al., 2013). Unfortunately, pre-excluding Interleukin-7 receptor (IL-7R) expressing cells from this precursor population (due to their B cell differentiation potential), masked the most potent population for pDC differentiation during this study. This issue was recently resolved in a study suggesting that pDCs mainly develop from IL-7R$^+$ lymphoid progenitors (Dress et al., 2019; Rodrigues et al., 2018). Within the IL-7R expressing CLPs, Ly6D$^+$Siglec-H$^+$ CLPs seemed to constitute committed pre-pDCs, whereas the Ly6D$^-$Siglec-H$^-$ subpopulation was still uncommitted. Depending on either Irf8 or Ebf1 expression in Ly6D$^+$ Siglec-H$^-$ CLPs, pDCs or B cells predominantly differentiate from this precursor population (Rodrigues et al., 2018). In general, expression of Irf8, E2-2 and Runx2 was found in Ly6D$^+$Siglec-H$^+$ progenitors and further increased in mature pDCs, while Spi-B and Irf7 expression was low or absent (Rodrigues et al., 2018). Expression of Ebf1 was restricted to a fraction of Ly6D^{+}Siglec-H^{-} progenitors allowing B cell lineage fate, whereas the other fraction expressed high levels of Irf8 and lacked Ebf1 expression. Finally, early Irf8 expression in Ly6D^{+}Siglec-H^{+} progenitors was a prerequisite for the development of pDCs, while later E2-2 expression was necessary to establish pDC identity and function (Rodrigues et al., 2018).

14.1.4 Spi-B is a fine tuner of the E2-2 to Id2 axis

While the E2-2 to Id2 axis depicts a dominant regulatory pathway separating pDC and cDC1 differentiation, other factors acting via this axis have been identified in pDCs. These transcription factors are direct targets of E2-2. For instance, the ETS-transcription factor Spi-B is important for both, pDC development and function, in mice and men (Sasaki et al., 2012; Schotte et al., 2004). In humans, the E2-2 target Spi-B synergistically acts with E2-2, thereby circumventing Id2-mediated repression of pDC development (Nagasawa et al., 2008). During this study, Spi-B alone was not sufficient to overcome Id2-dependent blockade of pDC development, but Id2 levels were reduced in Spi-B overexpressing mice implying that Spi-B is not a master-regulator of the pDC lineage, but rather a fine-tuner of E2-2 activity (Nagasawa et al., 2008). In addition to its developmental function, Spi-B supports the regulation of other E2-2 target genes in cooperation with IRFs. This includes the two nucleic acid sensors TLR7 and TLR9, since mice deficient for Spi-B exhibited an impaired IFN type I response to stimulation with the TLR7 agonist polyuridylic acid or the TLR9 agonist ODN1668 (Sasaki et al., 2012). In addition to its liaison with E2-2, Spi-B directly interacts with Bcl2, thereby regulating pDC survival in humans (Karrich et al., 2012).

14.1.5 The bona fide transcriptional repressor Bcl11a acts in E2-2 and Id2 regulation

Another transcription factor specific for pDCs and B cells in mice and men is the zinc finger protein and transcriptional repressor Bcl11a, which has been shown to be involved in E2-2 and Id2 regulation (Chistiakov et al., 2014; Marafioti et al., 2008; Pulford et al., 2006). Mice lacking Bcl11a display a considerable reduction in pDC numbers (Wu et al., 2013). This might not only be due to the regulation of E2-2 and Id2 transcription by Bcl11a, but also due to the essential role of Bcl11a for Flt3 expression in early progenitor cells (Wu et al., 2013). Furthermore, it was hypothesized that Bcl11a is a bona fide transcriptional repressor acting on genes important for cDC commitment, thereby favoring pDC generation (Ghosh et al., 2010).

14.1.6 Zeb2 facilitates cDC2 differentiation by repressing Id2 activity

Beside the already well described Zbtb46 and E2-2:Id2 transcription factors, recently, a new transcription factor was discovered to play a pivotal role in transcriptional regulation of DCs. The zinc finger transcription factor Zeb2 was shown to be present in pre-cDCs. Zeb2 displayed an increased expression in cDC2 and decreased during cDC1 development (Scott et al., 2016). While the E2-2:Id2 axis provides a platform for determination of lineage fate between cDC1 and pDCs, Zeb2 may rather provide the missing link explaining cDC2 differentiation within this axis. Conditional ablation of Zeb2 in CD11c expressing cells exhibited decreased cDC2 and pDC numbers, whereas cDC1 numbers were increased (Scott et al., 2016). On the other hand, forced Zeb2 overexpression in CD11c expressing cells led to decreased cDC1 numbers. Utilizing ChIP analyses, it was suggested that Id2 might be a direct target of Zeb2. This was in line with the upregulation of Id2 in Zeb2 deficient mice (Scott et al., 2016). Thus, Zeb2 is recognized as a regulator of DC development as it is preventing cDC1 lineage fate by repression of Id2.

14.1.7 Etv6 promotes cDC1-like programming of cDCs

The ETS family transcription factor Etv6 (also known as TEL), originally described as a transcription factor controlling hematopoietic stem cell and progenitor cell function as well as thrombopoiesis, plays a pivotal role in shaping the cDC1 lineage (Lau et al., 2018). Etv6 was found to be expressed in cDCs, while the expression of antagonists such as Ets1 expressed by pDCs, lymphocytes and neutrophils or Ehf expressed by cDC2 was lacking. In Flt3L-driven DC differentiation from bone marrow cells deficient in Etv6 expression, the differentiation of cDC1-like cells was perturbed. Additionally, the depletion of an floxed Etv6 in bone marrow cells by a tamoxifen-driven Cre recombinase revealed a development of primary cDC1 with an expression profile and chromatin structure reminiscent of pDCs, whereas cDC specific gene and chromatin signatures were downregulated (Lau et al., 2018). Hence, this transcription factor may act in further refining the pDC:cDC1 lineage commitment, as Etv6 deficient cDC1 were incapable of cross-priming tumor-specific T cell responses (Lau et al., 2018).

14.1.8 The PU.1 dependent transcriptional repressor DC-SCRIPT favors the development of cDC1 not pDCs

Only recently, PU.1, which also belongs to the ETS transcription factor family, was identified as another player acting in pushing cDC1 generation by inducing the transcriptional repressor DC-SCRIPT (Chopin et al., 2019). While the absence of PU.1 did not affect the expression of cDC1 hallmark transcription factors, such as Irf8, Id2, Irf4, and BatF3, expression of the zinc finger transcription factor DC-SCRIPT (gene Zfp366) decreased. The expression of DC-SCRIPT is switched on at the CDP stage and reaches its maximum in lineage-committed mature cDC1 (Chopin et al., 2019). By gain- and loss-of function experiments utilizing bone marrow cultures, DC-SCRIPT was shown to be involved in favoring cDC1 lineage fate, while repressing pDC development of progenitors. Of note, DC-SCRIPT did not regulate Nfil3, BatF3, or Id2 (Chopin et al., 2019). On the other hand, pDC progenitors express only low levels of PU.1, indicating the presence of a regulatory mechanism to possibly avoid transcriptional suppression by DC-SCRIPT in pDCs (Chopin et al., 2019).

14.2 Specific factors regulating transcription in cDC1

14.2.1 IRF8 in cDC1

In both, the tissue-resident as well as the migratory DC subpopulations, the major feature of cDC1 cells (or Irf8-dependent DCs) is their specialized capability to cross-present antigens and to induce strong cytotoxic $CD8^+$ T cell responses (Den Haan et al., 2000; Den Haan and Bevan, 2002; Dudziak et al., 2007; Hildner et al., 2008; Lehmann et al., 2017; Neubert et al., 2014).

The IRF transcription factor family comprises a total of nine members in mice and men. All of the IRF family members are exhibiting C-terminal IRF associated domains for protein-interaction and a conserved N-terminal DNA binding domain recognizing interferon responsive DNA elements (Tamura et al., 2008). The transcription factor Irf8 is of particular interest for the cDC1 differentiation. Following heterodimerization with other transcription factors, Irf8 can either act as transcriptional activator (in combination with PU.1, AP1 and BatF3) or repressor (Irf1 and Irf2) (Tamura et al., 2015). In addition to cMoP, Irf8 expression occurs as early as the CMP to CDP transition state and pursues in terminally differentiated cDCs and pDCs (Satpathy et al., 2012). Moreover, absence of Irf8 led to a reduced expression of several key transcription factors including Id2, Bach2, Bcl6, and Kfl4 (Becker et al., 2012). Further analyses revealed the necessity of Irf8 expression to escape neutrophil differentiation (Becker et al., 2012). Mice deficient for Irf8 developed a Chronic Myelogenous Leukemia-like disease, showed a decreased IL-12 production, and an impaired viral immunity (Holtschke et al., 1996). More detailed analyses of the total Irf8 knock out revealed that the ablation of Irf8 negatively influences several steps of cDC1 development (Sichien et al., 2016). Hence, Irf8 deficiency results in a severe developmental impairment of the cDC1 lineage in lymphatic organs and the periphery (Aliberti et al., 2003; Luda et al., 2016). Therefore, Irf8 may be considered as master regulator of cDC1 development even though its expression occurs in other immune cell populations as well (Netherby et al., 2017; Satpathy et al., 2012).

The activity of Irf8 is regulated on different levels. The transcription factor PU.1 was described to be involved in chromatin remodeling by inducing chromosomal looping. Subsequently, regulatory DNA elements, which are normally located far from each other, can interact allowing expression from the Irf8 locus, thereby contributing to DC lineage commitment (Schonheit et al., 2013). Only recently, the combination of PU.1 and Irf8 in early progenitors was found to foster a chromatin landscape, in which enhancers for DC developmental programs are activated (Kurotaki et al., 2018). In this study, lack of Irf8 led to an enhancer landscape, which would be rather reminiscent to an earlier progenitor cell stage (Kurotaki et al., 2018). Therefore, Irf8 may be important to establish chromatin configurations that are needed to induce DC specific transcriptional programs. Furthermore, BatF3 seems to sustain Irf8 auto-activation, which is necessary during the pre-cDC1 to cDC1 transition (Grajales-Reyes et al., 2015). Moreover, transgenic overexpression of Irf8 was able to overcome the developmental impairment of CD8α^+ DCs (cDC1) in BatF3 deficient mice (Grajales-Reyes et al., 2015). This data are in line with the initial characterization of the BatF3$^{-/-}$ mouse, in which the generation of CD8α^+ and CD103$^+$ cDC1s was hampered leading to a lack of cross presenting DCs and high susceptibility to viral infections and syngeneic tumors (Hildner et al., 2008). There, missing auto-activation of Irf8 by BatF3 might have potentially contributed to the observed phenotype. Furthermore, it was described that the combined action of Irf8, Id2 and BatF3 promotes cDC1 differentiation, whereas sole Irf8 activity seems to shift differentiation in favor of pDCs (Jaiswal et al., 2013). Utilizing chromatin immunoprecipitation DNA sequencing (ChiP-Seq), Grajales-Reyes et al. identified several transcription factor binding sites for BatF3 26 kilobases downstream (−26 kb) and 32 kb upstream (+32 kb) of the Irf8 transcriptional start site

(TSS) (Grajales-Reyes et al., 2015). Additionally, Irf8 binding sites were located at −26, −16, +32 and +41 kb from the Irf8 TSS (Grajales-Reyes et al., 2015). Both, cDC1 and pDCs, exhibited histone acetyltransferase binding at the +32 and +41 kb elements suggesting enhancer activity. On the other hand, Irf4 expressing cDC2 had no enhancer activity potentially implying that rather Irf8:BatF3 but not Irf4:BatF3 heterodimers are active at these elements (Grajales-Reyes et al., 2015). Moreover, the +32 enhancer, binding site for BatF3 and Irf8, was specifically activated in cDC1, whereas the +41 kb enhancer was specifically active in pDCs (Grajales-Reyes et al., 2015). Besides Irf8 binding, this enhancer exhibited several E-box motifs implying that E-proteins, such as E2-2, might control Irf8 expression in pDCs (Durai et al., 2019; Grajales-Reyes et al., 2015). During a follow up study, the group of Murphy recently investigated the importance of the +32 kb and + 41 kb enhancer in detail using DC chromatin profiling and CRISPR/Cas9 genome editing (Durai et al., 2019). Here, they demonstrated the necessity of the +32 kb enhancer in normal cDC1 (via BatF3) and compensatory cDC1 (via BatF and BatF2) development. Pre-cDC1 developed independently of this enhancer. Interestingly, assay for transposase-accessible chromatin using sequencing (ATAC seq) revealed +41 kb enhancer accessibility during the MDP to pre-cDC1 transition, while this locus was closed in mature cDC1 (Durai et al., 2019). CRISPR/Cas9 mediated deletion of the +41 kb enhancer not only decreased Irf8 expression in pDCs, but hindered Irf8 induction in CDP thereby preventing pre-cDC1 and cDC1 development (Durai et al., 2019). Hence, the developmental stage dependent activity of the +41 kb enhancer within the Irf8 super-enhancer is critical for cDC1 development. How this enhancer is controlled and potentially discriminates between pDC and cDC1 development was recently assessed (Bagadia et al., 2019). In $Zeb2^{high}Id2^{low}$ CDPs, Irf8 expression was found to be +41 kb enhancer dependent. Following an Nfil3 dependent transition of a fraction of $Zeb2^{high}Id2^{low}$ CDPs to $Zeb2^{low}Id2^{high}$ CDPs, the resulting cells exhibit exclusive cDC1 differentiation potential (Bagadia et al., 2019). Within the $Zeb2^{low}Id2^{high}$ pre-cDC1 population, Id2 may serve to inhibit E-proteins and their interaction with the +41 kb Irf8 enhancer previously described to be important for pDCs, while BatF3 maintains Irf8 activation at the +32 kb enhancer (Bagadia et al., 2019; Durai et al., 2019).

On a functional level, Irf8 dependent cDC1s are associated with the induction of CTL and Th1 T cell responses thereby promoting immunity against viruses and intracellular parasites (Tussiwand et al., 2015). In a previous study, mice resistant to phosphorylation-induced β-Catenin degradation, showed increased cDC1 numbers and a higher Irf8 promotor activity. This led to increased IL-12 production by splenic $CD8\alpha^+$ cDC1s following *Toxoplasma gondii* infection (Cohen et al., 2015). Thus, Irf8 is not only involved in cross-presentation and consequently the induction of cytotoxic $CD8^+$ T cell responses, but transcriptional regulation of Irf8 is also involved in supporting Th1 $CD4^+$ T cell immunity (Cohen et al., 2015).

14.2.2 BatF3 in cDC1

The transcription factor BatF3 was previously described to be highly important for cDC1 development in lymphoid organs and the periphery (Edelson et al., 2010; Hildner et al., 2008). BatF3 deficiency led to the ablation of cDC1, which was further associated with a loss of cytotoxic T cell cross-priming. Interestingly, during a study with *Toxoplasma gondii*, it was recognized that administration of murine IL-12 was able to restore cytotoxic $CD8^+$ T cell responses in BatF3 deficient mice, indicating a potential rescue mechanism for cDC1 in $BatF3^{-/-}$ mice (Mashayekhi et al., 2011). Furthermore, in another study, cDC1 ($CD8\alpha^+$ DCs) developed after $Id2^{-/-}$, $BatF3^{-/-}$ or $Nfil3^{-/-}$ bone marrow transfer into irradiated wild-type mice without additional stimuli in a solely Irf8 dependent manner (Seillet et al., 2013). In another study, BatF as well as BatF2 acted to compensate for BatF3 function in the development of cDC1 (Tussiwand et al., 2012). Moreover, this cDC1 recovery can be explained by the shared specificity of the leucine-zipper domain in BatF and BatF3 enabling these proteins to interact with Irf8 and Irf4 following heterodimerization with Jun (Tussiwand et al., 2012). In particular, the interaction of BatF with Irf8 would again allow for Irf8 auto-activation, which is essential during the differentiation of pre-cDC1 precursors to cDC1. Since BatF as well as BatF3 were found to be able to interact with Irf4 as well as Irf8, it was assumed that DC activity is rather dictated by the latter factors (Tussiwand et al., 2012). Owing to the fact that BatF is not expressed in pre-cDC1 or mature cDC1, the described rescue mechanism may only explain cDC1 recovery in BatF3 deficient mice in non-steady-state conditions (Grajales-Reyes et al., 2015).

14.2.3 Other transcription factors in cDC1: Nfil3

Besides the master regulator Irf8, other sets of transcription factors contribute to cDC1 lineage decision and function. Nfil3 is an important transcription factor during cDC1 development. Absence of this transcriptional repressor in $Nfil3^{-/-}$ mice led to a depletion of the cDC1 compartment, while cDC2 and pDCs remained unaffected (Kashiwada et al., 2011). This phenotype can be partially explained by a reduced BatF3 expression in Nfil3$^{-/-}$ mice (Kashiwada et al., 2011) and is in line with the finding that ectopic expression of BatF3 in Nfil3$^{-/-}$ mice restores the cDC1 compartment (Kashiwada et al., 2011). Interestingly, Nfil3 is also associated with the regulation of Id2 activity in a yet unidentified manner (Gascoyne et al., 2009; Murphy et al., 2016).

14.3 Specific factors for the transcriptional regulation of cDC2

cDC2 are the major subset involved in priming of CD4^{+} T cells (Den Haan and Bevan, 2002; Dudziak et al., 2007; Kamphorst et al., 2010). In the last couple of years, efforts were made to define transcriptional regulators of cDC2 development and function. Thereby Irf4, Notch/Rbp-J, Klf4, Irf2, RelB, Traf6 and Zeb2 have been identified (Caton et al., 2007; Ichikawa et al., 2004; Lewis et al., 2011; Scott et al., 2016; Suzuki et al., 2004; Wu et al., 1998). The role of these factors will be described in the following section.

14.3.1 Irf4 in cDC2

Irf4 has been described as one of the key transcriptional regulators in cDC2 development and function. Its importance for the cDC2 lineage in mice was underlined due to the loss of CD11b^{+}CD8^{-}CD103^{-} DCs after its conditional ablation in CD11c expressing cells (Vander Lugt et al., 2014). Similar to Irf8 for cDC1, Irf4 is assumed to be a master regulator of cDC2 development (Tamura et al., 2005). Examination of transcription factor expression in splenic DC subsets revealed Irf8 expression in cDC1 (CD8^{+}CD4^{-} DCs) and Irf4 expression in cDC2 (including CD8^{-}CD4^{+} DCs and CD8^{-}CD4^{-} DCs) as well as pDCs (Suzuki et al., 2004; Tamura et al., 2005). Utilizing Irf4, Irf8 and double deficient mice, the importance of Irf4 and Irf8 for these four subsets was demonstrated. After retroviral transduction, Irf4 or Irf8 expression was reconstituted in Irf4$^{-/-}$Irf8$^{-/-}$ bone marrow cells. Compared to double-deficient cells, both transcription factors promoted cDC and pDC development, reconstituted LPS sensitivity, and enhanced the expression of co-stimulatory molecules. On the other hand, the selective requirement of Irf4 for cDC2 and Irf8 for cDC1 was demonstrated (Tamura et al., 2005). Furthermore, GM-CSF-driven DC development was dependent on Irf4, whereas the Flt3 dependent DC development, however, rather required Irf8. Finally, selective requirement for Irf8 was shown for IL-12p40 and IFNα expression. IL-12 was previously demonstrated to be a key feature during type 1 $CD4^+$ T helper cell (Th1) differentiation (Manetti et al., 1994; Persky et al., 2005). This finding is in line with earlier studies indicating that cDC1 ($CD8^+$ DCs) drive Th1 responses, while cDC2 ($CD8^-$ DCs) induce Th2 responses (Maldonado-LÓpez et al., 1999; Pulendran et al., 1999; Williams et al., 2013; Yasumi et al., 2005). Taken together, this may imply a role of Irf8 in cDC1 and Irf4 in cDC2 in the generation of Th1 or Th2 effector responses, respectively.

With regard to functional modules, utilizing chromatin immunoprecipitation DNA sequencing (ChiP-Seq) combined with gene expression profiling an Irf4 dependent regulatory unit was identified (Vander Lugt et al., 2014). Irf4 expression positively correlated with CD11b, and proteases of the MHC II pathway, such as cathepsin S and legumain (AEP), which have been formerly shown to be expressed at higher levels in splenic cDC2 (Dudziak et al., 2007). On the other hand, Irf8 expression correlated with CD8α and CD103 as well as genes of the MHC I machinery, such as transporter associated with antigen processing (TAP) 1 and 2. Hence, Irf4 dependent DCs were found superior to Irf8 dependent DCs in the formation of pMHCII complexes. This superior MHC II machinery was found in both, resident and migratory, cDC2. Moreover, the described expression differences in Irf4, Irf8 and components of the MHC I and II presentation machinery seem to be conserved in the human counterparts of the murine DC subsets (Heidkamp et al., 2016b; Robbins et al., 2008).

Even though, Irf4 is widely accepted as master regulator of cDC2 development, one study claimed that it is rather the loss of the migratory capacity, but not the developmental potential, which changes in absence of Irf4 (Bajaña et al., 2012). During a follow up study, the same group described the importance of Irf4 for cDC2 in a tissue dependent manner (Bajana et al., 2016).

Besides the functional role of Irf4 as master regulator, Irf4 dependent cDC2 cells can be further subdivided into functionally different Notch2 and Klf4 dependent cDC2, which will be summarized in the following sections.

14.3.2 Notch2 in cDC2 cells

In mammals, Notch signaling is triggered via four Notch receptors - Notch1 to 4, combined with the binding of their ligands Delta, Serrate and Jagged (Bray, 2006). Following ligand recognition by the Notch receptor, the intracellular receptor domain (IRD) is cleaved, enabling its translocation to the nucleus. To orchestrate Notch target gene transcription, the IRD interacts with the DNA binding protein Rbp-j (also called CSL) and the co-factor Mastermind (Schweisguth, 2004). Because ligands of Notch are transmembrane proteins themselves, Notch signaling is predominantly controlled by neighboring cells (Bray, 2006). During a study, in which Rbp-j was selectively deleted in CD11c expressing cells, the cDC2 compartment in the spleen was split in two, whereas cDC1 remained unaffected (Caton et al., 2007). Interestingly, only cDC2 in the spleen seemed to be affected. Since it was described earlier (Dudziak et al., 2007) that splenic $DCIR2^+CD8^-$ cDC2 primarily reside within the marginal zone (MZ), location of cDC2 as well as expression of the Notch ligand Delta-like 1 (Dll1), which is a homolog of the notch ligand delta, were assessed by immunofluorescence analysis. The authors could demonstrate the co-localization of cDC2 cells and Dll1 expression within the marginal zone, while marginal zone cDC2 cells were absent in mice conditionally depleted of Rbp-j (Caton et al., 2007). Moreover, Notch1, 2 and 4, but not Notch3 expression was detected in cDC1 ($CD8^+$) and cDC2 ($CD8^-$) DCs. While conditional knockout of Rbp-j increased Notch1 expression in cDC2, Notch4 expression was reduced indicating a positive feedback loop between Rbp-j and Notch4. In summary, these studies provide evidence that Notch signaling is dependent on the tissue microenvironment and that only a fraction of cDC2s is dependent on Notch signaling.

Further characterization of the Notch-dependent $CX_3CR_1^{low}ESAM^{high}$ $CD4^+$ cDC2 fraction displayed that this population was solely dependent on Notch2 and no other notch receptors. Furthermore, this population was also dependent on lymphotoxin β receptor signaling and able to prime $CD4^+$ T cells (Lewis et al., 2011). The Notch2-independent $ESAM^{low}CD4^-$ cDC2 subset, however, displayed a rather monocyte-related gene expression and was found to be superior with respect to cytokine responses (TNFα and IL-12) following CpG stimulation (Lewis et al., 2011). Other studies of Cheng et al. reported that Notch1/2 signaling is also relevant during the development of DC precursors within the bone marrow (Cheng et al., 2003, 2007). There, bone marrow stroma cells expressed high levels of Jagged-1, which is supporting a DC precursor accumulation. On the other hand, stroma cells of the spleen expressed high levels of Delta-1, which is stimulating DC differentiation (Cheng et al., 2007). In addition to the role of Notch2 in the homeostasis of the splenic $ESAM^{high}$ cDC2, DC specific deletion of Notch2 using CD11c-Cre, revealed Notch2 dependency for resident $ESAM^+$ cDC2 within mesenteric and skin-draining lymph nodes. This is in line with a considerable reduction of migratory

$CD103^+CD11b^+$ cDC2 numbers within the lamina propria and the mesenteric lymph nodes in a CD11c-Cre x $Notch2^{fl/fl}$ model (in the mesenteric lymph nodes these cells are most probably identical to $ESAM^+$ cDC2) (Satpathy et al., 2013). These two examples are highlighting the CD11c-driven Notch2 deletion as an interesting murine model for the research of cDC2 function. Importantly, migratory dermal cDC2 and lung- and kidney resident cDC2, however, remained unchanged in this mouse model (Satpathy et al., 2013). Functional analyses of intestinal Notch2-dependent cDCs revealed them as the critical source of IL-23 and controllers of *Citrobacter rodentium* infection (Satpathy et al., 2013). Furthermore, lack of Notch2 and therefore of cDC2 dependent IL-23 was found to hamper ILC3 and Th17 cell activity (Lewis et al., 2011; Murphy et al., 2016; Satpathy et al., 2013).

14.3.3 Klf4 in cDC2

While Notch2 dependent cDC2 were described to be important drivers of Th17 T cell and ILC3 mediated immunity, Kruppel-like factor 4 (Klf4) was found to be important for the induction of Th2 T cell responses by cDC2 (Tussiwand et al., 2015). Kfl4 was originally identified as an important transcriptional regulator during hematopoiesis (Feinberg et al., 2007; Tussiwand et al., 2015). When utilizing cDC specific ablation (CD11c-Cre) of Klf4 during *Schistosoma mansoni* or house dust mite challenge, Th2 $CD4^+$ T cell responses were ablated and parasite control was diminished or allergy inhibited, respectively (Tussiwand et al., 2015). Moreover, cDC specific Klf4 ablation had no influence on CTL, Th17 and Th1 responses in appropriate infection models. Finally, mice conditionally deleted in Klf4 in DCs exhibited impaired Irf4 expression in pre-cDCs leading to the ablation of Klf4-dependent cDC2 across several tissues (Tussiwand et al., 2015).

Besides the importance of Klf4 in cDC2 function, Kfl4 was also demonstrated to play a role during monocyte development (Feinberg et al., 2007). Although not formerly proven, this would be in line with the finding that Notch2 independent $ESAM^{low}$ cDC2 display traces of a monocyte gene signature and that $ESAM^{low}CD4^+$ DCs are independent of Notch2, but rather dependent on Kfl4, whereas $ESAM^{high}CD4^+$ DCs are dependent on Notch2 (Feinberg et al., 2007; Lewis et al., 2011).

In summary, the transcription factors Notch2 and Kfl4 define two subsets with distinct functions accounting for at least some of the heterogeneity observed within the cDC2 compartment (Jaitin et al., 2014). This might also explain the different functional roles of the two subsets assessed in different infection models (see above). Notch2 and Kfl4 dependent cDC2 may further specialize functional modules downstream of Irf8 and Irf4. While Irf8 drives programs associated with the defense against intracellular pathogens, Irf4 acts in the defense against extracellular pathogens by priming of Th2 and Th17 responses. With respect to the previously described infection models (Lewis et al., 2011; Murphy et al., 2016; Satpathy et al., 2013; Tussiwand et al., 2015), Irf4 dependent cDC2 may be further subdivided into two specialized subsets. First, Notch2 dependent ESAM^{high} cDC2 might direct Th17 responses potentially via IL-6 and IL-23, whereas second, the Klf4 dependent ESAM^{low} cDC2 might direct Th2 responses (Lewis et al., 2011, Murphy et al., 2016, Satpathy et al., 2013, Tussiwand et al., 2015).

Based on the fact that deletion of either Notch2 or Klf4 only ablates one of two populations within the cDC2 subpopulation and since Klf4 deletion did not inhibit differentiation of CDPs into cDC1 or cDC2 (Tussiwand et al., 2015), it is likely that Notch2 and Klf4 act later than Irf4 and Irf8 during DC development or on a lower hierarchy level.

14.3.4 Other transcription factors in cDC2 cells: RelB, Traf6, and Irf2

Besides the master regulator Irf4 and the transcription factors Notch, Rbp-j, as well as Klf4, other transcription factors have been described to be important for cDC2 development. These include RelB, Irf2 and Traf6 (Caton et al., 2007; Ichikawa et al., 2004; Lewis et al., 2011; Scott et al., 2016; Wu et al., 1998).

The NF-κB/Rel family member RelB was found to be highly expressed in cDC2 (Burkly et al., 1995; Wu et al., 1998). In these studies, absence of RelB in the microenvironment of thymic DC was found to intervene in thymic DC development due to a defect in thymic structure. Furthermore, RelB was demonstrated to be a selective regulator of splenic cDC2 since mice deficient for RelB exclusively lack splenic cDC2, but not cDC1 (Wu et al., 1998). Recently, additional evidence for the dispensability of RelB expression for cDC development with the exception of Notch2 dependent ESAM$^+$CD4$^+$ splenic cDC2 could be provided. In contrast, RelB was required within the microenvironment of developmental niches (Briseno et al., 2017). Taken together, these results imply that RelB is an important factor for cDC2 development in specific microenvironments.

Upstream of RelB, the TNF receptor associated factor 6 (Traf6) acts within the NF-κB pathway. Traf6 deficiency was found to severely impair the development of splenic cDC2 (Kobayashi et al., 2003). Since NF-κB is a major target of Traf6, this effect on cDC2 development could also be explained by the missing RelB regulation in Traf6 deficient mice.

Another important transcription factor for the development of splenic cDC2 is Irf2. Deletion of Irf2 led to the ablation of splenic cDC2, a reduced number of LCs in the skin, and to an impaired NK cell development (Ichikawa et al., 2004). Interestingly, mice lacking both, Irf2 and the IFNα/β receptor, displayed an intact cDC2 compartment, while NK cell development was still affected (Ichikawa et al., 2004). Thus, Irf2 may act as a negative regulator of IFNα/β receptor signaling selectively controlling cDC2 fate.

14.4 Specific factors in transcriptional regulation of pDCs

As discussed before, the pDC master regulator E2-2 and its downstream targets are the driving forces of pDC commitment. Nevertheless, other transcription factors acting outside Id2 repression in pDCs, such as Runx2 and Ikaros contribute to pDC development and function.

14.4.1 Runx2 in peripheral pDCs homeostasis and function

The Runt family transcription factor Runx2 expressed in pDCs is a direct target of E2-2 (Sawai et al., 2013). Mice deficient for the transcription factor Runx2 exhibited normal pDC generation in the bone marrow, whereas numbers were significantly reduced in the periphery, indicating a responsibility of Runx2 for pDC migration into the periphery. Indeed, gene expression of several important pDC genes, such as Ccr2 and Ccr5, were found to be reduced in Runx2$^{-/-}$ mice. Hence, Runx2 was described as an important regulator of pDC migration (Sawai et al., 2013). During a later study (Chopin et al., 2016), Runx2 was also identified as a repressor of Cxcr4, which is important for the retention of cells in the bone marrow. Runx2 haploinsufficiency led to decreased pDC numbers in the periphery indicating that the homeostasis of pDCs is dose dependent on Runx2 (Chopin et al., 2016).

Furthermore, following LCMV infection, Runx2 deficient mice showed an impaired IFNα response and higher viral titers (Chopin et al., 2016). Since Irf7, which acts downstream of TLR signaling in pDCs and is thereby involved to induce IFN production, is a direct target of Runx2, lack of Runx2 in pDCs may explain the hampered IFN response in Runx2$^{-/-}$ mice (Chopin et al., 2016).

14.4.2 *Ikaros in pDCs*

The zinc finger transcription factor Ikaros, which is also critical for B cell differentiation and function, was suggested to be involved in pDC development (Allman et al., 2006; Kirstetter et al., 2002). In genetically engineered Ikaros-deficient mice (Ikaros exon 2 replaced by LacZ on both alleles), peripheral pDCs were missing, whereas cDCs remained unaffected (Allman et al., 2006). Furthermore, global gene expression profiling of pDCs from bone marrow of engineered mice showed an upregulation of non-pDC associated genes, thereby indicating that Ikaros contributes to pDC development by repression of non-pDC associated transcriptional programs (Allman et al., 2006). In humans, the Ikaros family zinc finger 1 (IKZF1) was recently found to be important during pDC and cDC development (Cytlak et al., 2018). Haploinsufficiency of patients in IKZF1 led to decreased pDC numbers, while cDC1 numbers were increased. Moreover, application of Lenalidomide, which initiates proteosomal degradation of IKZF1, shifted pDC differentiation to cDC1 in vitro. Finally, primary DCs from these IKZF1 haploinsufficient patients as well as in vitro treated DCs exhibited reduced secretion of IFNα and IL-12 in response to different TLR stimuli (Cytlak et al., 2018).

15. Impact of post-transcriptional regulation by microRNAs

While transcription factors directly control the transcription of target genes, microRNAs (miRNAs) act post-transcriptionally by repressing translation of target mRNAs. Hence, miRNAs appear to be predestined to more rapidly modify cellular protein levels. Therefore, miRNAs are usually postulated as fine-tuners of translational activity, rather than a simple on-off switch system (Lodish et al., 2008; Turner et al., 2011). miRNAs are transcribed by RNA polymerase II (Lee et al., 2004). These primary miRNAs are further processed by the nuclease Drosha and its co-factor DGCRB8 to form hairpin precursor miRNAs (pre-miRNAs) (Gregory et al., 2004). Subsequently, pre-miRNAs are exported from the nucleus into the cytoplasm via exportin 5, where pre-miRNAs are trimmed by the protein Dicer into about 22 nt long miRNA duplex structures (Grishok et al., 2001; Hutvagner et al., 2001; Lund et al., 2004). Finally, one single stranded microRNA is loaded into the RNA-induced silencing complex (RISC). In general, this loading procedure is directed since one of the two strands is clearly favored. The favored strand is referred to as miRNA, whereas the less frequently used strand is the star form miRNA (miRNA*) (Khvorova et al., 2003; Schwarz et al., 2003). Target mRNAs contain miRNA binding sites in their 3′ untranslated regions (3'UTR) and following recognition by miRNAs, RISC is guided to target the mRNA, which is leading to degradation or translational repression of the targeted mRNA (Baltimore et al., 2008; Bartel, 2009; Fabian et al., 2010). With respect to the observation that one mRNA regularly contains multiple binding sites for different miRNAs, miRNA-dependent regulation of transcription embraces another layer of regulatory capacity, since this allows for the combined action of agonistic or antagonistic miRNA species during the regulation of one specific mRNA (Gebert and Macrae, 2019). Moreover, one single miRNA species is able to recognize a diverse spectrum of mRNAs all displaying 3′ UTR binding sites (Gebert and Macrae, 2019). Following their discovery in 1993 (Lee et al., 1993; Wightman et al., 1993), important functions of miRNAs were demonstrated in hematopoiesis, oncogenesis or maintenance of tolerance (Chen et al., 2013a; Landgraf et al., 2007; Montagner et al., 2014; Peng and Croce, 2016). Beyond these functions, miRNAs are able to regulate transcription factor expression, hence directly fine-tuning the first layer of transcriptional regulation (Gebert and Macrae, 2019).

15.1 Systems to investigate miRNA function in specific cell types

First indications for the importance of miRNA within a specific cell type can be gained within comparative expression profiling studies followed up by miRNA level manipulation. This can either be achieved by miRNA overexpression, genetic ablation or posttranscriptional silencing (Pritchard et al., 2012; Turner et al., 2011). Nevertheless, studying the importance of miRNAs within primary DCs is not a simple task and is usually assessed by incorporating genetically engineered mice. First indications for the role of miRNAs in antigen presenting cell populations were found, when the miRNA duplex trimming protein Dicer was specifically deleted in CD11c expressing cells (CD11c-DICER mice). Interestingly, although CD11c is expressed on a variety of APCs, abnormalities were only found in LCs, which showed reduced cell numbers, exhibited maturation and antigen presentation defects, as well as a pro-apoptotic gene signature (Kuipers et al., 2010). In CD11c-DICER mice short-lived resident DCs seemed not to be influenced, which might partly be explained by the finding that the average miRNA half-life is about 119h. Therefore, DCs following

CD11c driven Dicer ablation may just not persist long enough to exhibit functional aberrations since 50% of the miRNAs are predicted to be still present 5 days after Dicer depletion (Gantier et al., 2011; Turner et al., 2011).

After the discovery of miRNAs, systems to manipulate miRNA expression, mainly utilizing the generation and application of short interfering RNAs (siRNAs), were established. These systems can be utilized to introduce RNA functioning as post-transcriptional expression silencer in cells (Kanasty et al., 2012). Furthermore, target cell specific carrier systems, such as antibody coated lipid particles, could be used to encapsulate siRNA and directly deliver them into target cells (Cruz et al., 2011; Kanasty et al., 2012; Liu et al., 2016; Su et al., 2019). Thereby, siRNAs could be applied to rapidly silence miRNA species within DCs to be independent on, e.g., CD11c promoter activity but rather be dependent on entry receptor expression, which might be important to circumvent the issue of short lifespans of primary DCs in future studies (Cruz et al., 2011).

15.2 Important functions of miRNAs in murine BMDCs and human MoDCs

15.2.1 *miRNA-21, 34a, 99b and 511 delay the development of MoDCs*

Most of the knowledge acquired about the function of miRNAs in DCs is derived from data in BMDCs or MoDCs. miRNA profiling during human MoDC differentiation displayed distinct expression of 20 different miRNAs, such as miRNA-21 and miRNA-34a (Hashimi et al., 2009). Following usage of specific anti-miRNA oligonucleotides against miRNA-21 or miRNA-34a, MoDC development was delayed as demonstrated by CD14/CD209 expression (Hashimi et al., 2009). This effect was boosted following simultaneous block of both miRNAs, fostering the notion that these miRNAs orchestrate MoDC differentiation. The mRNAs for the proteins WNT1 and JAG1 were ascertained as targets of miRNA-21 and miRNA-34a, and furthermore, WNT1 and JAG1 were found to inhibit MoDC development. These results indicate that miRNA-21 and miRNA-34a negatively regulate the two repressors WNT1 and JAG1. The addition of anti-miRNA oligonucleotides or exogenous addition of WNT1 and JAG1 were shown to decrease endocytic activity, thereby also influencing MoDC function (Hashimi et al., 2009). How these miRNAs influence primary DC endocytic activity is not known yet. Another study demonstrated reduced CD209 levels on MoDCs following miRNA-99b and miRNA-511 inhibition (Tserel et al., 2011). In addition to the described role of miRNA-21 in reducing endocytic activity in human MoDCs, BMDCs from miRNA-21$^{-/-}$ mice produced more IL-12 following LPS treatment and skewed the Th1/Th2 balance in favor of Th1 cells (Lu et al., 2011b).

15.2.2 miRNA-34a is involved in the generation of pro-inflammatory immune responses in human inflammatory and murine BMDCs

In addition to the above described functions of miRNA-34a in MoDC development, it was shown that miRNA-34a negatively regulates the inhibitory DC protein AXL in CD1c^{+} human MoDCs. Following treatment of MoDCs with miRNA inhibitors, production of pro-inflammatory cytokines was reduced (Kurowska-Stolarska et al., 2017). In mice exhibiting a miRNA-34a deficiency, BMDC-T cell interactions were impaired, thus negatively influencing the Th17 compartment. Moreover, miRNA-34a mice were resistant to collagen induced arthritis (Kurowska-Stolarska et al., 2017). Interestingly, miRNA-34a was even found upregulated in CD1c^{+} DCs, including CD1c^{+} inflammatory DCs in patients suffering from rheumatoid arthritis (Kurowska-Stolarska et al., 2017).

15.2.3 miRNA-221 and miRNA-155 negatively and positively regulate apoptosis in human MoDCs

By performing expression profiling in human monocytes and immature and mature monocyte derived DCs (MoDCs), 27 diverging miRNAs were identified between monocytes and MoDCs (Ceppi et al., 2009; Lu et al., 2011a; Zhan and Wu, 2012). Two miRNAs of particular importance were miRNA-155 and miRNA-221. Both were found to be involved in apoptosis regulation since miRNA-221 silencing or miRNA-155 overexpression were associated with increased apoptosis potentially caused by p27^{kip1} enrichment in MoDCs (Lu et al., 2011a). The p27^{kip1} protein is a known cell cycle inhibitor exhibiting miRNA-155 and miRNA-221 binding sites in its 3'UTR (Lu et al., 2011a).

15.2.4 miRNA-155 prevents continuous cytokine production following TLR stimulation in MoDCs

Furthermore, miRNA-155 was involved in regulating IL-12 production by negatively regulating the negative regulator suppressor of cytokine signaling 1 (SOCS-1) (Lu et al., 2011a). In addition, miRNA-155 expression was highly upregulated in human MoDCs subsequent to lipopolysaccharide (LPS) stimulation directly acting in the inflammatory TLR/IL-1 pathway. This is implying that miRNA-155 is involved in forestalling continuous inflammatory cytokine production after TLR signaling (Ceppi et al., 2009). This hypothesis is also supported by another study, in which concomitant upregulation of CD80, CD83, CD86, HLA class II, and miRNA-155 was observed after LPS or IFNγ stimulation of immature MoDCs (Jin et al., 2010).

15.2.5 miRNA-155 controls the maturation status of murine BMDCs

miRNA-155 was demonstrated to be essential for DC maturation and function potentially by inhibition of c-Fos (Dunand-Sauthier et al., 2011). In this study, $CD4^+$ T cell activation by LPS matured miRNA-$155^{-/-}$ BMDCs was found impaired, while DCs exhibited lower levels of MHC II, the co-stimulatory molecules CD40 and CD86, as well as a reduced cytokine response regarding IL-12 and TNFα. In contrast, another study reported a similar lack of T cell stimulatory capacity of LPS matured miRNA-$155^{-/-}$ BMDCs, while the levels of MHC II and CD86 were not altered (Rodriguez et al., 2007). How exactly miRNA-155 regulates DC functions remains unclear. Nevertheless, miRNA-155 appears to be essential for DC maturation and function (Rodriguez et al., 2007), whereas the development of cDCs and pDCs is not hampered by miRNA-155 depletion (Dunand-Sauthier et al., 2011).

15.2.6 miRNA-24, 30b and 142-3p are negative regulators of antigen presentation in human MoDCs

Together with miRNA-24 and 30b, miRNA-142-3p was demonstrated to interfere with antigen presentation of human MoDCs (Naqvi et al., 2016). Overexpression of these miRNA species resulted in impaired antigen uptake and processing. Furthermore, antigen presentation to T cells was negatively affected, since $CD4^+$ T cell proliferation was reduced (Naqvi et al., 2016). Moreover, miRNA overexpression was found to result in expression of Programmed Death-Ligand 1 (PD-L1), thereby inhibiting T cell proliferation. In line with the results, checkpoint blockade of PD-L1 was able to overcome the perturbance of T cell priming (Naqvi et al., 2016).

15.2.7 miRNA-148a, 148b and 152 are negative regulators of antigen presentation in murine BMDCs

miRNA-148a, 148b and 152 are suggested to be negative regulators of antigen presentation activity in murine BMDCs (Liu et al., 2010). Subsequent to Tlr3, 4 or 9 triggered DC activation, miRNA-148/152 expression increased, whereas Calcium/calmodulin dependent protein kinase II

(CaMKII) expression decreased. CaMKII was previously described to be important during antigen presentation by human DCs (Herrmann et al., 2005). In this case, targeting of CaMKII by miRNA-148/152 commanded inhibition of IL-12, IL-6, TNFα, IFNβ, upregulation of MHC II, and the induction of antigen-specific $CD4^+$ T cell proliferation (Liu et al., 2010). Hence, the miRNA-148/152 family might be considered as fine tuner of antigen presentation.

15.3 miRNAs involved in murine and human primary DC development

15.3.1 miRNA-142 is important for the development of cDC2 cells

miRNA pattern analysis of bone marrow progenitor cells, monocytes and splenic DCs revealed differential miRNA profiles indicating specific roles of different miRNA species during development (Mildner et al., 2013). Among them, miRNA-142 was found to disturb homeostasis of splenic Flt3-ligand dependent $CD4^+$ cDC2 (Mildner et al., 2013). This was associated with an impaired $CD4^+$ T cell priming (Mildner et al., 2013). Since classical cDC2 ($CD4^+$ DCs) were not present in the spleen of miRNA-$142^{-/-}$ mice, Flt3-ligand dependent bone marrow cultures containing cDC2 equivalents were analyzed (Naik et al., 2005). Interestingly, these cells exhibited upregulation of the cDC1 master regulator Irf8 (Mildner et al., 2013) and might suggest that miRNA-142 exhibits functions during cDC1:cDC2 lineage fate decisions.

15.3.2 miRNA expression patterns show substantial differences between cDCs and pDCs

Furthermore, a differential expression pattern of miRNA clusters between cDCs and pDCs indicates that miRNAs are also involved in DC lineage decisions during cDC:pDC development (Kuipers et al., 2010; Mildner et al., 2013; Smyth et al., 2015). In contrast to cDCs, the pDC miRNA expression profile was suggested to share more characteristics with lymphoid cells, especially with T cells (Kuipers et al., 2010). In line with this, pDCs exhibit high expression of the miRNA-17-92 cluster, which is expressed in lymphocyte progenitors, too (Xiao et al., 2008). Compared to pDCs, cDCs display higher expression of the miRNA-23a/24/27a cluster (Kuipers et al., 2010). Previously, this cluster was described as inhibitor of lymphopoiesis with miRNA-23a as a downstream target of the PU.1 promotor, and PU.1 as a transcription factor originally identified as regulator of myeloid cell differentiation (Kong et al., 2010; Kuipers et al., 2010).

Beside the cluster of miRNA-17-92 and 23a/24/27a, miRNA-221 and miRNA-222 may be involved in the regulation of the cDC/pDC lineages since application of antisense oligonucleotides in DC differentiation cultures from bone marrow cells fostered the generation of pDCs (Kuipers et al., 2010).

The before-mentioned miRNA cluster revealed a function in lineage decision between pDCs and cDCs. Recently, a set of miRNAs were described that display cDC subset specific functions. One example is miRNA-22, which was pointed out to be expressed at high levels in cDC2 (Li et al., 2012). Overexpression of miRNA-22 resulted in increased cDC2 and reduced pDCs numbers, while knockdown led to decreased cDC numbers (Li et al., 2012). It may be assumed that the repressive function of miRNA-22 might be mediated by knocking down Irf8 mRNA, as Irf8 mRNA contains miRNA-22 binding sides in its 3′ untranslated mRNA region (Li et al., 2012).

15.4 miRNAs regulating DC function

15.4.1 miRNA-155 and its star form regulate type I IFN production following TLR stimuli in human pDCs

miRNA profiling of human pDCs revealed a strong increase in miRNA-155 and miRNA155* following Tlr7 stimulation (Zhou et al., 2010). Interestingly, both miRNA forms comprised regulatory functions within the same signaling pathway. In contrast to miRNA-155* amplifying type I IFN production, miRNA-155 counteracted IFN production. Interestingly, a kinetic analysis demonstrated that both miRNA forms appear in a time-dependent fashion: miRNA-155* was found to be induced during the early phase of TLR signaling, whereas miRNA-155 acted during a later stage in attenuating TLR-dependent activation (Zhou et al., 2010). Furthermore, miRNA-155 and miRNA-155* regulated themselves on the post-transcriptional level, thereby adding another layer of regulation within this complex network (Zhou et al., 2010).

15.4.2 miRNA-126 substantially contributes to anti-viral responses of murine pDCs

Expression profiling of miRNA-126 revealed a specific and strong expression in pDCs, where it was found to be important during the response to viral infections (Agudo et al., 2014). Furthermore, miRNA-$126^{-/-}$ mice exhibited great susceptibility for viral infection, since it controls the function of the genes for Tlr7, Tlr9, Nfkb1, and Kdr, which is coding for the Vascular endothelial growth factor receptor 2 (VEGFR2) (Agudo et al., 2014).

Similar to miRNA deletion, ablation of VEGFR2 resulted in an impaired type I IFN response indicating a connection of VEGFR2 and miRNA-126 in the regulation of TLR-mediated type I IFN responses (Agudo et al., 2014).

15.4.3 miRNA-146a negatively regulates pDC activation following TLR signaling

After TLR signaling, it was observed that miRNA-146a is directly induced and functions as a negative feedback-loop, thereby avoiding pDC over-activation as strict measurement against autoimmunity. This was demonstrated during a study with human pDCs, in which miRNA-146a overexpression resulted in aggrieved induction of NF-κB following TLR signaling and consequently in a decreased survival and reduced cytokine production (Karrich et al., 2013).

15.4.4 miRNA-10a expression is reduced by inflammatory stimuli thereby boosting cytokine responses

In intestinal DCs, miRNA-10a expression was found to be reduced in mice housed under specific pathogen-free (SPF) conditions compared to germ-free (GF) housed mice (Xue et al., 2011). Likewise, treatment of BMDCs with lysates of different intestinal lumen occupying bacteria strains reduced miRNA-10a expression. This outcome was reproducible by direct usage of ligands for the TLRs 1/2, 4, 5 and 9 all resulting in decreased miRNA-10a levels (Xue et al., 2011). Utilizing binding site prediction followed by a luciferase reporter assay in RAW264.7 macrophages, IL-12/IL-23p40 downregulation was shown to be miR-10a dependent (Xue et al., 2011). Furthermore, overexpression of miRNA 10a in BMDCs was suggested to reduce the mRNA or proteins levels of IL-12/23p40, IL-12p70, and IL-23. Interestingly, these findings might therefore resemble the situation in mice suffering from colitis, as miRNA-10a displayed a low level expression and correlating with high levels of IL-12/IL-23p40 in this pathology (Xue et al., 2011).

15.4.5 Activation of DCs following exogenous and endogenous danger signals is tightly controlled by miRNAs

Most of the aforementioned studies focus on miRNA dependent actions following exogenous stimuli triggering pathogen sensors, such as TLRs. But also endogenous danger signals, such as high mobility group box-1 (HMGB1), can stimulate primary DCs to foster T cell responses and induce cytokine secretion. Only recently, miRNA-181a-5p was found to be upregulated following DC maturation and activation with HMGB1. Luciferase assays validated a predicted interaction of miRNA-181a-5p with TNFα on mRNA level (Zhu et al., 2017). Moreover, overexpression or inhibition of miRNA-181a-5p commanded reduced or elevated TNFα levels, respectively. Therefore, miRNA-181a-5p constitutes a negative regulator of damage associated responses driven by HMGB1 (Zhu et al., 2017). In combination with other studies, demonstrating anti-inflammatory activity of miRNAs, this may render miRNAs as interesting targets for rapid and transient manipulation of DC pathways.

Another miRNA species with anti-inflammatory potential is miRNA-142-3p. This miRNA was highly expressed during steady-state conditions (without TLR stimuli) in primary murine cDC1 ($CD8^+$ DCs) and cDC2 ($CD8^-$ DCs) and human peripheral mononuclear blood cells (PBMCs) derived MoDCs (Sun et al., 2011). Inhibition of miRNA-142-3p fostered IL-6 production in immature and LPS matured BMDCs in vitro (Sun et al., 2011). Finally, application of an oligonucleotide complementary to miRNA-142-3p resulted only in wild type, but not $IL\text{-}6^{-/-}$ mice in reduced endotoxin dependent septic shock following high dose LPS treatment (Sun et al., 2011). This is in line with a previous study demonstrating anti-inflammatory capacity of IL-6 during local and systemic inflammatory immune responses (Xing et al., 1998). However, IL-6 deficiency as well as miRNA silencing was not specific to DCs within this study (Sun et al., 2011).

15.4.6 miRNAs potentially contribute to cell-cell communication

Besides cell intrinsic functional roles, there is evidence for the transfer of miRNAs from cell to cell either in a free form or via exosomes contributing to cell-cell communication (Montecalvo et al., 2012; Wang et al., 2010). Furthermore, for BMDCs and the DC-like cell line DC2.4 transfer, uptake and functional manipulation induced by transferred miRNAs have been described (Montecalvo et al., 2012).

Taken together, miRNAs appear as fine-tunes of DC differentiation and function. Interestingly, many studies dedicate anti-inflammatory properties to different miRNA species. Hence, miRNAs may operate as regulatory unit, thereby avoiding DC over-activation. Furthermore, miRNA function during antigen presentation may comprise a platform for rapid regulation of adaptive immune responses.

16. Epigenetic regulation of DCs

Hematopoietic stem cells share a common genome. Utilizing an identical toolbox to start with, cells different in phenotype and function emerge (Adams et al., 2012). Obviously, many factors contribute to this phenomenon. Transcription factors play a special role during differentiation and function since they directly control transcription of target gene programs. At the forefront though, epigenetic programming enables the establishment of cell specific gene expression. This is achieved by chromatin remodeling, a process that is allowing or denying access to specific regions of the genomic DNA (Adams et al., 2012; Tian et al., 2017). While the genome is stable under most circumstances, chromatin is a highly dynamic, modifiable structure, but its state is also inheritable to daughter cells (Moazed, 2011). Together with manipulations at cytosines of the DNA, the code of modifications at histones and nucleosomes constitute the epigenome (Adams et al., 2012, Tian et al., 2017).

16.1 Organization of the epigenome

The epigenetic code is organized by protein complexes consisting of readers and writers, which are able to recognize and manipulate histone modifications (Rothbart and Strahl, 2014). On DNA level, cytosine residues are either methylated or demethylated by DNA methyltransferases (DNMTs) or DNA demethylases (e.g., TET) usually leading to transcriptional repression or activation, respectively (Boukhaled et al., 2019; Jones, 2012). The major modifications taking place at histones comprise (de-)acetylation, (de-)methylation and ubiquitylation. These modifications are performed by histone deacetylases (HDACs), histone acetyltransferases (HATs), lysine demethylases (KDM) and histone methyltransferases (HMTs) (Boukhaled et al., 2019; Jones, 2012; Rothbart and Strahl, 2014). Ubiquitination by E3 ligases was primarily studied at histone H2A (Boukhaled et al., 2019). In general, the modifications $H_3K_36me_3$, H_3K_27Ac and $H_3K_3me_3$ are associated with open chromatin and gene activity, while $H_3K_27me_3$ and $H_3K_9me_3$ act in gene suppression (Boukhaled et al., 2019). In combination, histone modifications orchestrate the recruitment of the chromatin remodeling complexes and determine the accessibility of the DNA for the transcriptional machinery (Boukhaled et al., 2019; Yadav et al., 2018).

16.2 The epigenetic code of DCs

Only recently, an atlas for cis-regulatory elements in an array of murine immune cells was released (Yoshida et al., 2019). By combining the results from assay for transposase-accessible chromatin using sequencing (ATACSeq) and RNASeq, this study provides an overview of the chromatin status of different cellular lineages and precursors and links chromatin accessibility with expression profiling. For instance, in splenic DCs a new open chromatin region within the CD8 locus specific for cDC1 ($CD8^+$ DCs) as well as one specific to pDCs were identified, while these open chromatin regions were closed in $CD4^+$ cDC2 (CD4+ DCs) (Yoshida et al., 2019). Nevertheless, most open chromatin regions in cells of one lineage clustered together. Interestingly, many open chromatin regions in pDCs were found to be more distant from cDCs than from T cells. This finding further strengthens the notion that pDCs may arise from lymphoid progenitors (Musumeci et al., 2019; Reizis et al., 2011; Rodrigues et al., 2018). Furthermore, combining transcription factor binding site mapping with open chromatin region data, several binding motifs related to prominent, cell-specific transcription factors were identified in the open chromatin regions of certain cell types. This was the case for the E2-2 binding motif, which was found highly enriched in pDCs. On the other hand, transcription factor binding sites in open chromatin regions for Irf8 and BatF3 were found to be enriched in splenic cDC1 ($CD8^+$ DCs). Overall, profiles of intestinal $CD103^+CD11b^-$ cDC1 resembled splenic cDC1 (Yoshida et al., 2019). Open chromatin regions of both cDC1 populations presented multiple transcription factor binding sites for Stat2 and Irf9, which are involved in type I IFN signaling. This chromatin landscape may be designed to sensitize steady-state cDC1 to respond to viral infections by cross-presentation (Diamond et al., 2011; Helft et al., 2012). In contrast, splenic $CD4^+$ cDC2 exhibited enriched transcription factor binding sites for CCAAT/enhancer (C/EBP) family members. C/EBPs are known to be involved in inflammation and hematopoiesis and C/EBP binding sites have been previously identified in many important genes, such as TNFα, IL-1β, IL-6, or IL-12, and in the promoter region of the GM-CSF receptor (Tsukada et al., 2011; Yoshida et al., 2019). Utilizing transcription factors experimentally proven to be important during development as a reference, the approach of this study may enable the identification of new, bona fide transcription factors potentially involved in DC lineage commitment.

17. Conclusions

Single cell RNA sequencing, miRNA sequencing, analyses of the open chromatin structures combined with cutting-edge phenotypical characterization of surface markers and transcription factors pushed the understanding of DC development and function tremendously forward. By generation of mouse models deficient in transcription factors, the role of newly identified transcription factors, such as Zeb2, was proven important for the function of cDC2. However, newly generated data on known candidates, such as Notch2 and Klf4, also identified further possible specialized subpopulations of cDC2 with respect to the induction of Th2 and Th17 $CD4^+$ T cell responses. In a concerted manner, Flt3L and GM-CSF seem to work together in the differentiation from their bone marrow progenitor cells to lineage committed DCs. How this concerted action might be achieved in vivo is not understood. This knowledge and the understanding, which factors might play in a timely defined manner, will also allow for the establishment of culture protocols for the generation of primary DC subsets from precursor cells of patient material that are more sophisticated and thereby advance options for DC therapies. Furthermore, there are open questions regarding the pDC development, which is more unclear than some years ago. In the future, fate-mapping approaches might clarify, if there are two or more developmental routes for pDC development. A point that is only started to be addressed very recently is the understanding of the role of the tissue microenvironment, in which a (pre)-DC is arriving or settled in, on their overall immunological function. This will be important for a better determination of critical factors that can be used in immunomodulatory aspects (e.g., for the function of DCs in tumor environment or in autoinflammatory diseases).

Acknowledgments

This work was supported by the German Research Foundation to D.D. (CRC1181-A7, RTG1962-A6, RTG2504-B2, DU548/5-1), and intramural funding by the IZKF (A65) and the Emerging Fields Initiative (EFI) "BIG-THERA." Ch.L. received intramural funding by the IZKF (J54) and the German Research Foundation (RTG2504-B2). The authors have no conflicting financial interests.

Statement

Parts of this review are also parts of the PhD theses of Ch.L., L.A., J.S., A.B., L.H.

References

Adams, D., Altucci, L., Antonarakis, S.E., Ballesteros, J., Beck, S., Bird, A., Bock, C., Boehm, B., Campo, E., Caricasole, A., Dahl, F., Dermitzakis, E.T., Enver, T., Esteller, M., Estivill, X., Ferguson-Smith, A., Fitzgibbon, J., Flicek, P., Giehl, C., Graf, T., Grosveld, F., Guigo, R., Gut, I., Helin, K., Jarvius, J., KÜppers, R., Lehrach, H., Lengauer, T., Lernmark, Å., Leslie, D., Loeffler, M., Macintyre, E., Mai, A., Martens, J.H.A., Minucci, S., Ouwehand, W.H., Pelicci, P.G., Pendeville, H., Porse, B., Rakyan, V., Reik, W., Schrappe, M., SchÜbeler, D., Seifert, M., Siebert, R., Simmons, D., Soranzo, N., Spicuglia, S., Stratton, M., Stunnenberg, H.G., Tanay, A., Torrents, D., Valencia, A., Vellenga, E., Vingron, M., Walter, J., Willcocks, S., 2012. BLUEPRINT to decode the epigenetic signature written in blood. Nat. Biotechnol. 30, 224.

Adolfsson, J., Mansson, R., Buza-Vidas, N., Hultquist, A., Liuba, K., Jensen, C.T., Bryder, D., Yang, L., Borge, O.J., Thoren, L.A., Anderson, K., Sitnicka, E., Sasaki, Y., Sigvardsson, M., Jacobsen, S.E., 2005. Identification of Flt3+ lympho-myeloid stem cells lacking erythro-megakaryocytic potential a revised road map for adult blood lineage commitment. Cell 121, 295–306.

Agudo, J., Ruzo, A., Tung, N., Salmon, H., Leboeuf, M., Hashimoto, D., Becker, C., Garrett-Sinha, L.A., Baccarini, A., Merad, M., Brown, B.D., 2014. The miR-126-VEGFR2 axis controls the innate response to pathogen-associated nucleic acids. Nat. Immunol. 15, 54–62.

Akira, S., Uematsu, S., Takeuchi, O., 2006. Pathogen recognition and innate immunity. Cell 124, 783–801.

Alcantara-Hernandez, M., Leylek, R., Wagar, L.E., Engleman, E.G., Keler, T., Marinkovich, M.P., Davis, M.M., Nolan, G.P., Idoyaga, J., 2017. High-dimensional phenotypic mapping of human dendritic cells reveals interindividual variation and tissue specialization. Immunity 47, 1037–1050 e6.

Ali, S., Mann-Nüttel, R., Schulze, A., Richter, L., Alferink, J., Scheu, S., 2019. Sources of type I interferons in infectious immunity: plasmacytoid dendritic cells not always in the driver's seat. Front. Immunol. 10 (778).

Aliberti, J., Schulz, O., Pennington, D.J., Tsujimura, H., Reis, E., Sousa, C., Ozato, K., Sher, A., 2003. Essential role for ICSBP in the in vivo development of murine CD8alpha + dendritic cells. Blood 101, 305–310.

Allman, D., Dalod, M., Asselin-Paturel, C., Delale, T., Robbins, S.H., Trinchieri, G., Biron, C.A., Kastner, P., Chan, S., 2006. Ikaros is required for plasmacytoid dendritic cell differentiation. Blood 108, 4025–4034.

Alloatti, A., Rookhuizen, D.C., Joannas, L., Carpier, J.M., Iborra, S., Magalhaes, J.G., Yatim, N., Kozik, P., Sancho, D., Albert, M.L., Amigorena, S., 2017. Critical role for Sec22b-dependent antigen cross-presentation in antitumor immunity. J. Exp. Med. 214, 2231–2241.

Angkasekwinai, P., Park, H., Wang, Y.H., Wang, Y.H., Chang, S.H., Corry, D.B., Liu, Y.J., Zhu, Z., Dong, C., 2007. Interleukin 25 promotes the initiation of proallergic type 2 responses. J. Exp. Med. 204, 1509–1517.

Asselin-Paturel, C., Trinchieri, G., 2005. Production of type I interferons: plasmacytoid dendritic cells and beyond. J. Exp. Med. 202, 461–465.

Asselin-Paturel, C., Boonstra, A., Dalod, M., Durand, I., Yessaad, N., Dezutter-Dambuyant, C., Vicari, A., O'Garra, A., Biron, C., Briere, F., Trinchieri, G., 2001. Mouse type I Ifn-producing cells are immature APCs with plasmacytoid morphology. Nat. Immunol. 2, 1144–1150.

Asseman, C., Mauze, S., Leach, M.W., Coffman, R.L., Powrie, F., 1999. An essential role for interleukin 10 in the function of regulatory T cells that inhibit intestinal inflammation. J. Exp. Med. 190, 995.

Audiger, C., Rahman, M.J., Yun, T.J., Tarbell, K.V., Lesage, S., 2017. The importance of dendritic cells in maintaining immune tolerance. J. Immunol. 198, 2223–2231.

Auffray, C., Fogg, D.K., Narni-Mancinelli, E., Senechal, B., Trouillet, C., Saederup, N., Leemput, J., Bigot, K., Campisi, L., Abitbol, M., Molina, T., Charo, I., Hume, D.A., Cumano, A., Lauvau, G., Geissmann, F., 2009. CX3CR1+ CD115+ CD135+ common macrophage/DC precursors and the role of CX3CR1 in their response to inflammation. J. Exp. Med. 206, 595.

Bachem, A., Güttler, S., Hartung, E., Ebstein, F., Schaefer, M., Tannert, A., Salama, A., Movassaghi, K., Opitz, C., Mages, H.W., Henn, V., Kloetzel, P.-M., Gurka, S., Kroczek, R.A., 2010. Superior antigen cross-presentation and XCR1 expression define human CD11c+CD141+ cells as homologues of mouse CD8+ dendritic cells. J. Exp. Med. 207, 1273.

Bachem, A., Hartung, E., Güttler, S., Mora, A., Zhou, X., Hegemann, A., Plantinga, M., Mazzini, E., Stoitzner, P., Gurka, S., Henn, V., Mages, H., Kroczek, R., 2012. Expression of XCR1 characterizes the Batf3-dependent lineage of dendritic cells capable of antigen cross-presentation. Front. Immunol. 3, 214.

Bagadia, P., Huang, X., Liu, T.-T., Durai, V., Grajales-Reyes, G.E., Nitschké, M., Modrusan, Z., Granja, J.M., Satpathy, A.T., BriseÑO, C.G., Gargaro, M., Iwata, A., Kim, S., Chang, H.Y., Shaw, A.S., Murphy, T.L., Murphy, K.M., 2019. An Nfil3–Zeb2–Id2 pathway imposes Irf8 enhancer switching during cDC1 development. Nat. Immunol. 20, 1174–1185.

Bajaña, S., Roach, K., Turner, S., Paul, J., Kovats, S., 2012. IRF4 promotes cutaneous dendritic cell migration to lymph nodes during homeostasis and inflammation. J. Immunol. 189, 3368.

Bajana, S., Turner, S., Paul, J., Ainsua-Enrich, E., Kovats, S., 2016. IRF4 and IRF8 act in CD11c+ cells to regulate terminal differentiation of lung tissue dendritic cells. J. Immunol. 196, 1666–1677.

Bakri, Y., Sarrazin, S., Mayer, U.P., Tillmanns, S., Nerlov, C., Boned, A., Sieweke, M.H., 2005. Balance of MafB and PU.1 specifies alternative macrophage or dendritic cell fate. Blood 105, 2707–2716.

Ballesteros-Tato, A., León, B., Lund, F.E., Randall, T.D., 2010. Temporal changes in dendritic cell subsets, cross-priming and costimulation via CD70 control CD8+ T cell responses to influenza. Nat. Immunol. 11, 216.

Baltimore, D., Boldin, M.P., O'connell, R.M., Rao, D.S., Taganov, K.D., 2008. MicroRNAs: new regulators of immune cell development and function. Nat. Immunol. 9, 839–845.

Bamboat, Z.M., Stableford, J.A., Plitas, G., Burt, B.M., Nguyen, H.M., Welles, A.P., Gonen, M., Young, J.W., Dematteo, R.P., 2009. Human liver dendritic cells promote T cell hyporesponsiveness. J. Immunol. 182, 1901–1911.

Banchereau, J., Steinman, R.M., 1998. Dendritic cells and the control of immunity. Nature 392, 245–252.

Barclay, A.N., Brown, M.H., 2006. The SIRP family of receptors and immune regulation. Nat. Rev. Immunol. 6, 457–464.

Bartel, D.P., 2009. MicroRNAs: target recognition and regulatory functions. Cell 136, 215–233.

Beaulieu, A.M., Sant'angelo, D.B., 2011. The BTB-ZF family of transcription factors: key regulators of lineage commitment and effector function development in the immune system. J. Immunol. 187, 2841–2847.

Becher, B., Schlitzer, A., Chen, J., Mair, F., Sumatoh, H.R., Teng, K.W.W., Low, D., Ruedl, C., Riccardi-Castagnoli, P., Poidinger, M., Greter, M., Ginhoux, F., Newell, E.W., 2014. High-dimensional analysis of the murine myeloid cell system. Nat. Immunol. 15, 1181.

Becker, M.R., Choi, Y.S., Millar, S.E., Udey, M.C., 2011. Wnt signaling influences the development of murine epidermal Langerhans cells. J. Invest. Dermatol. 131, 1861–1868.

Becker, A.M., Michael, D.G., Satpathy, A.T., Sciammas, R., Singh, H., Bhattacharya, D., 2012. IRF-8 extinguishes neutrophil production and promotes dendritic cell lineage commitment in both myeloid and lymphoid mouse progenitors. Blood 119, 2003–2012.

Becker-Herman, S., Lantner, F., Shachar, I., 2002. Id2 negatively regulates B cell differentiation in the spleen. J. Immunol. 168, 5507–5513.

Belkaid, Y., Piccirillo, C.A., Mendez, S., Shevach, E.M., Sacks, D.L., 2002. CD4+CD25+ regulatory T cells control Leishmania major persistence and immunity. Nature 420, 502–507.

Benvenuti, F., Lagaudriere-Gesbert, C., Grandjean, I., Jancic, C., Hivroz, C., Trautmann, A., Lantz, O., Amigorena, S., 2004. Dendritic cell maturation controls adhesion, synapse formation, and the duration of the interactions with naive T lymphocytes. J. Immunol. 172, 292–301.

Bettelli, E., Carrier, Y., Gao, W., Korn, T., Strom, T.B., Oukka, M., Weiner, H.L., Kuchroo, V.K., 2006. Reciprocal developmental pathways for the generation of pathogenic effector TH17 and regulatory T cells. Nature 441, 235–238.

Bevan, M.J., 1976. Cross-priming for a secondary cytotoxic response to minor H antigens with H-2 congenic cells which do not cross-react in the cytotoxic assay. J. Exp. Med. 143, 1283–1288.

Bigley, V., McGovern, N., Milne, P., Dickinson, R., Pagan, S., Cookson, S., Haniffa, M., Collin, M., 2015. Langerin-expressing dendritic cells in human tissues are related to CD1c+ dendritic cells and distinct from Langerhans cells and CD141high XCR1+ dendritic cells. J. Leukoc. Biol. 97, 627–634.

Blander, J.M., 2007. Signalling and phagocytosis in the orchestration of host defence. Cell. Microbiol. 9, 290–299.

Blander, J.M., Nair-Gupta, P., 2013. An updated view of the intracellular mechanisms regulating cross-presentation. Front. Immunol. 4, 401.

Blasius, A.L., Cella, M., Maldonado, J., Takai, T., Colonna, M., 2006. Siglec-H is an IPC-specific receptor that modulates type I IFN secretion through DAP12. Blood 107, 2474–2476.

Blum, J.S., Wearsch, P.A., Cresswell, P., 2013. Pathways of antigen processing. Annu. Rev. Immunol. 31, 443–473.

Bogunovic, M., Ginhoux, F., Helft, J., Shang, L., Hashimoto, D., Greter, M., Liu, K., Jakubzick, C., Ingersoll, M.A., Leboeuf, M., Stanley, E.R., Nussenzweig, M., Lira, S.A., Randolph, G.J., Merad, M., 2009. Origin of the lamina propria dendritic cell network. Immunity 31, 513–525.

Boiocchi, L., Lonardi, S., Vermi, W., Fisogni, S., Facchetti, F., 2013. BDCA-2 (CD303): a highly specific marker for normal and neoplastic plasmacytoid dendritic cells. Blood 122, 296–297.

Bol, K.F., Schreibelt, G., Gerritsen, W.R., De Vries, I.J., Figdor, C.G., 2016. Dendritic cell-based immunotherapy: state of the art and beyond. Clin. Cancer Res. 22, 1897–1906.

Bonifaz, L., Bonnyay, D., Mahnke, K., Rivera, M., Nussenzweig, M.C., Steinman, R.M., 2002. Efficient targeting of protein antigen to the dendritic cell receptor DEC-205 in the steady state leads to antigen presentation on major histocompatibility complex class I products and peripheral CD8+ T cell tolerance. J. Exp. Med. 196, 1627–1638.

Bonifaz, L.C., Bonnyay, D.P., Charalambous, A., Darguste, D.I., Fujii, S., Soares, H., Brimnes, M.K., Moltedo, B., Moran, T.M., Steinman, R.M., 2004. In vivo targeting of antigens to maturing dendritic cells via the DEC-205 receptor improves T cell vaccination. J. Exp. Med. 199, 815–824.

Borkowski, T.A., Letterio, J.J., Farr, A.G., Udey, M.C., 1996. A role for endogenous transforming growth factor β1 in langerhans cell biology: the skin of transforming growth factor β1 null mice is devoid ofepidermal langerhans cells. J. Exp. Med. 184, 2417.
Boscardin, S.B., Hafalla, J.C.R., Masilamani, R.F., Kamphorst, A.O., Zebroski, H.A., Rai, U., Morrot, A., Zavala, F., Steinman, R.M., Nussenzweig, R.S., Nussenzweig, M.C., 2006. Antigen targeting to dendritic cells elicits long-lived T cell help for antibody responses. J. Exp. Med. 203, 599.
Bosschaerts, T., Guilliams, M., Stijlemans, B., Morias, Y., Engel, D., Tacke, F., Hérin, M., De Baetselier, P., Beschin, A., 2010. Tip-DC development during parasitic infection is regulated by IL-10 and requires CCL2/CCR2, IFN-γ and MyD88 signaling. PLoS Pathog. 6, e1001045.
Boukhaled, G.M., Corrado, M., Guak, H., Krawczyk, C.M., 2019. Chromatin architecture as an essential determinant of dendritic cell function. Front. Immunol. 10, 1119.
Bozzacco, L., Trumpfheller, C., Siegal, F.P., Mehandru, S., Markowitz, M., Carrington, M., Nussenzweig, M.C., Piperno, A.G., Steinman, R.M., 2007. DEC-205 receptor on dendritic cells mediates presentation of HIV gag protein to CD8+ T cells in a spectrum of human MHC I haplotypes. Proc. Natl. Acad. Sci. 104, 1289.
Brasel, K., McKenna, H.J., Morrissey, P.J., Charrier, K., Morris, A.E., Lee, C.C., Williams, D.E., Lyman, S.D., 1996. Hematologic effects of flt3 ligand in vivo in mice. Blood 88, 2004.
Bray, S.J., 2006. Notch signalling: a simple pathway becomes complex. Nat. Rev. Mol. Cell Biol. 7, 678–689.
Breton, G., Lee, J., Zhou, Y.J., Schreiber, J.J., Keler, T., Puhr, S., Anandasabapathy, N., Schlesinger, S., Caskey, M., Liu, K., Nussenzweig, M.C., 2015. Circulating precursors of human CD1c+ and CD141+ dendritic cells. J. Exp. Med. 212, 401–413.
Breton, G., Zheng, S., Valieris, R., Tojal Da Silva, I., Satija, R., Nussenzweig, M.C., 2016. Human dendritic cells (DCs) are derived from distinct circulating precursors that are precommitted to become CD1c+ or CD141+ DCs. J. Exp. Med. 213, 2861–2870
Briseno, C.G., Gargaro, M., Durai, V., Davidson, J.T., Theisen, D.J., Anderson 3rd, D.A., Novack, D.V., Murphy, T.L., Murphy, K.M., 2017. Deficiency of transcription factor RelB perturbs myeloid and DC development by hematopoietic-extrinsic mechanisms. Proc. Natl. Acad. Sci. U. S. A. 114, 3957–3962.
BriseñO, C.G., Satpathy, A.T., Davidson, J.T., Ferris, S.T., Durai, V., Bagadia, P., O'connor, K.W., Theisen, D.J., Murphy, T.L., Murphy, K.M., 2018. Notch2 dependent DC2s mediate splenic germinal center responses. Proc. Natl. Acad. Sci. 115, 10726.
Burkly, L., Hession, C., Ogata, L., Reilly, C., Marconi, L.A., Olson, D., Tizard, R., Cate, R., Lo, D., 1995. Expression of relB is required for the development of thymic medulla and dendritic cells. Nature 373, 531–536.
Cabeza-Cabrerizo, M., Van Blijswijk, J., Wienert, S., Heim, D., Jenkins, R.P., Chakravarty, P., Rogers, N., Frederico, B., Acton, S., Beerling, E., Van Rheenen, J., Clevers, H., Schraml, B.U., Bajénoff, M., Gerner, M., Germain, R.N., Sahai, E., Klauschen, F., Reis, E., Sousa, C., 2019. Tissue clonality of dendritic cell subsets and emergency DCpoiesis revealed by multicolor fate mapping of DC progenitors. Sci. Immunol. 4, eaaw1941.
Calabro, S., Gallman, A., Gowthaman, U., Liu, D., Chen, P., Liu, J., Krishnaswamy, J.K., Nascimento, M.S.L., Xu, L., Patel, S.R., Williams, A., Tormey, C.A., Hod, E.A., Spitalnik, S.L., Zimring, J.C., Hendrickson, J.E., Stowell, S.R., Eisenbarth, S.C., 2016. Bridging channel dendritic cells induce immunity to transfused red blood cells. J. Exp. Med. 213, 887.

Caminschi, I., Proietto, A.I., Ahmet, F., Kitsoulis, S., Shin Teh, J., Lo, J.C., Rizzitelli, A., Wu, L., Vremec, D., Van Dommelen, S.L., Campbell, I.K., Maraskovsky, E., Braley, H., Davey, G.M., Mottram, P., Van De Velde, N., Jensen, K., Lew, A.M., Wright, M.D., Heath, W.R., Shortman, K., Lahoud, M.H., 2008. The dendritic cell subtype-restricted C-type lectin Clec9A is a target for vaccine enhancement. Blood 112, 3264–3273.

Caton, M.L., Smith-Raska, M.R., Reizis, B., 2007. Notch-RBP-J signaling controls the homeostasis of CD8-dendritic cells in the spleen. J. Exp. Med. 204, 1653–1664.

Cebrian, I., Visentin, G., Blanchard, N., Jouve, M., Bobard, A., Moita, C., Enninga, J., Moita, L.F., Amigorena, S., Savina, A., 2011. Sec22b regulates phagosomal maturation and antigen crosspresentation by dendritic cells. Cell 147, 1355–1368.

Cella, M., Jarrossay, D., Facchetti, F., Alebardi, O., Nakajima, H., Lanzavecchia, A., Colonna, M., 1999. Plasmacytoid monocytes migrate to inflamed lymph nodes and produce large amounts of type I interferon. Nat. Med. 5, 919–923.

Ceppi, M., Pereira, P.M., Dunand-Sauthier, I., Barras, E., Reith, W., Santos, M.A., Pierre, P., 2009. MicroRNA-155 modulates the interleukin-1 signaling pathway in activated human monocyte-derived dendritic cells. Proc. Natl. Acad. Sci. 106, 2735.

Chen, W., Jin, W., Hardegen, N., Lei, K.J., Li, L., Marinos, N., Mcgrady, G., Wahl, S.M., 2003. Conversion of peripheral CD4+CD25-naive T cells to CD4+CD25+ regulatory T cells by TGF-beta induction of transcription factor Foxp3. J. Exp. Med. 198, 1875–1886.

Chen, C.Z., Schaffert, S., Fragoso, R., Loh, C., 2013a. Regulation of immune responses and tolerance: the microRNA perspective. Immunol. Rev. 253, 112–128.

Chen, Y.L., Chen, T.T., Pai, L.M., Wesoly, J., Bluyssen, H.A., Lee, C.K., 2013b. A type I IFN-Flt3 ligand axis augments plasmacytoid dendritic cell development from common lymphoid progenitors. J. Exp. Med. 210, 2515–2522.

Cheng, P., Nefedova, Y., Miele, L., Osborne, B.A., Gabrilovich, D., 2003. Notch signaling is necessary but not sufficient for differentiation of dendritic cells. Blood 102, 3980–3988.

Cheng, P., Nefedova, Y., Corzo, C.A., Gabrilovich, D.I., 2007. Regulation of dendritic-cell differentiation by bone marrow stroma via different Notch ligands. Blood 109, 507–515.

Chistiakov, D.A., Orekhov, A.N., Sobenin, I.A., Bobryshev, Y.V., 2014. Plasmacytoid dendritic cells: development, functions, and role in atherosclerotic inflammation. Front. Physiol. 5, 279.

Chopin, M., Allan, R., Belz, G., 2012. Transcriptional regulation of dendritic cell diversity. Front. Immunol. 3, 26.

Chopin, M., Seillet, C., Chevrier, S., Wu, L., Wang, H., Morse, H.C., Belz, G.T., Nutt, S.L., 2013. Langerhans cells are generated by two distinct PU.1-dependent transcriptional networks. J. Exp. Med. 210, 2967.

Chopin, M., Preston, S.P., Lun, A.T.L., Tellier, J., Smyth, G.K., Pellegrini, M., Belz, G.T., Corcoran, L.M., Visvader, J.E., Wu, L., Nutt, S.L., 2016. RUNX2 mediates plasmacytoid dendritic cell egress from the bone marrow and controls viral immunity. Cell Rep. 15, 866–878.

Chopin, M., Lun, A.T., Zhan, Y., Schreuder, J., Coughlan, H., D'amico, A., Mielke, L.A., Almeida, F.F., Kueh, A.J., Dickins, R.A., Belz, G.T., Naik, S.H., Lew, A.M., Bouillet, P., Herold, M.J., Smyth, G.K., Corcoran, L.M., Nutt, S.L., 2019. Transcription factor PU.1 promotes conventional dendritic cell identity and function via induction of transcriptional regulator DC-SCRIPT. Immunity 50, 77–90 e5.

Chu, C.C., Ali, N., Karagiannis, P., Di Meglio, P., Skowera, A., Napolitano, L., Barinaga, G., Grys, K., Sharif-Paghaleh, E., Karagiannis, S.N., Peakman, M., Lombardi, G., Nestle, F.O., 2012. Resident CD141 (BDCA3)+ dendritic cells in human skin produce IL-10 and induce regulatory T cells that suppress skin inflammation. J. Exp. Med. 209, 935–945.

Cisse, B., Caton, M.L., Lehner, M., Maeda, T., Scheu, S., Locksley, R., Holmberg, D., Zweier, C., Den Hollander, N.S., Kant, S.G., Holter, W., Rauch, A., Zhuang, Y., Reizis, B., 2008. Transcription factor E2-2 is an essential and specific regulator of plasmacytoid dendritic cell development. Cell 135, 37–48.

Clausen, B.E., Burkhardt, C., Reith, W., Renkawitz, R., Forster, I., 1999. Conditional gene targeting in macrophages and granulocytes using LysMcre mice. Transgenic Res. 8, 265–277.

Clayton, K., Vallejo, A.F., Davies, J., Sirvent, S., Polak, M.E., 2017. Langerhans cells—programmed by the epidermis. Front. Immunol. 8, 1676.

Cohen, S.B., Smith, N.L., McDougal, C., Pepper, M., Shah, S., Yap, G.S., Acha-Orbea, H., Jiang, A., Clausen, B.E., Rudd, B.D., Denkers, E.Y., 2015. Beta-catenin signaling drives differentiation and proinflammatory function of IRF8-dependent dendritic cells. J. Immunol. 194, 210–222.

Cohn, L., Chatterjee, B., Esselborn, F., Smed-Sorensen, A., Nakamura, N., Chalouni, C., Lee, B.C., Vandlen, R., Keler, T., Lauer, P., Brockstedt, D., Mellman, I., Delamarre, L., 2013. Antigen delivery to early endosomes eliminates the superiority of human blood BDCA3+ dendritic cells at cross presentation. J. Exp. Med. 210, 1049–1063.

Collin, M., Bigley, V., 2018. Human dendritic cell subsets: an update. Immunology 154, 3–20.

Collin, M., Milne, P., 2016. Langerhans cell origin and regulation. Curr. Opin. Hematol. 23, 28–35.

Collison, L.W., Workman, C.J., Kuo, T.T., Boyd, K., Wang, Y., Vignali, K.M., Cross, R., Sehy, D., Blumberg, R.S., Vignali, D.A., 2007. The inhibitory cytokine IL-35 contributes to regulatory T-cell function. Nature 450, 566–569.

Colonna, M., Trinchieri, G., Liu, Y.J., 2004. Plasmacytoid dendritic cells in immunity. Nat. Immunol. 5, 1219–1226.

Contreras, V., Urien, C., Guiton, R., Alexandre, Y., Vu Manh, T.-P., Andrieu, T., Crozat, K., Jouneau, L., Bertho, N., Epardaud, M., Hope, J., Savina, A., Amigorena, S., Bonneau, M., Dalod, M., Schwartz-Cornil, I., 2010. Existence of CD8α-like dendritic cells with a conserved functional specialization and a common molecular signature in distant mammalian species. J. Immunol. 185, 3313.

Coombes, J.L., Siddiqui, K.R., Arancibia-Carcamo, C.V., Hall, J., Sun, C.M., Belkaid, Y., Powrie, F., 2007. A functionally specialized population of mucosal CD103+ DCs induces Foxp3+ regulatory T cells via a TGF-beta and retinoic acid-dependent mechanism. J. Exp. Med. 204, 1757–1764.

Coutant, F., Miossec, P., 2016. Altered dendritic cell functions in autoimmune diseases: distinct and overlapping profiles. Nat. Rev. Rheumatol. 12, 703.

Crotzer, V.L., Blum, J.S., 2010. Autophagy and adaptive immunity. Immunology 131, 9–17.

Crozat, K., Guiton, R., Contreras, V., Feuillet, V., Dutertre, C.A., Ventre, E., Vu Manh, T.P., Baranek, T., Storset, A.K., Marvel, J., Boudinot, P., Hosmalin, A., Schwartz-Cornil, I., Dalod, M., 2010. The XC chemokine receptor 1 is a conserved selective marker of mammalian cells homologous to mouse CD8alpha+ dendritic cells. J. Exp. Med. 207, 1283–1292.

Crozat, K., Tamoutounour, S., Vu Manh, T.P., Fossum, E., Luche, H., Ardouin, L., Guilliams, M., Azukizawa, H., Bogen, B., Malissen, B., Henri, S., Dalod, M., 2011. Cutting edge: expression of XCR1 defines mouse lymphoid-tissue resident and migratory dendritic cells of the CD8alpha+ type. J. Immunol. 187, 4411–4415.

Cruz, L.J., Tacken, P.J., Bonetto, F., Buschow, S.I., Croes, H.J., Wijers, M., De Vries, I.J., Figdor, C.G., 2011. Multimodal imaging of nanovaccine carriers targeted to human dendritic cells. Mol. Pharm. 8, 520–531.

Cytlak, U., Resteu, A., Bogaert, D., Kuehn, H.S., Altmann, T., Gennery, A., Jackson, G., Kumanovics, A., Voelkerding, K.V., Prader, S., Dullaers, M., Reichenbach, J., Hill, H., Haerynck, F., Rosenzweig, S.D., Collin, M., Bigley, V., 2018. Ikaros family zinc finger 1 regulates dendritic cell development and function in humans. Nat. Commun. 9, 1239.

Czakai, K., Leonhardt, I., Dix, A., Bonin, M., Linde, J., Einsele, H., Kurzai, O., Loeffler, J., 2016. Krüppel-like factor 4 modulates interleukin-6 release in human dendritic cells after in vitro stimulation with Aspergillus fumigatus and Candida albicans. Sci. Rep. 6, 27990.

Darrah, P.A., Patel, D.T., De Luca, P.M., Lindsay, R.W.B., Davey, D.F., Flynn, B.J., Hoff, S.T., Andersen, P., Reed, S.G., Morris, S.L., Roederer, M., Seder, R.A., 2007. Multifunctional TH1 cells define a correlate of vaccine-mediated protection against Leishmania major. Nat. Med. 13, 843–850.

Davies, L.C., Jenkins, S.J., Allen, J.E., Taylor, P.R., 2013. Tissue-resident macrophages. Nat. Immunol. 14, 986–995.

Delamarre, L., Holcombe, H., Mellman, I., 2003. Presentation of exogenous antigens on major histocompatibility complex (MHC) class I and MHC class II molecules is differentially regulated during dendritic cell maturation. J. Exp. Med. 198, 111–122.

Demedts, I.K., Brusselle, G.G., Vermaelen, K.Y., Pauwels, R.A., 2005. Identification and characterization of human pulmonary dendritic cells. Am. J. Respir. Cell Mol. Biol. 32, 177–184.

de Mingo Pulido, A., Gardner, A., Hiebler, S., Soliman, H., Rugo, H.S., Krummel, M.F., Coussens, L.M., Ruffell, B., 2018. TIM-3 regulates CD103(+) dendritic cell function and response to chemotherapy in breast cancer. Cancer Cell, 33, 60.e6–74.e6.

Den Haan, J.M., Bevan, M.J., 2002. Constitutive versus activation-dependent cross-presentation of immune complexes by CD8(+) and CD8(-) dendritic cells in vivo. J. Exp. Med. 196, 817–827.

Den Haan, J.M., Lehar, S.M., Bevan, M.J., 2000. CD8(+) but not CD8(-) dendritic cells cross-prime cytotoxic T cells in vivo. J. Exp. Med. 192, 1685–1696.

Denzin, L.K., 2013. Inhibition of HLA-DM Mediated MHC class II Peptide Loading by HLA-DO promotes self tolerance. Front. Immunol. 4, 465.

Denzin, L.K., Hammond, C., Cresswell, P., 1996. HLA-DM interactions with intermediates in HLA-DR maturation and a role for HLA-DM in stabilizing empty HLA-DR molecules. J. Exp. Med. 184, 2153–2165.

Desch, A.N., Gibbings, S.L., Clambey, E.T., Janssen, W.J., Slansky, J.E., Kedl, R.M., Henson, P.M., Jakubzick, C., 2014. Dendritic cell subsets require cis-activation for cytotoxic CD8 T-cell induction. Nat. Commun. 5, 4674.

De Sousa Linhares, A., Leitner, J., Grabmeier-Pfistershammer, K., Steinberger, P., 2018. Not all immune checkpoints are created equal. Front. Immunol. 9, 1909.

De Trez, C., Magez, S., Akira, S., Ryffel, B., Carlier, Y., Muraille, E., 2009. iNOS-producing inflammatory dendritic cells constitute the major infected cell type during the chronic Leishmania major infection phase of C57BL/6 resistant mice. PLoS Pathog. 5, e1000494.

De Vries, H.E., Hendriks, J.J., Honing, H., De Lavalette, C.R., Van Der Pol, S.M., Hooijberg, E., Dijkstra, C.D., Van Den Berg, T.K., 2002. Signal-regulatory protein alpha-CD47 interactions are required for the transmigration of monocytes across cerebral endothelium. J. Immunol. 168, 5832–5839.

Diamond, M.S., Kinder, M., Matsushita, H., Mashayekhi, M., Dunn, G.P., Archambault, J.M., Lee, H., Arthur, C.D., White, J.M., Kalinke, U., Murphy, K.M., Schreiber, R.D., 2011. Type I interferon is selectively required by dendritic cells for immune rejection of tumors. J. Exp. Med. 208, 1989–2003.

Dillon, S., Agrawal, A., Van Dyke, T., Landreth, G., McCauley, L., Koh, A., Maliszewski, C., Akira, S., Pulendran, B., 2004. A toll-like receptor 2 ligand stimulates Th2 responses in vivo, via induction of extracellular signal-regulated kinase mitogen-activated protein kinase and c-Fos in dendritic cells. J. Immunol. 172, 4733.

Dorner, B.G., Dorner, M.B., Zhou, X., Opitz, C., Mora, A., Guttler, S., Hutloff, A., Mages, H.W., Ranke, K., Schaefer, M., Jack, R.S., Henn, V., Kroczek, R.A., 2009. Selective expression of the chemokine receptor XCR1 on cross-presenting dendritic cells determines cooperation with CD8+ T cells. Immunity 31, 823–833.

Draheim, M., Wlodarczyk, M.F., Crozat, K., Saliou, J.M., Alayi, T.D., Tomavo, S., Hassan, A., Salvioni, A., Demarta-Gatsi, C., Sidney, J., Sette, A., Dalod, M., Berry, A., Silvie, O., Blanchard, N., 2017. Profiling MHC II immunopeptidome of blood-stage malaria reveals that cDC1 control the functionality of parasite-specific CD4 T cells. EMBO Mol. Med. 9, 1605–1621.

Dress, R.J., Dutertre, C.-A., Giladi, A., Schlitzer, A., Low, I., Shadan, N.B., Tay, A., Lum, J., Kairi, M.F.B.M., Hwang, Y.Y., Becht, E., Cheng, Y., Chevrier, M., Larbi, A., Newell, E.W., Amit, I., Chen, J., Ginhoux, F., 2019. Plasmacytoid dendritic cells develop from Ly6D+ lymphoid progenitors distinct from the myeloid lineage. Nat. Immunol. 20, 852–864.

Dudziak, D., Kamphorst, A.O., Heidkamp, G.F., Buchholz, V.R., Trumpfheller, C., Yamazaki, S., Cheong, C., Liu, K., Lee, H.-W., Park, C.G., Steinman, R.M., Nussenzweig, M.C., 2007. Differential antigen processing by dendritic cell subsets in vivo. Science 315, 107.

Dunand-Sauthier, I., Santiago-Raber, M.L., Capponi, L., Vejnar, C.E., Schaad, O., Irla, M., Seguin-Estevez, Q., Descombes, P., Zdobnov, E.M., Acha-Orbea, H., Reith, W., 2011. Silencing of c-Fos expression by microRNA-155 is critical for dendritic cell maturation and function. Blood 117, 4490–4500.

Durai, V., Bagadia, P., Granja, J.M., Satpathy, A.T., Kulkarni, D.H., Davidson, J.T., Wu, R., Patel, S.J., Iwata, A., Liu, T.-T., Huang, X., BriseÑO, C.G., Grajales-Reyes, G.E., WÖhner, M., Tagoh, H., Kee, B.L., Newberry, R.D., Busslinger, M., Chang, H.Y., Murphy, T.L., Murphy, K.M., 2019. Cryptic activation of an Irf8 enhancer governs cDC1 fate specification. Nat. Immunol. 20, 1161–1173.

Dursun, E., Endele, M., Musumeci, A., Failmezger, H., Wang, S.-H., Tresch, A., Schroeder, T., Krug, A.B., 2016. Continuous single cell imaging reveals sequential steps of plasmacytoid dendritic cell development from common dendritic cell progenitors. Sci. Rep. 6, 37462.

Dzionek, A., Fuchs, A., Schmidt, P., Cremer, S., Zysk, M., Miltenyi, S., Buck, D.W., Schmitz, J., 2000. BDCA-2, BDCA-3, and BDCA-4: three markers for distinct subsets of dendritic cells in human peripheral blood. J. Immunol. 165, 6037–6046.

Dzionek, A., Sohma, Y., Nagafune, J., Cella, M., Colonna, M., Facchetti, F., Günther, G., Johnston, I., Lanzavecchia, A., Nagasaka, T., Okada, T., Vermi, W., Winkels, G., Yamamoto, T., Zysk, M., Yamaguchi, Y., Schmitz, J., 2001. BDCA-2, a novel plasmacytoid dendritic cell-specific type II C type lectin, mediates antigen capture and is a potent inhibitor of interferon α/β induction. J. Exp. Med. 194, 1823.

Edelson, B.T., Kc, W., Juang, R., Kohyama, M., Benoit, L.A., Klekotka, P.A., Moon, C., Albring, J.C., Ise, W., Michael, D.G., Bhattacharya, D., Stappenbeck, T.S., Holtzman, M.J., Sung, S.S., Murphy, T.L., Hildner, K., Murphy, K.M., 2010. Peripheral CD103+ dendritic cells form a unified subset developmentally related to CD8alpha + conventional dendritic cells. J. Exp. Med. 207, 823–836.

Edelson, B.T., Bradstreet, T.R., Kc, W., Hildner, K., Herzog, J.W., Sim, J., Russell, J.H., Murphy, T.L., Unanue, E.R., Murphy, K.M., 2011. Batf3-dependent CD11b(low/-) peripheral dendritic cells are GM-CSF-independent and are not required for Th cell priming after subcutaneous immunization. PLoS One 6, e25660.

Eisenbarth, S.C., 2019. Dendritic cell subsets in T cell programming: location dictates function. Nat. Rev. Immunol. 19, 89–103.

Embgenbroich, M., Burgdorf, S., 2018. Current concepts of antigen cross-presentation. Front. Immunol. 9, 1643.

Fabian, M.R., Sonenberg, N., Filipowicz, W., 2010. Regulation of mRNA translation and stability by microRNAs. Annu. Rev. Biochem. 79, 351–379.
Fallarino, F., Grohmann, U., Hwang, K.W., Orabona, C., Vacca, C., Bianchi, R., Belladonna, M.L., Fioretti, M.C., Alegre, M.-L., Puccetti, P., 2003. Modulation of tryptophan catabolism by regulatory T cells. Nat. Immunol. 4, 1206.
Farrand, K.J., Dickgreber, N., Stoitzner, P., Ronchese, F., Petersen, T.R., Hermans, I.F., 2009. Langerin+CD8α+ dendritic cells are critical for cross-priming and IL-12 production in response to systemic antigens. J. Immunol. 183, 7732.
Feinberg, M.W., Wara, A.K., Cao, Z., Lebedeva, M.A., Rosenbauer, F., Iwasaki, H., Hirai, H., Katz, J.P., Haspel, R.L., Gray, S., Akashi, K., Segre, J., Kaestner, K.H., Tenen, D.G., Jain, M.K., 2007. The Kruppel-like factor KLF4 is a critical regulator of monocyte differentiation. EMBO J. 26, 4138–4148.
Figdor, C.G., Van Kooyk, Y., Adema, G.J., 2002. C-type lectin receptors on dendritic cells and langerhans cells. Nat. Rev. Immunol. 2, 77–84.
Fiorentino, D.F., Bond, M.W., Mosmann, T.R., 1989. Two types of mouse T helper cell. IV. Th2 clones secrete a factor that inhibits cytokine production by Th1 clones. J. Exp. Med. 170, 2081.
Flacher, V., Tripp, C.H., Mairhofer, D.G., Steinman, R.M., Stoitzner, P., Idoyaga, J., Romani, N., 2014. Murine langerin+ dermal dendritic cells prime CD8+ T cells while langerhans cells induce cross-tolerance. EMBO Mol. Med. 6, 1191–1204.
Fogg, D.K., Sibon, C., Miled, C., Jung, S., Aucouturier, P., Littman, D.R., Cumano, A., Geissmann, F., 2006. A clonogenic bone marrow progenitor specific for macrophages and dendritic cells. Science 311, 83–87.
Fontenot, J.D., Gavin, M.A., Rudensky, A.Y., 2003. Foxp3 programs the development and function of CD4+CD25+ regulatory T cells. Nat. Immunol. 4, 330–336.
Fort, M.M., Cheung, J., Yen, D., Li, J., Zurawski, S.M., Lo, S., Menon, S., Clifford, T., Hunte, B., Lesley, R., Muchamuel, T., Hurst, S.D., Zurawski, G., Leach, M.W., Gorman, D.M., Rennick, D.M., 2001. IL-25 induces IL-4, IL-5, and IL-13 and Th2-associated pathologies in vivo. Immunity 15, 985–995.
Fu, C., Jiang, A., 2018. Dendritic cells and CD8 T cell immunity in tumor microenvironment. Front. Immunol. 9, 3059.
Fuertes, M.B., Kacha, A.K., Kline, J., Woo, S.R., Kranz, D.M., Murphy, K.M., Gajewski, T.F., 2011. Host type I IFN signals are required for antitumor CD8+ T cell responses through CD8{alpha}+ dendritic cells. J. Exp. Med. 208, 2005–2016.
Gaiser, M.R., LÄmmermann, T., Feng, X., Igyarto, B.Z., Kaplan, D.H., Tessarollo, L., Germain, R.N., Udey, M.C., 2012. Cancer-associated epithelial cell adhesion molecule (EpCAM; CD326) enables epidermal Langerhans cell motility and migration in vivo. Proc. Natl. Acad. Sci. 109, E889.
Gantier, M.P., McCoy, C.E., Rusinova, I., Saulep, D., Wang, D., Xu, D., Irving, A.T., Behlke, M.A., Hertzog, P.J., Mackay, F., Williams, B.R., 2011. Analysis of microRNA turnover in mammalian cells following Dicer1 ablation. Nucleic Acids Res. 39, 5692–5703.
Gao, Y., Nish, S.A., Jiang, R., Hou, L., Licona-Limon, P., Weinstein, J.S., Zhao, H., Medzhitov, R., 2013. Control of T helper 2 responses by transcription factor IRF4-dependent dendritic cells. Immunity 39, 722–732.
Garlisi, C.G., Falcone, A., Kung, T.T., Stelts, D., Pennline, K.J., Beavis, A.J., Smith, S.R., Egan, R.W., Umland, S.P., 1995. T cells are necessary for Th2 cytokine production and eosinophil accumulation in airways of antigen-challenged allergic mice. Clin. Immunol. Immunopathol. 75, 75–83.
Garnache-Ottou, F., Jacob, M.-C., Biichle, S., Chaperot, L., Trimoreau, F., Plumas, J., Tiberghien, P., Saas, P., Feuillard, J., 2005. Identification of BDCA-2 and high levels of CD123 expression as useful markers for the diagnosis of plasmacytoid dendritic cell leukemia. Blood 106, 3269.

Garstka, M.A., Neefjes, J., 2013. How to target MHC class II into the MIIC compartment. Mol. Immunol. 55, 162–165.
Gascoyne, D.M., Long, E., Veiga-Fernandes, H., De Boer, J., Williams, O., Seddon, B., Coles, M., Kioussis, D., Brady, H.J., 2009. The basic leucine zipper transcription factor E4BP4 is essential for natural killer cell development. Nat. Immunol. 10, 1118–1124.
Gebert, L.F.R., Macrae, I.J., 2019. Regulation of microRNA function in animals. Nat. Rev. Mol. Cell Biol. 20, 21–37.
Geissmann, F., Manz, M.G., Jung, S., Sieweke, M.H., Merad, M., Ley, K., 2010. Development of monocytes, macrophages, and dendritic cells. Science 327, 656–661.
Gerner, M.Y., Kastenmuller, W., Ifrim, I., Kabat, J., Germain, R.N., 2012. Histo-cytometry: a method for highly multiplex quantitative tissue imaging analysis applied to dendritic cell subset microanatomy in lymph nodes. Immunity 37, 364–376.
Gerner, M.Y., Casey, K.A., Kastenmuller, W., Germain, R.N., 2017. Dendritic cell and antigen dispersal landscapes regulate T cell immunity. J. Exp. Med. 214, 3105–3122.
Ghosh, H.S., Cisse, B., Bunin, A., Lewis, K.L., Reizis, B., 2010. Continuous expression of the transcription factor e2-2 maintains the cell fate of mature plasmacytoid dendritic cells. Immunity 33, 905–916.
Ginhoux, F., Jung, S., 2014. Monocytes and macrophages: developmental pathways and tissue homeostasis. Nat. Rev. Immunol. 14, 392.
Ginhoux, F., Tacke, F., Angeli, V., Bogunovic, M., Loubeau, M., Dai, X.-M., Stanley, E.R., Randolph, G.J., Merad, M., 2006. Langerhans cells arise from monocytes in vivo. Nat. Immunol. 7, 265–273.
Ginhoux, F., Collin, M.P., Bogunovic, M., Abel, M., Leboeuf, M., Helft, J., Ochando, J., Kissenpfennig, A., Malissen, B., Grisotto, M., Snoeck, H., Randolph, G., Merad, M., 2007. Blood-derived dermal langerin+ dendritic cells survey the skin in the steady state. J. Exp. Med. 204, 3133–3146.
Ginhoux, F., Liu, K., Helft, J., Bogunovic, M., Greter, M., Hashimoto, D., Price, J., Yin, N., Bromberg, J., Lira, S.A., Stanley, E.R., Nussenzweig, M., Merad, M., 2009. The origin and development of nonlymphoid tissue CD103+ DCs. J. Exp. Med. 206, 3115–3130.
Gomez De Agüero, M., Vocanson, M., Hacini-Rachinel, F., Taillardet, M., Sparwasser, T., Kissenpfennig, A., Malissen, B., Kaiserlian, D., Dubois, B., 2012. Langerhans cells protect from allergic contact dermatitis in mice by tolerizing CD8+ T cells and activating Foxp3+ regulatory T cells. J. Clin. Invest. 122, 1700–1711.
Gomez Perdiguero, E., Klapproth, K., Schulz, C., Busch, K., Azzoni, E., Crozet, L., Garner, H., Trouillet, C., De Bruijn, M.F., Geissmann, F., Rodewald, H.-R., 2014. Tissue-resident macrophages originate from yolk-sac-derived erythro-myeloid progenitors. Nature 518, 547.
Gottschalk, C., Mettke, E., Kurts, C., 2015. The role of invariant natural killer T cells in dendritic cell licensing, cross-priming, and memory CD8(+) T cell generation. Front. Immunol. 6, 379.
Grajales-Reyes, G.E., Iwata, A., Albring, J., Wu, X., Tussiwand, R., Kc, W., Kretzer, N.M., Briseno, C.G., Durai, V., Bagadia, P., Haldar, M., Schonheit, J., Rosenbauer, F., Murphy, T.L., Murphy, K.M., 2015. Batf3 maintains autoactivation of Irf8 for commitment of a CD8alpha(+) conventional DC clonogenic progenitor. Nat. Immunol. 16, 708–717.
Granot, T., Senda, T., Carpenter, D.J., Matsuoka, N., Weiner, J., Gordon, C.L., Miron, M., Kumar, B.V., Griesemer, A., Ho, S.H., Lerner, H., Thome, J.J.C., Connors, T., Reizis, B., Farber, D.L., 2017. Dendritic cells display subset and tissue-specific maturation dynamics over human life. Immunity 46, 504–515.
Gregory, R.I., Yan, K.P., Amuthan, G., Chendrimada, T., Doratotaj, B., Cooch, N., Shiekhattar, R., 2004. The microprocessor complex mediates the genesis of microRNAs. Nature 432, 235–240.

Greter, M., Helft, J., Chow, A., Hashimoto, D., Mortha, A., Agudo-Cantero, J., Bogunovic, M., Gautier, E.L., Miller, J., Leboeuf, M., Lu, G., Aloman, C., Brown, B.D., Pollard, J.W., Xiong, H., Randolph, G.J., Chipuk, J.E., Frenette, P.S., Merad, M., 2012. GM-CSF controls nonlymphoid tissue dendritic cell homeostasis but is dispensable for the differentiation of inflammatory dendritic cells. Immunity 36, 1031–1046.

Grishok, A., Pasquinelli, A.E., Conte, D., Li, N., Parrish, S., Ha, I., Baillie, D.L., Fire, A., Ruvkun, G., Mello, C.C., 2001. Genes and mechanisms related to RNA interference regulate expression of the small temporal RNAs that control *C. elegans* developmental timing. Cell 106, 23–34.

Gros, M., Amigorena, S., 2019. Regulation of antigen export to the cytosol during cross-presentation. Front. Immunol. 10, 41.

Guermonprez, P., Valladeau, J., Zitvogel, L., Thery, C., Amigorena, S., 2002. Antigen presentation and T cell stimulation by dendritic cells. Annu. Rev. Immunol. 20, 621–667.

Guilliams, M., Crozat, K., Henri, S., Tamoutounour, S., Grenot, P., Devilard, E., De Bovis, B., Alexopoulou, L., Dalod, M., Malissen, B., 2010. Skin-draining lymph nodes contain dermis-derived CD103(-) dendritic cells that constitutively produce retinoic acid and induce Foxp3(+) regulatory T cells. Blood 115, 1958–1968.

Guilliams, M., Ginhoux, F., Jakubzick, C., Naik, S.H., Onai, N., Schraml, B.U., Segura, E., Tussiwand, R., Yona, S., 2014. Dendritic cells, monocytes and macrophages: a unified nomenclature based on ontogeny. Nat. Rev. Immunol. 14, 571–578.

Guilliams, M., Dutertre, C.A., Scott, C.L., McGovern, N., Sichien, D., Chakarov, S., Van Gassen, S., Chen, J., Poidinger, M., De Prijck, S., Tavernier, S.J., Low, I., Irac, S.E., Mattar, C.N., Sumatoh, H.R., Low, G.H.L., Chung, T.J.K., Chan, D.K.H., Tan, K.K., Hon, T.L.K., Fossum, E., Bogen, B., Choolani, M., Chan, J.K.Y., Larbi, A., Luche, H., Henri, S., Saeys, Y., Newell, E.W., Lambrecht, B.N., Malissen, B., Ginhoux, F., 2016. Unsupervised high-dimensional analysis aligns dendritic cells across tissues and species. Immunity 45, 669–684.

Guram, K., Kim, S.S., Wu, V., Sanders, P.D., Patel, S., Schoenberger, S.P., Cohen, E.E.W., Chen, S.-Y., Sharabi, A.B., 2019. A threshold model for T-cell activation in the era of checkpoint blockade immunotherapy. Front. Immunol. 10, 491.

Gurka, S., Hartung, E., Becker, M., Kroczek, R.A., 2015. Mouse conventional dendritic cells can be universally classified based on the mutually exclusive expression of XCR1 and SIRPalpha. Front. Immunol. 6, 35.

Gutierrez-Martinez, E., Planès, R., Anselmi, G., Reynolds, M., Menezes, S., Adiko, A.C., Saveanu, L., Guermonprez, P., 2015. Cross-presentation of cell-associated antigens by MHC class I in dendritic cell subsets. Front. Immunol. 6, 363.

Guttman-Yassky, E., Lowes, M.A., Fuentes-Duculan, J., Whynot, J., Novitskaya, I., Cardinale, I., Haider, A., Khatcherian, A., Carucci, J.A., Bergman, R., Krueger, J.G., 2007. Major differences in inflammatory dendritic cells and their products distinguish atopic dermatitis from psoriasis. J. Allergy Clin. Immunol. 119, 1210–1217.

Hacker, C., Kirsch, R.D., Ju, X.S., Hieronymus, T., Gust, T.C., Kuhl, C., Jorgas, T., Kurz, S.M., Rose-John, S., Yokota, Y., Zenke, M., 2003. Transcriptional profiling identifies Id2 function in dendritic cell development. Nat. Immunol. 4, 380–386.

Hadeiba, H., Butcher, E.C., 2013. Thymus-homing dendritic cells in central tolerance. Eur. J. Immunol. 43, 1425–1429.

Hambleton, S., Salem, S., Bustamante, J., Bigley, V., Boisson-Dupuis, S., Azevedo, J., Fortin, A., Haniffa, M., Ceron-Gutierrez, L., Bacon, C.M., Menon, G., Trouillet, C., McDonald, D., Carey, P., Ginhoux, F., Alsina, L., Zumwalt, T.J., Kong, X.F., Kumararatne, D., Butler, K., Hubeau, M., Feinberg, J., Al-Muhsen, S., Cant, A., Abel, L., Chaussabel, D., Doffinger, R., Talesnik, E., Grumach, A., Duarte, A., Abarca, K., Moraes-Vasconcelos, D., Burk, D., Berghuis, A., Geissmann, F., Collin, M., Casanova, J.L., Gros, P., 2011. IRF8 mutations and human dendritic-cell immunodeficiency. N. Engl. J. Med. 365, 127–138.

Hammad, H., Plantinga, M., Deswarte, K., Pouliot, P., Willart, M.A.M., Kool, M., Muskens, F., Lambrecht, B.N., 2010. Inflammatory dendritic cells—not basophils—are necessary and sufficient for induction of Th2 immunity to inhaled house dust mite allergen. J. Exp. Med. 207, 2097.

Haniffa, M., Shin, A., Bigley, V., McGovern, N., Teo, P., See, P., Wasan, P.S., Wang, X.N., Malinarich, F., Malleret, B., Larbi, A., Tan, P., Zhao, H., Poidinger, M., Pagan, S., Cookson, S., Dickinson, R., Dimmick, I., Jarrett, R.F., Renia, L., Tam, J., Song, C., Connolly, J., Chan, J.K., Gehring, A., Bertoletti, A., Collin, M., Ginhoux, F., 2012. Human tissues contain CD141hi cross-presenting dendritic cells with functional homology to mouse CD103+ nonlymphoid dendritic cells. Immunity 37, 60–73.

Haniffa, M., Collin, M., Ginhoux, F., 2013. Ontogeny and functional specialization of dendritic cells in human and mouse. Adv. Immunol. 120, 1–49.

Harpur, C.M., Kato, Y., Dewi, S.T., Stankovic, S., Johnson, D.N., Bedoui, S., Whitney, P.G., Lahoud, M.H., Caminschi, I., Heath, W.R., Brooks, A.G., Gebhardt, T., 2019. Classical type 1 dendritic cells dominate priming of Th1 responses to herpes simplex virus type 1 skin infection. J. Immunol. 202, 653.

Harrington, L.E., Hatton, R.D., Mangan, P.R., Turner, H., Murphy, T.L., Murphy, K.M., Weaver, C.T., 2005. Interleukin 17-producing CD4+ effector T cells develop via a lineage distinct from the T helper type 1 and 2 lineages. Nat. Immunol. 6, 1123–1132.

Hashimi, S.T., Fulcher, J.A., Chang, M.H., Gov, L., Wang, S., Lee, B., 2009. MicroRNA profiling identifies miR-34a and miR-21 and their target genes JAG1 and WNT1 in the coordinate regulation of dendritic cell differentiation. Blood 114, 404–414.

Hashimoto-Hill, S., Friesen, L., Park, S., Im, S., Kaplan, M.H., Kim, C.H., 2018. RARα supports the development of Langerhans cells and langerin-expressing conventional dendritic cells. Nat. Commun. 9, 3896.

Hawiger, D., Inaba, K., Dorsett, Y., Guo, M., Mahnke, K., Rivera, M., Ravetch, J.V., Steinman, R.M., Nussenzweig, M.C., 2001. Dendritic cells induce peripheral T cell unresponsiveness under steady state conditions in vivo. J. Exp. Med. 194, 769.

Heath, W.R., Belz, G.T., Behrens, G.M., Smith, C.M., Forehan, S.P., Parish, I.A., Davey, G.M., Wilson, N.S., Carbone, F.R., Villadangos, J.A., 2004. Cross-presentation, dendritic cell subsets, and the generation of immunity to cellular antigens. Immunol. Rev. 199, 9–26.

Heger, L., Balk, S., Lühr, J.J., Heidkamp, G.F., Lehmann, C.H.K., Hatscher, L., Purbojo, A., Hartmann, A., Garcia-Martin, F., Nishimura, S.-I., Cesnjevar, R., Nimmerjahn, F., Dudziak, D., 2018. CLEC10A is a specific marker for human CD1c+ dendritic cells and enhances their toll-like receptor 7/8-induced cytokine secretion. Front. Immunol. 744.

Heidkamp, G.F., Lehmann, C.H.K., Heger, L., Baransk, A., Hoffmann, A., Lühr, J., Dudziak, D., 2016a. Functional Specialization of Dendritic Cell Subsets. Encyclopedia of Cell Biology.

Heidkamp, G.F., Sander, J., Lehmann, C.H.K., Heger, L., Eissing, N., Baranska, A., Lühr, J.J., Hoffmann, A., Reimer, K.C., Lux, A., Söder, S., Hartmann, A., Zenk, J., Ulas, T., McGovern, N., Alexiou, C., Spriewald, B., Mackensen, A., Schuler, G., Schauf, B., Forster, A., Repp, R., Fasching, P.A., Purbojo, A., Cesnjevar, R., Ullrich, E., Ginhoux, F., Schlitzer, A., Nimmerjahn, F., Schultze, J.L., Dudziak, D., 2016b. Human lymphoid organ dendritic cell identity is predominantly dictated by ontogeny, not tissue microenvironment. Sci. Immunol. 1, eaai7677.

Heinz, L.X., Platzer, B., Reisner, P.M., Jorgl, A., Taschner, S., Gobel, F., Strobl, H., 2006. Differential involvement of PU.1 and Id2 downstream of TGF-beta1 during Langerhans-cell commitment. Blood 107, 1445–1453.

Helft, J., Manicassamy, B., Guermonprez, P., Hashimoto, D., Silvin, A., Agudo, J., Brown, B.D., Schmolke, M., Miller, J.C., Leboeuf, M., Murphy, K.M., Garcia-Sastre, A., Merad, M., 2012. Cross-presenting CD103+ dendritic cells are protected from influenza virus infection. J. Clin. Invest. 122, 4037–4047.

Helft, J., Bottcher, J., Chakravarty, P., Zelenay, S., Huotari, J., Schraml, B.U., Goubau, D., Reis, E., Sousa, C., 2015. GM-CSF mouse bone marrow cultures comprise a heterogeneous population of CD11c(+)MHCII(+) macrophages and dendritic cells. Immunity 42, 1197–1211.

Hemmi, H., Yoshino, M., Yamazaki, H., Naito, M., Iyoda, T., Omatsu, Y., Shimoyama, S., Letterio, J.J., Nakabayashi, T., Tagaya, H., Yamane, T., Ogawa, M., Nishikawa, S., Ryoke, K., Inaba, K., Hayashi, S., Kunisada, T., 2001. Skin antigens in the steady state are trafficked to regional lymph nodes by transforming growth factor-beta1-dependent cells. Int. Immunol. 13, 695–704.

Henri, S., Vremec, D., Kamath, A., Waithman, J., Williams, S., Benoist, C., Burnham, K., Saeland, S., Handman, E., Shortman, K., 2001. The dendritic cell populations of mouse lymph nodes. J. Immunol. 167, 741–748.

Henri, S., Poulin, L.F., Tamoutounour, S., Ardouin, L., Guilliams, M., De Bovis, B., Devilard, E., Viret, C., Azukizawa, H., Kissenpfennig, A., Malissen, B., 2010. CD207+ CD103+ dermal dendritic cells cross-present keratinocyte-derived antigens irrespective of the presence of Langerhans cells. J. Exp. Med. 207, 189–206.

Henry, C.J., Ornelles, D.A., Mitchell, L.M., Brzoza-Lewis, K.L., Hiltbold, E.M., 2008. IL-12 produced by dendritic cells augments CD8+ T cell activation through the production of the chemokines CCL1 and CCL17. J. Immunol. (Baltimore, MD: 1950) 181, 8576–8584.

Herman, J.S., Sagar, GrüN, D., 2018. FateID infers cell fate bias in multipotent progenitors from single-cell RNA-seq data. Nat. Methods 15, 379–386.

Herrmann, T.L., Morita, C.T., Lee, K., Kusner, D.J., 2005. Calmodulin kinase II regulates the maturation and antigen presentation of human dendritic cells. J. Leukoc. Biol. 78, 1397–1407.

Hespel, C., Moser, M., 2012. Role of inflammatory dendritic cells in innate and adaptive immunity. Eur. J. Immunol. 42, 2535–2543.

Hettinger, J., Richards, D.M., Hansson, J., Barra, M.M., Joschko, A.C., Krijgsveld, J., Feuerer, M., 2013. Origin of monocytes and macrophages in a committed progenitor. Nat. Immunol. 14, 821–830.

Hildner, K., Edelson, B.T., Purtha, W.E., Diamond, M., Matsushita, H., Kohyama, M., Calderon, B., Schraml, B.U., Unanue, E.R., Diamond, M.S., Schreiber, R.D., Murphy, T.L., Murphy, K.M., 2008. Batf3 deficiency reveals a critical role for CD8alpha+ dendritic cells in cytotoxic T cell immunity. Science 322, 1097–1100.

Hoeffel, G., Wang, Y., Greter, M., See, P., Teo, P., Malleret, B., Leboeuf, M., Low, D., Oller, G., Almeida, F., Choy, S.H.Y., Grisotto, M., Renia, L., Conway, S.J., Stanley, E.R., Chan, J.K.Y., Ng, L.G., Samokhvalov, I.M., Merad, M., Ginhoux, F., 2012. Adult LANGERHANS cells derive predominantly from embryonic fetal liver monocytes with a minor contribution of yolk sac–derived macrophages. J. Exp. Med. 209, 1167.

Hoeffel, G., Chen, J., Lavin, Y., Low, D., Almeida, F.F., See, P., Beaudin, A.E., Lum, J., Low, I., Forsberg, E.C., Poidinger, M., Zolezzi, F., Larbi, A., Ng, L.G., Chan, J.K.Y., Greter, M., Becher, B., Samokhvalov, I.M., Merad, M., Ginhoux, F., 2015. C-Myb+ erythro-myeloid progenitor-derived fetal monocytes give rise to adult tissue-resident macrophages. Immunity 42, 665–678.

Hohl, T.M., Rivera, A., Lipuma, L., Gallegos, A., Shi, C., Mack, M., Pamer, E.G., 2009. Inflammatory monocytes facilitate adaptive CD4 T cell responses during respiratory fungal infection. Cell Host Microbe 6, 470–481.

Holtschke, T., Lohler, J., Kanno, Y., Fehr, T., Giese, N., Rosenbauer, F., Lou, J., Knobeloch, K.P., Gabriele, L., Waring, J.F., Bachmann, M.F., Zinkernagel, R.M., Morse 3rd, H.C., Ozato, K., Horak, I., 1996. Immunodeficiency and chronic myelogenous leukemia-like syndrome in mice with a targeted mutation of the ICSBP gene. Cell 87, 307–317.

Hoving, J.C., Wilson, G.J., Brown, G.D., 2014. Signalling C-type lectin receptors, microbial recognition and immunity. Cell. Microbiol. 16, 185–194.

Hume, D.A., Robinson, A.P., Macpherson, G.G., Gordon, S., 1983. The mononuclear phagocyte system of the mouse defined by immunohistochemical localization of antigen F4/80. Relationship between macrophages, Langerhans cells, reticular cells, and dendritic cells in lymphoid and hematopoietic organs. J. Exp. Med. 158, 1522–1536.

Hunger, R.E., Sieling, P.A., Ochoa, M.T., Sugaya, M., Burdick, A.E., Rea, T.H., Brennan, P.J., Belisle, J.T., Blauvelt, A., Porcelli, S.A., Modlin, R.L., 2004. Langerhans cells utilize CD1a and langerin to efficiently present nonpeptide antigens to T cells. J. Clin. Invest. 113, 701–708.

Hutvagner, G., McLachlan, J., Pasquinelli, A.E., Balint, E., Tuschl, T., Zamore, P.D., 2001. A cellular function for the RNA-interference enzyme dicer in the maturation of the let-7 small temporal RNA. Science 293, 834–838.

Ichikawa, E., Hida, S., Omatsu, Y., Shimoyama, S., Takahara, K., Miyagawa, S., Inaba, K., Taki, S., 2004. Defective development of splenic and epidermal CD4+ dendritic cells in mice deficient for IFN regulatory factor-2. Proc. Natl. Acad. Sci. U. S. A. 101, 3909.

Idoyaga, J., Cheong, C., Suda, K., Suda, N., Kim, J.Y., Lee, H., Park, C.G., Steinman, R.M., 2008. Cutting edge: langerin/CD207 receptor on dendritic cells mediates efficient antigen presentation on MHC I and II products in vivo. J. Immunol. 180, 3647–3650.

Idoyaga, J., Fiorese, C., Zbytnuik, L., Lubkin, A., Miller, J., Malissen, B., Mucida, D., Merad, M., Steinman, R.M., 2013. Specialized role of migratory dendritic cells in peripheral tolerance induction. J. Clin. Invest. 123, 844–854.

Igyarto, B.Z., Haley, K., Ortner, D., Bobr, A., Gerami-Nejad, M., Edelson, B.T., Zurawski, S.M., Malissen, B., Zurawski, G., Berman, J., Kaplan, D.H., 2011. Skin-resident murine dendritic cell subsets promote distinct and opposing antigen-specific T helper cell responses. Immunity 35, 260–272.

Iijima, N., Mattei, L.M., Iwasaki, A., 2011. Recruited inflammatory monocytes stimulate antiviral Th1 immunity in infected tissue. Proc. Natl. Acad. Sci. 108, 284.

Ikutani, M., Yanagibashi, T., Ogasawara, M., Tsuneyama, K., Yamamoto, S., Hattori, Y., Kouro, T., Itakura, A., Nagai, Y., Takaki, S., Takatsu, K., 2012. Identification of innate IL-5-producing cells and their role in lung eosinophil regulation and antitumor immunity. J. Immunol. 188, 703.

Ippolito, G.C., Dekker, J.D., Wang, Y.H., Lee, B.K., Shaffer, A.L., 3rd, Lin, J., Wall, J.K., Lee, B.S., Staudt, L.M., Liu, Y.J., Iyer, V.R., Tucker, H.O., 2014. Dendritic cell fate is determined by BCL11A. Proc. Natl. Acad. Sci. U.S.A. 111, E998–E1006.

Irla, M., Küpfer, N., Suter, T., Lissilaa, R., Benkhoucha, M., Skupsky, J., Lalive, P.H., Fontana, A., Reith, W., Hugues, S., 2010. MHC class II-restricted antigen presentation by plasmacytoid dendritic cells inhibits T cell-mediated autoimmunity. J. Exp. Med. 207, 1891.

Ivanov, I.I., McKenzie, B.S., Zhou, L., Tadokoro, C.E., Lepelley, A., Lafaille, J.J., Cua, D.J., Littman, D.R., 2006. The orphan nuclear receptor RORgammat directs the differentiation program of proinflammatory IL-17+ T helper cells. Cell 126, 1121–1133.

Iwama, A., Osawa, M., Hirasawa, R., Uchiyama, N., Kaneko, S., Onodera, M., Shibuya, K., Shibuya, A., Vinson, C., Tenen, D.G., Nakauchi, H., 2002. Reciprocal roles for CCAAT/enhancer binding protein (C/EBP) and PU.1 transcription factors in langerhans cell commitment. J. Exp. Med. 195, 547.

Jaiswal, H., Kaushik, M., Sougrat, R., Gupta, M., Dey, A., Verma, R., Ozato, K., Tailor, P., 2013. Batf3 and Id2 have a synergistic effect on Irf8-directed classical CD8alpha+ dendritic cell development. J. Immunol. 191, 5993–6001.

Jaitin, D.A., Kenigsberg, E., Keren-Shaul, H., Elefant, N., Paul, F., Zaretsky, I., Mildner, A., Cohen, N., Jung, S., Tanay, A., Amit, I., 2014. Massively parallel single-cell RNA-seq for marker-free decomposition of tissues into cell types. Science 343, 776.

Janas, M.L., Groves, P., Kienzle, N., Kelso, A., 2005. IL-2 regulates perforin and granzyme gene expression in CD8+ T cells independently of its effects on survival and proliferation. J. Immunol. 175, 8003–8010.

Jiang, A., Bloom, O., Ono, S., Cui, W., Unternaehrer, J., Jiang, S., Whitney, J.A., Connolly, J., Banchereau, J., Mellman, I., 2007. Disruption of E-cadherin-mediated adhesion induces a functionally distinct pathway of dendritic cell maturation. Immunity 27, 610–624.

Jin, P., Han, T.H., Ren, J., Saunders, S., Wang, E., Marincola, F.M., Stroncek, D.F., 2010. Molecular signatures of maturing dendritic cells: implications for testing the quality of dendritic cell therapies. J. Transl. Med. 8, 4.

Joffre, O.P., Segura, E., Savina, A., Amigorena, S., 2012. Cross-presentation by dendritic cells. Nat. Rev. Immunol. 12, 557.

Johnson, B.J., Costelloe, E.O., Fitzpatrick, D.R., Haanen, J.B., Schumacher, T.N., Brown, L.E., Kelso, A., 2003. Single-cell perforin and granzyme expression reveals the anatomical localization of effector CD8+ T cells in influenza virus-infected mice. Proc. Natl. Acad. Sci. U. S. A. 100, 2657–2662.

Johnston, R.J., Poholek, A.C., Ditoro, D., Yusuf, I., Eto, D., Barnett, B., Dent, A.L., Craft, J., Crotty, S., 2009. Bcl6 and Blimp-1 are reciprocal and antagonistic regulators of T follicular helper cell differentiation. Science 325, 1006–1010.

Jones, A., Bourque, J., Kuehm, L., Opejin, A., Teague, R.M., Gross, C., Hawiger, D., 2016. Immunomodulatory functions of BTLA and HVEM govern induction of extrathymic regulatory T cells and tolerance by dendritic cells. Immunity 45, 1066–1077.

Jones, P.A., 2012. Functions of DNA methylation: islands, start sites, gene bodies and beyond. Nat. Rev. Genet. 13, 484–492.

Jongbloed, S.L., Kassianos, A.J., McDonald, K.J., Clark, G.J., Ju, X., Angel, C.E., Chen, C.J., Dunbar, P.R., Wadley, R.B., Jeet, V., Vulink, A.J., Hart, D.N., Radford, K.J., 2010. Human CD141+ (BDCA-3)+ dendritic cells (DCs) represent a unique myeloid DC subset that cross-presents necrotic cell antigens. J. Exp. Med. 207, 1247–1260.

Kamata, T., Giblett, S., Pritchard, C., 2017. KRAS(G12D) expression in lung-resident myeloid cells promotes pulmonary LCH-like neoplasm sensitive to statin treatment. Blood 130, 514–526.

Kamphorst, A.O., Guermonprez, P., Dudziak, D., Nussenzweig, M.C., 2010. Route of antigen uptake differentially impacts presentation by dendritic cells and activated monocytes. J. Immunol. 185, 3426–3435.

Kanasty, R.L., Whitehead, K.A., Vegas, A.J., Anderson, D.G., 2012. Action and reaction: the biological response to siRNA and its delivery vehicles. Mol. Ther. 20, 513–524.

Kano, S., Sato, K., Morishita, Y., Vollstedt, S., Kim, S., Bishop, K., Honda, K., Kubo, M., Taniguchi, T., 2008. The contribution of transcription factor IRF1 to the interferon-gamma-interleukin 12 signaling axis and TH1 versus TH-17 differentiation of CD4+ T cells. Nat. Immunol. 9, 34–41.

Kaplan, D.H., 2017. Ontogeny and function of murine epidermal Langerhans cells. Nat. Immunol. 18, 1068.

Karrich, J.J., Balzarolo, M., Schmidlin, H., Libouban, M., Nagasawa, M., Gentek, R., Kamihira, S., Maeda, T., Amsen, D., Wolkers, M.C., Blom, B., 2012. The transcription factor Spi-B regulates human plasmacytoid dendritic cell survival through direct induction of the antiapoptotic gene BCL2-A1. Blood 119, 5191–5200.

Karrich, J.J., Jachimowski, L.C., Libouban, M., Iyer, A., Brandwijk, K., Taanman-Kueter, E.W., Nagasawa, M., De Jong, E.C., Uittenbogaart, C.H., Blom, B., 2013. MicroRNA-146a regulates survival and maturation of human plasmacytoid dendritic cells. Blood 122, 3001–3009.

Kashem, S.W., Haniffa, M., Kaplan, D.H., 2017. Antigen-presenting cells in the skin. Annu. Rev. Immunol. 35, 469–499.

Kashiwada, M., Pham, N.L., Pewe, L.L., Harty, J.T., Rothman, P.B., 2011. NFIL3/E4BP4 is a key transcription factor for CD8alpha(+) dendritic cell development. Blood 117, 6193–6197.

Kastenmuller, K., Wille-Reece, U., Lindsay, R.W., Trager, L.R., Darrah, P.A., Flynn, B.J., Becker, M.R., Udey, M.C., Clausen, B.E., Igyarto, B.Z., Kaplan, D.H., Kastenmuller, W., Germain, R.N., Seder, R.A., 2011. Protective T cell immunity in mice following protein-TLR7/8 agonist-conjugate immunization requires aggregation, type I IFN, and multiple DC subsets. J. Clin. Invest. 121, 1782–1796.

Kaufmann, S.H., 2007. The contribution of immunology to the rational design of novel antibacterial vaccines. Nat. Rev. Microbiol. 5, 491–504.

Kee, B.L., 2009. E and ID proteins branch out. Nat. Rev. Immunol. 9, 175–184.

Khvorova, A., Reynolds, A., Jayasena, S.D., 2003. Functional siRNAs and miRNAs exhibit strand bias. Cell 115, 209–216.

Kim, T.S., Braciale, T.J., 2009. Respiratory dendritic cell subsets differ in their capacity to support the induction of virus-specific cytotoxic CD8+ T cell responses. PLoS One 4, e4204.

Kim, S., Kaiser, V., Beier, E., Bechheim, M., Guenthner-Biller, M., Ablasser, A., Berger, M., Endres, S., Hartmann, G., Hornung, V., 2014. Self-priming determines high type I IFN production by plasmacytoid dendritic cells. Eur. J. Immunol. 44, 807–818.

Kingston, D., Schmid, M.A., Onai, N., Obata-Onai, A., Baumjohann, D., Manz, M.G., 2009. The concerted action of GM-CSF and Flt3-ligand on in vivo dendritic cell homeostasis. Blood 114, 835–843.

Kirkling, M.E., Cytlak, U., Lau, C.M., Lewis, K.L., Resteu, A., Khodadadi-Jamayran, A., Siebel, C.W., Salmon, H., Merad, M., Tsirigos, A., Collin, M., Bigley, V., Reizis, B., 2018. Notch signaling facilitates in vitro generation of cross-presenting classical dendritic cells. Cell Rep. 23, 3658.e6–3672.e6.

Kirstetter, P., Thomas, M., Dierich, A., Kastner, P., Chan, S., 2002. Ikaros is critical for B cell differentiation and function. Eur. J. Immunol. 32, 720–730.

Kisselev, A.F., Akopian, T.N., Woo, K.M., Goldberg, A.L., 1999. The sizes of peptides generated from protein by mammalian 26 and 20 S proteasomes: implications For understanding the degradative mechanism and antigen presentation. J. Biol. Chem. 274, 3363–3371.

Kitajima, M., Ziegler, S.F., 2013. Cutting edge: identification of the thymic stromal lymphopoietin-responsive dendritic cell subset critical for initiation of type 2 contact hypersensitivity. J. Immunol. 191, 4903–4907.

Kloetzel, P.M., 2001. Antigen processing by the proteasome. Nat. Rev. Mol. Cell Biol. 2, 179–187.

Kobayashi, T., Walsh, P.T., Walsh, M.C., Speirs, K.M., Chiffoleau, E., King, C.G., Hancock, W.W., Caamano, J.H., Hunter, C.A., Scott, P., Turka, L.A., Choi, Y., 2003. TRAF6 is a critical factor for dendritic cell maturation and development. Immunity 19, 353–363.

Kong, K.Y., Owens, K.S., Rogers, J.H., Mullenix, J., Velu, C.S., Grimes, H.L., Dahl, R., 2010. MIR-23A microRNA cluster inhibits B-cell development. Exp. Hematol. 38 (629–640) e1.

Konkel, J.E., Zhang, D., Zanvit, P., Chia, C., Zangarle-Murray, T., Jin, W., Wang, S., Chen, W., 2017. Transforming growth factor-beta signaling in regulatory T cells controls T helper-17 cells and tissue-specific immune responses. Immunity 46, 660–674.

Kool, M., SoulliÉ, T., Van Nimwegen, M., Willart, M.A.M., Muskens, F., Jung, S., Hoogsteden, H.C., Hammad, H., Lambrecht, B.N., 2008. Alum adjuvant boosts adaptive immunity by inducing uric acid and activating inflammatory dendritic cells. J. Exp. Med. 205, 869.

Kopf, M., Brombacher, F., Köhler, G., Kienzle, G., Widmann, K.H., Lefrang, K., Humborg, C., Ledermann, B., Solbach, W., 1996. IL-4-deficient Balb/c mice resist infection with Leishmania major. J. Exp. Med. 184, 1127.

Korn, T., Bettelli, E., Gao, W., Awasthi, A., Jager, A., Strom, T.B., Oukka, M., Kuchroo, V.K., 2007. IL-21 initiates an alternative pathway to induce proinflammatory T(H)17 cells. Nature 448, 484–487.

Kretschmer, K., Apostolou, I., Hawiger, D., Khazaie, K., Nussenzweig, M.C., Von Boehmer, H., 2005. Inducing and expanding regulatory T cell populations by foreign antigen. Nat. Immunol. 6, 1219–1227.

Kroczek, R.A., Henn, V., 2012. The role of XCR1 and its ligand XCL1 in antigen cross-presentation by murine and human dendritic cells. Front. Immunol. 3, 14.

Kueh, H.Y., Champhekar, A., Nutt, S.L., Elowitz, M.B., Rothenberg, E.V., 2013. Positive feedback between PU.1 and the cell cycle controls myeloid differentiation. Science 341, 670–673.

Kuipers, H., Schnorfeil, F.M., Brocker, T., 2010. Differentially expressed microRNAs regulate plasmacytoid vs. conventional dendritic cell development. Mol. Immunol. 48, 333–340.

Kumagai, Y., Kumar, H., Koyama, S., Kawai, T., Takeuchi, O., Akira, S., 2009. Cutting edge: TLR-dependent viral recognition along with type I IFN positive feedback signaling masks the requirement of viral replication for IFN-{alpha} production in plasmacytoid dendritic cells. J. Immunol. 182, 3960–3964.

Kumamoto, Y., Linehan, M., Weinstein, J.S., Laidlaw, B.J., Craft, J.E., Iwasaki, A., 2013. CD301b(+) dermal dendritic cells drive T helper 2 cell-mediated immunity. Immunity 39, 733–743.

Kurata, H., Lee, H.J., O'Garra, A., Arai, N., 1999. Ectopic expression of activated Stat6 induces the expression of Th2-specific cytokines and transcription factors in developing Th1 cells. Immunity 11, 677–688.

Kurotaki, D., Nakabayashi, J., Nishiyama, A., Sasaki, H., Kawase, W., Kaneko, N., Ochiai, K., Igarashi, K., Ozato, K., Suzuki, Y., Tamura, T., 2018. Transcription factor IRF8 governs enhancer landscape dynamics in mononuclear phagocyte progenitors. Cell Rep. 22, 2628–2641.

Kurowska-Stolarska, M., Alivernini, S., Melchor, E.G., Elmesmari, A., Tolusso, B., Tange, C., Petricca, L., Gilchrist, D.S., Di Sante, G., Keijzer, C., Stewart, L., Di Mario, C., Morrison, V., Brewer, J.M., Porter, D., Milling, S., Baxter, R.D., Mccarey, D., Gremese, E., Lemke, G., Ferraccioli, G., Mcsharry, C., Mcinnes, I.B., 2017. MicroRNA-34a dependent regulation of AXL controls the activation of dendritic cells in inflammatory arthritis. Nat. Commun. 8, 15877.

Kurts, C., Robinson, B.W., Knolle, P.A., 2010. Cross-priming in health and disease. Nat. Rev. Immunol. 10, 403–414.

Kusunoki, T., Sugai, M., Katakai, T., Omatsu, Y., Iyoda, T., Inaba, K., Nakahata, T., Shimizu, A., Yokota, Y., 2003. TH2 dominance and defective development of a CD8+ dendritic cell subset in Id2-deficient mice. J. Allergy Clin. Immunol. 111, 136–142.

Landgraf, P., Rusu, M., Sheridan, R., Sewer, A., Iovino, N., Aravin, A., Pfeffer, S., Rice, A., Kamphorst, A.O., Landthaler, M., Lin, C., Socci, N.D., Hermida, L., Fulci, V., Chiaretti, S., FoÀ, R., Schliwka, J., Fuchs, U., Novosel, A., MÜller, R.-U., Schermer, B., Bissels, U., Inman, J., Phan, Q., Chien, M., Weir, D.B., Choksi, R., De Vita, G., Frezzetti, D., Trompeter, H.-I., Hornung, V., Teng, G., Hartmann, G., Palkovits, M., Di Lauro, R., Wernet, P., Macino, G., Rogler, C.E., Nagle, J.W.,

Ju, J., Papavasiliou, F.N., Benzing, T., Lichter, P., Tam, W., Brownstein, M.J., Bosio, A., Borkhardt, A., Russo, J.J., Sander, C., Zavolan, M., Tuschl, T., 2007. A mammalian microRNA expression atlas based on small RNA library sequencing. Cell 129, 1401–1414.

Langlet, C., Tamoutounour, S., Henri, S., Luche, H., Ardouin, L., Gregoire, C., Malissen, B., Guilliams, M., 2012. CD64 expression distinguishes monocyte-derived and conventional dendritic cells and reveals their distinct role during intramuscular immunization. J. Immunol. 188, 1751–1760.

Lau, C.M., Tiniakou, I., Perez, O.A., Kirkling, M.E., Yap, G.S., Hock, H., Reizis, B., 2018. Transcription factor Etv6 regulates functional differentiation of cross-presenting classical dendritic cells. J. Exp. Med. 215, 2265–2278.

Laustsen, A., Bak, R.O., Krapp, C., KjÆR, L., Egedahl, J.H., Petersen, C.C., Pillai, S., Tang, H.Q., Uldbjerg, N., Porteus, M., Roan, N.R., Nyegaard, M., Denton, P.W., Jakobsen, M.R., 2018. Interferon priming is essential for human CD34+ cell-derived plasmacytoid dendritic cell maturation and function. Nat. Commun. 9, 3525.

Lauterbach, H., Bathke, B., Gilles, S., Traidl-Hoffmann, C., Luber, C.A., Fejer, G., Freudenberg, M.A., Davey, G.M., Vremec, D., Kallies, A., Wu, L., Shortman, K., Chaplin, P., Suter, M., O'keeffe, M., Hochrein, H., 2010. Mouse CD8alpha+ DCs and human BDCA3+ DCs are major producers of IFN-lambda in response to poly IC. J. Exp. Med. 207, 2703–2717.

Lee, R.C., Feinbaum, R.L., Ambros, V., 1993. The C. elegans heterochronic gene lin-4 encodes small RNAs with antisense complementarity to lin-14. Cell 75, 843–854.

Lee, Y., Kim, M., Han, J., Yeom, K.H., Lee, S., Baek, S.H., Kim, V.N., 2004. MicroRNA genes are transcribed by RNA polymerase II. EMBO J. 23, 4051–4060.

Lee, J., Breton, G., Aljoufi, A., Zhou, Y.J., Puhr, S., Nussenzweig, M.C., Liu, K., 2015a. Clonal analysis of human dendritic cell progenitor using a stromal cell culture. J. Immunol. Methods 425, 21–26.

Lee, J., Breton, G., Oliveira, T.Y.K., Zhou, Y.J., Aljoufi, A., Puhr, S., Cameron, M.J., SÉkaly, R.-P., Nussenzweig, M.C., Liu, K., 2015b. Restricted dendritic cell and monocyte progenitors in human cord blood and bone marrow. J. Exp. Med. 212, 385.

Lehmann, C.H., Heger, L., Heidkamp, G.F., Baranska, A., Luhr, J.J., Hoffmann, A., Dudziak, D., 2016. Direct delivery of antigens to dendritic cells via antibodies specific for endocytic receptors as a promising strategy for future therapies. Vaccines 4, 8.

Lehmann, C.H.K., Baranska, A., Heidkamp, G.F., Heger, L., Neubert, K., Lühr, J.J., Hoffmann, A., Reimer, K.C., BrÜckner, C., Beck, S., Seeling, M., Kießling, M., Soulat, D., Krug, A.B., Ravetch, J.V., Leusen, J.H.W., Nimmerjahn, F., Dudziak, D., 2017. DC subset–specific induction of T cell responses upon antigen uptake via Fcγ receptors in vivo. J. Exp. Med. 214, 1509.

Lehtonen, A., Veckman, V., Nikula, T., Lahesmaa, R., Kinnunen, L., Matikainen, S., Julkunen, I., 2005. Differential expression of IFN regulatory factor 4 gene in human monocyte-derived dendritic cells and macrophages. J. Immunol. 175, 6570–6579.

Leibundgut-Landmann, S., Waldburger, J.M., Reis, E., Sousa, C., Acha-Orbea, H., Reith, W., 2004. MHC class II expression is differentially regulated in plasmacytoid and conventional dendritic cells. Nat. Immunol. 5, 899–908.

Leon, B., Lopez-Bravo, M., Ardavin, C., 2007. Monocyte-derived dendritic cells formed at the infection site control the induction of protective T helper 1 responses against Leishmania. Immunity 26, 519–531.

Levin, C., Bonduelle, O., Nuttens, C., Primard, C., Verrier, B., Boissonnas, A., Combadiere, B., 2017. Critical role for skin-derived migratory DCs and langerhans cells in TFH and GC responses after intradermal immunization. J. Invest. Dermatol. 137, 1905–1913.

Lewis, K.L., Caton, M.L., Bogunovic, M., Greter, M., Grajkowska, L.T., Ng, D., Klinakis, A., Charo, I.F., Jung, S., Gommerman, J.L., Ivanov, I., Liu, K., Merad, M., Reizis, B., 2011. Notch2 receptor signaling controls functional differentiation of dendritic cells in the spleen and intestine. Immunity 35, 780–791.

Li, M.O., Wan, Y.Y., Flavell, R.A., 2007. T cell-produced transforming growth factor-beta1 controls T cell tolerance and regulates Th1- and Th17-cell differentiation. Immunity 26, 579–591.

Li, J., Park, J., Foss, D., Goldschneider, I., 2009. Thymus-homing peripheral dendritic cells constitute two of the three major subsets of dendritic cells in the steady-state thymus. J. Exp. Med. 206, 607–622.

Li, H.S., Greeley, N., Sugimoto, N., Liu, Y.J., Watowich, S.S., 2012. miR-22 controls Irf8 mRNA abundance and murine dendritic cell development. PLoS One 7, e52341.

Liang, S.C., Tan, X.-Y., Luxenberg, D.P., Karim, R., Dunussi-Joannopoulos, K., Collins, M., Fouser, L.A., 2006. Interleukin (IL)-22 and IL-17 are coexpressed by Th17 cells and cooperatively enhance expression of antimicrobial peptides. J. Exp. Med. 203, 2271.

Lin, M.L., Zhan, Y., Proietto, A.I., Prato, S., Wu, L., Heath, W.R., Villadangos, J.A., Lew, A.M., 2008. Selective suicide of cross-presenting CD8+ dendritic cells by cytochrome c injection shows functional heterogeneity within this subset. Proc. Natl. Acad. Sci. U. S. A. 105, 3029–3034.

Lindquist, R.L., Shakhar, G., Dudziak, D., Wardemann, H., Eisenreich, T., Dustin, M.L., Nussenzweig, M.C., 2004. Visualizing dendritic cell networks in vivo. Nat. Immunol. 5, 1243–1250.

Liu, K., Nussenzweig, M.C., 2010. Development and homeostasis of dendritic cells. Eur. J. Immunol. 40, 2099–2102.

Liu, K., Iyoda, T., Saternus, M., Kimura, Y., Inaba, K., Steinman, R.M., 2002. Immune tolerance after delivery of dying cells to dendritic cells in situ. J. Exp. Med. 196, 1091–1097.

Liu, K., Idoyaga, J., Charalambous, A., Fujii, S., Bonito, A., Mordoh, J., Wainstok, R., Bai, X.F., Liu, Y., Steinman, R.M., 2005. Innate NKT lymphocytes confer superior adaptive immunity via tumor-capturing dendritic cells. J. Exp. Med. 202, 1507–1516.

Liu, X., Zhan, Z., Xu, L., Ma, F., Li, D., Guo, Z., Li, N., Cao, X., 2010. MicroRNA-148/152 impair innate response and antigen presentation of TLR-triggered dendritic cells by targeting CaMKIIalpha. J. Immunol. 185, 7244–7251.

Liu, L., Yi, H., Wang, C., He, H., Li, P., Pan, H., Sheng, N., Ji, M., Cai, L., Ma, Y., 2016. Integrated nanovaccine with microRNA-148a inhibition reprograms tumor-associated dendritic cells by modulating miR-148a/DNMT1/SOCS1 axis. J. Immunol. 197, 1231.

Lodish, H.F., Zhou, B., Liu, G., Chen, C.Z., 2008. Micromanagement of the immune system by microRNAs. Nat. Rev. Immunol. 8, 120–130.

Loschko, J., Schlitzer, A., Dudziak, D., Drexler, I., Sandholzer, N., Bourquin, C., Reindl, W., Krug, A.B., 2011. Antigen delivery to plasmacytoid dendritic cells via BST2 induces protective T cell-mediated immunity. J. Immunol. 186, 6718.

Loschko, J., Rieke, G.J., Schreiber, H.A., Meredith, M.M., Yao, K.H., Guermonprez, P., Nussenzweig, M.C., 2016. Inducible targeting of cDCs and their subsets in vivo. J. Immunol. Methods 434, 32–38.

Lu, C., Huang, X., Zhang, X., Roensch, K., Cao, Q., Nakayama, K.I., Blazar, B.R., Zeng, Y., Zhou, X., 2011a. miR-221 and miR-155 regulate human dendritic cell development, apoptosis, and IL-12 production through targeting of p27kip1, KPC1, and SOCS-1. Blood 117, 4293–4303.

Lu, T.X., Hartner, J., Lim, E.-J., Fabry, V., Mingler, M.K., Cole, E.T., Orkin, S.H., Aronow, B.J., Rothenberg, M.E., 2011b. MicroRNA-21 limits in vivo immune response-mediated activation of the IL-12/IFN-γ pathway, Th1 polarization, and the severity of delayed-type hypersensitivity. J. Immunol. 187, 3362.

Luckashenak, N., Schroeder, S., Endt, K., Schmidt, D., Mahnke, K., Bachmann, M.F., Marconi, P., Deeg, C.A., Brocker, T., 2008. Constitutive crosspresentation of tissue antigens by dendritic cells controls CD8+ T cell tolerance in vivo. Immunity 28, 521–532.
Luda, K.M., Joeris, T., Persson, E.K., Rivollier, A., Demiri, M., Sitnik, K.M., Pool, L., Holm, J.B., Melo-Gonzalez, F., Richter, L., Lambrecht, B.N., Kristiansen, K., Travis, M.A., Svensson-Frej, M., Kotarsky, K., Agace, W.W., 2016. IRF8 transcription-factor-dependent classical dendritic cells are essential for intestinal T cell homeostasis. Immunity 44, 860–874.
Lund, E., Guttinger, S., Calado, A., Dahlberg, J.E., Kutay, U., 2004. Nuclear export of microRNA precursors. Science 303, 95–98.
Ma, W., Lee, J., Backenroth, D., Zhou, Y.J., Bush, E., Sims, P., Liu, K., Shen, Y., 2019. Single cell RNA-Seq reveals pre-cDCs fate determined by transcription factor combinatorial dose. BMC Mol. Cell Biol. 20, 20.
Mabbott, N.A., Baillie, J.K., Brown, H., Freeman, T.C., Hume, D.A., 2013. An expression atlas of human primary cells: inference of gene function from coexpression networks. BMC Genomics 14, 632.
Macdonald, K.P., Munster, D.J., Clark, G.J., Dzionek, A., Schmitz, J., Hart, D.N., 2002. Characterization of human blood dendritic cell subsets. Blood 100, 4512–4520.
Mackarehtschian, K., Hardin, J.D., Moore, K.A., Boast, S., Goff, S.P., Lemischka, I.R., 1995. Targeted disruption of the flk2/flt3 gene leads to deficiencies in primitive hematopoietic progenitors. Immunity 3, 147–161.
Maldonado-LÓpez, R., De Smedt, T., Michel, P., Godfroid, J., Pajak, B., Heirman, C., Thielemans, K., Leo, O., Urbain, J., Moser, M., 1999. CD8alpha+ and CD8alpha- subclasses of dendritic cells direct the development of distinct T helper cells in vivo. J. Exp. Med. 189, 587–592.
Malissen, B., Tamoutounour, S., Henri, S., 2014. The origins and functions of dendritic cells and macrophages in the skin. Nat. Rev. Immunol. 14, 417–428.
Manetti, R., Gerosa, F., Giudizi, M.G., Biagiotti, R., Parronchi, P., Piccinni, M.P., Sampognaro, S., Maggi, E., Romagnani, S., Trinchieri, G., 1994. Interleukin 12 induces stable priming for interferon gamma (IFN-gamma) production during differentiation of human T helper (Th) cells and transient IFN-gamma production in established Th2 cell clones. J. Exp. Med. 179, 1273.
Manicassamy, S., Pulendran, B., 2011. Dendritic cell control of tolerogenic responses. Immunol. Rev. 241, 206–227.
Marafioti, T., Paterson, J.C., Ballabio, E., Reichard, K.K., Tedoldi, S., Hollowood, K., Dictor, M., Hansmann, M.L., Pileri, S.A., Dyer, M.J., Sozzani, S., Dikic, I., Shaw, A.S., Petrella, T., Stein, H., Isaacson, P.G., Facchetti, F., Mason, D.Y., 2008. Novel markers of normal and neoplastic human plasmacytoid dendritic cells. Blood 111, 3778–3792.
Maraskovsky, E., Brasel, K., Teepe, M., Roux, E.R., Lyman, S.D., Shortman, K., McKenna, H.J., 1996. Dramatic increase in the numbers of functionally mature dendritic cells in Flt3 ligand-treated mice: multiple dendritic cell subpopulations identified. J. Exp. Med. 184, 1953–1962.
Maraskovsky, E., Daro, E., Roux, E., Teepe, M., Maliszewski, C.R., Hoek, J., Caron, D., Lebsack, M.E., McKenna, H.J., 2000. In vivo generation of human dendritic cell subsets by Flt3 ligand. Blood 96, 878–884.
Martinez-Lopez, M., Iborra, S., Conde-Garrosa, R., Sancho, D., 2015. Batf3-dependent CD103+ dendritic cells are major producers of IL-12 that drive local Th1 immunity against Leishmania major infection in mice. Eur. J. Immunol. 45, 119–129.
Martin-Gayo, E., Gonzalez-Garcia, S., Garcia-Leon, M.J., Murcia-Ceballos, A., Alcain, J., Garcia-Peydro, M., Allende, L., De Andres, B., Gaspar, M.L., Toribio, M.L., 2017. Spatially restricted JAG1-Notch signaling in human thymus provides suitable DC developmental niches. J. Exp. Med. 214, 3361–3379.

Mashayekhi, M., Sandau, M.M., Dunay, I.R., Frickel, E.M., Khan, A., Goldszmid, R.S., Sher, A., Ploegh, H.L., Murphy, T.L., Sibley, L.D., Murphy, K.M., 2011. CD8alpha(+) dendritic cells are the critical source of interleukin-12 that controls acute infection by Toxoplasma gondii tachyzoites. Immunity 35, 249–259.
Mathers, A.R., Morelli, A.E., Janelsins, B., Rubin, J.P., Tkacheva, O.A., Falo, L.D., Larregina, A.T., 2007. Human skin migratory dendritic cells that do not secrete IL-12p70 stimulate both Th17 and Th1 allogeneic responses (36.20). J. Immunol. 178, S16.
Matsui, T., Connolly, J.E., Michnevitz, M., Chaussabel, D., Yu, C.-I., Glaser, C., Tindle, S., Pypaert, M., Freitas, H., Piqueras, B., Banchereau, J., Palucka, A.K., 2009. CD2 distinguishes two subsets of human plasmacytoid dendritic cells with distinct phenotype and functions. J. Immunol. 182, 6815.
Matthews, W., Jordan, C.T., Gavin, M., Jenkins, N.A., Copeland, N.G., Lemischka, I.R., 1991. A receptor tyrosine kinase cDNA isolated from a population of enriched primitive hematopoietic cells and exhibiting close genetic linkage to c-kit. Proc. Natl. Acad. Sci. 88, 9026.
Mayer, C.T., Ghorbani, P., Nandan, A., Dudek, M., Arnold-Schrauf, C., Hesse, C., Berod, L., Stuve, P., Puttur, F., Merad, M., Sparwasser, T., 2014. Selective and efficient generation of functional Batf3-dependent CD103+ dendritic cells from mouse bone marrow. Blood 124, 3081–3091.
Mayer-Barber, K.D., Andrade, B.B., Barber, D.L., Hieny, S., Feng, C.G., Caspar, P., Oland, S., Gordon, S., Sher, A., 2011. Innate and adaptive interferons suppress IL-1alpha and IL-1beta production by distinct pulmonary myeloid subsets during Mycobacterium tuberculosis infection. Immunity 35, 1023–1034.
McGovern, N., Shin, A., Low, G., Low, D., Duan, K., Yao, L.J., Msallam, R., Low, I., Shadan, N.B., Sumatoh, H.R., Soon, E., Lum, J., Mok, E., Hubert, S., See, P., Kunxiang, E.H., Lee, Y.H., Janela, B., Choolani, M., Mattar, C.N.Z., Fan, Y., Lim, T.K.H., Chan, D.K.H., Tan, K.K., Tam, J.K.C., Schuster, C., Elbe-Burger, A., Wang, X.N., Bigley, V., Collin, M., Haniffa, M., Schlitzer, A., Poidinger, M., Albani, S., Larbi, A., Newell, E.W., Chan, J.K.Y., Ginhoux, F., 2017. Human fetal dendritic cells promote prenatal T-cell immune suppression through arginase-2. Nature 546, 662–666.
McKenna, H.J., Stocking, K.L., Miller, R.E., Brasel, K., De Smedt, T., Maraskovsky, E., Maliszewski, C.R., Lynch, D.H., Smith, J., Pulendran, B., Roux, E.R., Teepe, M., Lyman, S.D., Peschon, J.J., 2000. Mice lacking flt3 ligand have deficient hematopoiesis affecting hematopoietic progenitor cells, dendritic cells, and natural killer cells. Blood 95, 3489–3497.
Medzhitov, R., 2001. Toll-like receptors and innate immunity. Nat. Rev. Immunol. 1, 135.
Mellman, I., 2013. Dendritic cells: master regulators of the immune response. Cancer Immunol. Res. 1, 145–149.
Mellor, A.L., Munn, D.H., 2004. Ido expression by dendritic cells: tolerance and tryptophan catabolism. Nat. Rev. Immunol. 4, 762–774.
Menezes, S., Melandri, D., Anselmi, G., Perchet, T., Loschko, J., Dubrot, J., Patel, R., Gautier, E.L., Hugues, S., Longhi, M.P., Henry, J.Y., Quezada, S.A., Lauvau, G., Lennon-Dumenil, A.M., Gutierrez-Martinez, E., Bessis, A., Gomez-Perdiguero, E., Jacome-Galarza, C.E., Garner, H., Geissmann, F., Golub, R., Nussenzweig, M.C., Guermonprez, P., 2016. The heterogeneity of Ly6C(hi) monocytes controls their differentiation into iNOS(+) macrophages or monocyte-derived dendritic cells. Immunity 45, 1205–1218.
Merad, M., Ginhoux, F., 2007. Dendritic cell genealogy: a new stem or just another branch? Nat. Immunol. 8, 1199.
Merad, M., Manz, M.G., 2009. Dendritic cell homeostasis. Blood 113, 3418–3427.

Merad, M., Manz, M.G., Karsunky, H., Wagers, A., Peters, W., Charo, I., Weissman, I.L., Cyster, J.G., Engleman, E.G., 2002. Langerhans cells renew in the skin throughout life under steady-state conditions. Nat. Immunol. 3, 1135–1141.

Merad, M., Ginhoux, F., Collin, M., 2008. Origin, homeostasis and function of Langerhans cells and other langerin-expressing dendritic cells. Nat. Rev. Immunol. 8, 935.

Merad, M., Sathe, P., Helft, J., Miller, J., Mortha, A., 2013. The dendritic cell lineage: ontogeny and function of dendritic cells and their subsets in the steady state and the inflamed setting. Annu. Rev. Immunol. 31, 563–604.

Meredith, M.M., Liu, K., Darrasse-Jeze, G., Kamphorst, A.O., Schreiber, H.A., Guermonprez, P., Idoyaga, J., Cheong, C., Yao, K.H., Niec, R.E., Nussenzweig, M.C., 2012a. Expression of the zinc finger transcription factor zDC (Zbtb46, Btbd4) defines the classical dendritic cell lineage. J. Exp. Med. 209, 1153–1165.

Meredith, M.M., Liu, K., Kamphorst, A.O., Idoyaga, J., Yamane, A., Guermonprez, P., Rihn, S., Yao, K.H., Silva, I.T., Oliveira, T.Y., Skokos, D., Casellas, R., Nussenzweig, M.C., 2012b. Zinc finger transcription factor zDC is a negative regulator required to prevent activation of classical dendritic cells in the steady state. J. Exp. Med. 209, 1583–1593.

Mildner, A., Jung, S., 2014. Development and function of dendritic cell subsets. Immunity 40, 642–656.

Mildner, A., Chapnik, E., Manor, O., Yona, S., Kim, K.W., Aychek, T., Varol, D., Beck, G., Itzhaki, Z.B., Feldmesser, E., Amit, I., Hornstein, E., Jung, S., 2013. Mononuclear phagocyte miRNome analysis identifies miR-142 as critical regulator of murine dendritic cell homeostasis. Blood 121, 1016–1027.

Min, J., Yang, D., Kim, M., Haam, K., Yoo, A., Choi, J.-H., Schraml, B.U., Kim, Y.S., Kim, D., Kang, S.-J., 2018. Publisher correction: inflammation induces two types of inflammatory dendritic cells in inflamed lymph nodes. Exp. Mol. Med. 50, 33.

Mittag, D., Proietto, A.I., Loudovaris, T., Mannering, S.I., Vremec, D., Shortman, K., Wu, L., Harrison, L.C., 2011. Human dendritic cell subsets from spleen and blood are similar in phenotype and function but modified by donor health status. J. Immunol. 186, 6207–6217.

Moazed, D., 2011. Mechanisms for the inheritance of chromatin states. Cell 146, 510–518.

Mohan, J.F., Unanue, E.R., 2012. Unconventional recognition of peptides by T cells and the implications for autoimmunity. Nat. Rev. Immunol. 12, 721.

Montagner, S., DehÓ, L., Monticelli, S., 2014. MicroRNAs in hematopoietic development. BMC Immunol. 15, 14.

Montealegre, S., Van Endert, P.M., 2019. Endocytic recycling of MHC class I molecules in non-professional antigen presenting and dendritic cells. Front. Immunol. 9, 3098.

Montecalvo, A., Larregina, A.T., Shufesky, W.J., Stolz, D.B., Sullivan, M.L., Karlsson, J.M., Baty, C.J., Gibson, G.A., Erdos, G., Wang, Z., Milosevic, J., Tkacheva, O.A., Divito, S.J., Jordan, R., Lyons-Weiler, J., Watkins, S.C., Morelli, A.E., 2012. Mechanism of transfer of functional microRNAs between mouse dendritic cells via exosomes. Blood 119, 756–766.

Monteiro, J.T., Lepenies, B., 2017. Myeloid C-type lectin receptors in viral recognition and antiviral immunity. Viruses 9.

Murphy, K.M., Reiner, S.L., 2002. The lineage decisions of helper T cells. Nat. Rev. Immunol. 2, 933–944.

Murphy, T.L., Grajales-Reyes, G.E., Wu, X., Tussiwand, R., Briseno, C.G., Iwata, A., Kretzer, N.M., Durai, V., Murphy, K.M., 2016. Transcriptional control of dendritic cell development. Annu. Rev. Immunol. 34, 93–119.

Musumeci, A., Lutz, K., Winheim, E., Krug, A.B., 2019. What makes a pDC: recent advances in understanding plasmacytoid DC development and heterogeneity. Front. Immunol. 10, 1222.

Na, H., Cho, M., Chung, Y., 2016. Regulation of Th2 cell immunity by dendritic cells. Immune Netw. 16, 1–12.

Nagasawa, M., Schmidlin, H., Hazekamp, M.G., Schotte, R., Blom, B., 2008. Development of human plasmacytoid dendritic cells depends on the combined action of the basic helix-loop-helix factor E2-2 and the Ets factor Spi-B. Eur. J. Immunol. 38, 2389–2400.

Naik, S.H., Proietto, A.I., Wilson, N.S., Dakic, A., Schnorrer, P., Fuchsberger, M., Lahoud, M.H., O'keeffe, M., Shao, Q.X., Chen, W.F., Villadangos, J.A., Shortman, K., Wu, L., 2005. Cutting edge: generation of splenic CD8+ and CD8- dendritic cell equivalents in Fms-like tyrosine kinase 3 ligand bone marrow cultures. J. Immunol. 174, 6592–6597.

Naik, S.H., Metcalf, D., Van Nieuwenhuijze, A., Wicks, I., Wu, L., O'keeffe, M., Shortman, K., 2006. Intrasplenic steady-state dendritic cell precursors that are distinct from monocytes. Nat. Immunol. 7, 663–671.

Naik, S.H., Sathe, P., Park, H.Y., Metcalf, D., Proietto, A.I., Dakic, A., Carotta, S., O'keeffe, M., Bahlo, M., Papenfuss, A., Kwak, J.Y., Wu, L., Shortman, K., 2007. Development of plasmacytoid and conventional dendritic cell subtypes from single precursor cells derived in vitro and in vivo. Nat. Immunol. 8, 1217–1226.

Naik, S.H., Perie, L., Swart, E., Gerlach, C., Van Rooij, N., De Boer, R.J., Schumacher, T.N., 2013. Diverse and heritable lineage imprinting of early haematopoietic progenitors. Nature 496, 229–232.

Nair-Gupta, P., Baccarini, A., Tung, N., Seyffer, F., Florey, O., Huang, Y., Banerjee, M., Overholtzer, M., Roche, P.A., Tampe, R., Brown, B.D., Amsen, D., Whiteheart, S.W., Blander, J.M., 2014. TLR signals induce phagosomal MHC-I delivery from the endosomal recycling compartment to allow cross-presentation. Cell 158, 506–521.

Nakano, H., Yanagita, M., Gunn, M.D., 2001. CD11c(+)B220(+)Gr-1(+) cells in mouse lymph nodes and spleen display characteristics of plasmacytoid dendritic cells. J. Exp. Med. 194, 1171–1178.

Nakano, H., Lin, K.L., Yanagita, M., Charbonneau, C., Cook, D.N., Kakiuchi, T., Gunn, M.D., 2009. Blood-derived inflammatory dendritic cells in lymph nodes stimulate acute T helper type 1 immune responses. Nat. Immunol. 10, 394–402.

Naqvi, A.R., Fordham, J.B., Ganesh, B., Nares, S., 2016. miR-24, miR-30b and miR-142-3p interfere with antigen processing and presentation by primary macrophages and dendritic cells. Sci. Rep. 6, 32925.

Nedjic, J., Aichinger, M., Emmerich, J., Mizushima, N., Klein, L., 2008. Autophagy in thymic epithelium shapes the T-cell repertoire and is essential for tolerance. Nature 455, 396–400.

Neefjes, J., Jongsma, M.L., Paul, P., Bakke, O., 2011. Towards a systems understanding of MHC class I and MHC class II antigen presentation. Nat. Rev. Immunol. 11, 823–836.

Netherby, C.S., Messmer, M.N., Burkard-Mandel, L., Colligan, S., Miller, A., Cortes Gomez, E., Wang, J., Nemeth, M.J., Abrams, S.I., 2017. The granulocyte progenitor stage is a key target of IRF8-mediated regulation of myeloid-derived suppressor cell production. J. Immunol. 198, 4129–4139.

Neubert, K., Lehmann, C.H., Heger, L., Baranska, A., Staedtler, A.M., Buchholz, V.R., Yamazaki, S., Heidkamp, G.F., Eissing, N., Zebroski, H., Nussenzweig, M.C., Nimmerjahn, F., Dudziak, D., 2014. Antigen delivery to CD11c+CD8- dendritic cells induces protective immune responses against experimental melanoma in mice in vivo. J. Immunol. 192, 5830–5838.

Nimmerjahn, F., Kobelt, D., Steinkasserer, A., Menke, A., Hobom, G., Behrends, U., Bornkamm, G.W., Mautner, J., 2003. Efficient generation and expansion of antigen-specific CD4+ T cells by recombinant influenza viruses. Eur. J. Immunol. 33, 3331–3341.

Niu, H., Cattoretti, G., Dalla-Favera, R., 2003. BCL6 controls the expression of the B7-1/CD80 costimulatory receptor in germinal center B cells. J. Exp. Med. 198, 211–221.

Ohnmacht, C., Pullner, A., King, S.B., Drexler, I., Meier, S., Brocker, T., Voehringer, D., 2009. Constitutive ablation of dendritic cells breaks self-tolerance of CD4 T cells and results in spontaneous fatal autoimmunity. J. Exp. Med. 206, 549–559.

Ohtsuka, H., Sakamoto, A., Pan, J., Inage, S., Horigome, S., Ichii, H., Arima, M., Hatano, M., Okada, S., Tokuhisa, T., 2011. Bcl6 is required for the development of mouse CD4+ and CD8alpha+ dendritic cells. J. Immunol. 186, 255–263.

Onai, N., Obata-Onai, A., Schmid, M.A., Ohteki, T., Jarrossay, D., Manz, M.G., 2007. Identification of clonogenic common Flt3+M-CSFR+ plasmacytoid and conventional dendritic cell progenitors in mouse bone marrow. Nat. Immunol. 8, 1207–1216.

Onai, N., Kurabayashi, K., Hosoi-Amaike, M., Toyama-Sorimachi, N., Matsushima, K., Inaba, K., Ohteki, T., 2013. A clonogenic progenitor with prominent plasmacytoid dendritic cell developmental potential. Immunity 38, 943–957.

Ouchi, T., Kubo, A., Yokouchi, M., Adachi, T., Kobayashi, T., Kitashima, D.Y., Fujii, H., Clausen, B.E., Koyasu, S., Amagai, M., Nagao, K., 2011. Langerhans cell antigen capture through tight junctions confers preemptive immunity in experimental staphylococcal scalded skin syndrome. J. Exp. Med. 208, 2607.

Ouchi, T., Nakato, G., Udey, M.C., 2016. EpCAM expressed by murine epidermal langerhans cells modulates immunization to an epicutaneously applied protein antigen. J. Invest. Dermatol. 136, 1627–1635.

Palucka, K., Banchereau, J., 2012. Cancer immunotherapy via dendritic cells. Nat. Rev. Cancer 12, 265.

Palucka, K., Banchereau, J., Mellman, I., 2010. Designing vaccines based on biology of human dendritic cell subsets. Immunity 33, 464–478.

Paludan, C., Schmid, D., Landthaler, M., Vockerodt, M., Kube, D., Tuschl, T., Munz, C., 2005. Endogenous MHC class II processing of a viral nuclear antigen after autophagy. Science 307, 593–596.

Park, H., Li, Z., Yang, X.O., Chang, S.H., Nurieva, R., Wang, Y.H., Wang, Y., Hood, L., Zhu, Z., Tian, Q., Dong, C., 2005. A distinct lineage of CD4 T cells regulates tissue inflammation by producing interleukin 17. Nat. Immunol. 6, 1133–1141.

Paul, W.E., Zhu, J., 2010. How are TH2-type immune responses initiated and amplified? Nat. Rev. Immunol. 10, 225.

Paul, F., Arkin, Y., Giladi, A., Jaitin, D.A., Kenigsberg, E., Keren-Shaul, H., Winter, D., Lara-Astiaso, D., Gury, M., Weiner, A., David, E., Cohen, N., Lauridsen, F.K., Haas, S., Schlitzer, A., Mildner, A., Ginhoux, F., Jung, S., Trumpp, A., Porse, B.T., Tanay, A., Amit, I., 2015. Transcriptional heterogeneity and lineage commitment in myeloid progenitors. Cell 163, 1663–1677.

Peiser, M., Wanner, R., Kolde, G., 2004. Human epidermal Langerhans cells differ from monocyte-derived Langerhans cells in CD80 expression and in secretion of IL-12 after CD40 cross-linking. J. Leukoc. Biol. 76, 616–622.

Peng, Y., Croce, C.M., 2016. The role of microRNAs in human cancer. Signal Transduct. Target. Ther. 1, 15004.

Persky, M.E., Murphy, K.M., Farrar, J.D., 2005. IL-12, but not IFN-alpha, promotes STAT4 activation and Th1 development in murine CD4+ T cells expressing a chimeric murine/human Stat2 gene. J. Immunol. 174, 294–301.

Pietras, E.M., Reynaud, D., Kang, Y.A., Carlin, D., Calero-Nieto, F.J., Leavitt, A.D., Stuart, J.M., Gottgens, B., Passegue, E., 2015. Functionally distinct subsets of lineage-biased multipotent progenitors control blood production in normal and regenerative conditions. Cell Stem Cell 17, 35–46.

Plantinga, M., Guilliams, M., Vanheerswynghels, M., Deswarte, K., Branco-Madeira, F., Toussaint, W., Vanhoutte, L., Neyt, K., Killeen, N., Malissen, B., Hammad, H., Lambrecht, B.N., 2013. Conventional and monocyte-derived CD11b(+) dendritic cells initiate and maintain T helper 2 cell-mediated immunity to house dust mite allergen. Immunity 38, 322–335.

Platzer, B., Stout, M., Fiebiger, E., 2014. Antigen cross-presentation of immune complexes. Front. Immunol. 5, 140.

Poulin, L.F., Salio, M., Griessinger, E., Anjos-Afonso, F., Craciun, L., Chen, J.L., Keller, A.M., Joffre, O., Zelenay, S., Nye, E., Le Moine, A., Faure, F., Donckier, V., Sancho, D., Cerundolo, V., Bonnet, D., Reis, E., Sousa, C., 2010. Characterization of human DNGR-1+ BDCA3+ leukocytes as putative equivalents of mouse CD8alpha+ dendritic cells. J. Exp. Med. 207, 1261–1271.

Poulin, L.F., Reyal, Y., Uronen-Hansson, H., Schraml, B.U., Sancho, D., Murphy, K.M., Hakansson, U.K., Moita, L.F., Agace, W.W., Bonnet, D., Reis, E., Sousa, C., 2012. DNGR-1 is a specific and universal marker of mouse and human Batf3-dependent dendritic cells in lymphoid and nonlymphoid tissues. Blood 119, 6052–6062.

Pritchard, C.C., Cheng, H.H., Tewari, M., 2012. MicroRNA profiling: approaches and considerations. Nat. Rev. Genet. 13, 358.

Pulendran, B., Smith, J.L., Caspary, G., Brasel, K., Pettit, D., Maraskovsky, E., Maliszewski, C.R., 1999. Distinct dendritic cell subsets differentially regulate the class of immune response in vivo. Proc. Natl. Acad. Sci. U. S. A. 96, 1036–1041.

Pulford, K., Banham, A.H., Lyne, L., Jones, M., Ippolito, G.C., Liu, H., Tucker, P.W., Roncador, G., Lucas, E., Ashe, S., Stockwin, L., Walewska, R., Karran, L., Gascoyne, R.D., Mason, D.Y., Dyer, M.J., 2006. The BCL11AXL transcription factor: its distribution in normal and malignant tissues and use as a marker for plasmacytoid dendritic cells. Leukemia 20, 1439–1441.

Qiu, C.H., Miyake, Y., Kaise, H., Kitamura, H., Ohara, O., Tanaka, M., 2009. Novel subset of CD8{alpha}+ dendritic cells localized in the marginal zone is responsible for tolerance to cell-associated antigens. J. Immunol. 182, 4127–4136.

Reizis, B., Bunin, A., Ghosh, H.S., Lewis, K.L., Sisirak, V., 2011. Plasmacytoid dendritic cells: recent progress and open questions. Annu. Rev. Immunol. 29, 163–183.

Reizis, B., 2019. Plasmacytoid dendritic cells: development, regulation, and function. Immunity 50, 37–50.

Riedhammer, C., Weissert, R., 2015. Antigen presentation, autoantigens, and immune regulation in multiple sclerosis and other autoimmune diseases. Front. Immunol. 6, 322.

Rissoan, M.-C., Duhen, T., Bridon, J.-M., Bendriss-Vermare, N., Péronne, C., Vis, B.D.S., Brière, F., Bates, E.E.M., 2002. Subtractive hybridization reveals the expression of immunoglobulinlike transcript 7, Eph-B1, granzyme B, and 3 novel transcripts in human plasmacytoid dendritic cells. Blood 100, 3295–3303.

Robbins, S.H., Walzer, T., Dembele, D., Thibault, C., Defays, A., Bessou, G., Xu, H., Vivier, E., Sellars, M., Pierre, P., Sharp, F.R., Chan, S., Kastner, P., Dalod, M., 2008. Novel insights into the relationships between dendritic cell subsets in human and mouse revealed by genome-wide expression profiling. Genome Biol. 9, R17.

Roberts, E.W., Broz, M.L., Binnewies, M., Headley, M.B., Nelson, A.E., Wolf, D.M., Kaisho, T., Bogunovic, D., Bhardwaj, N., Krummel, M.F., 2016. Critical role for CD103(+)/CD141(+) dendritic cells bearing CCR7 for tumor antigen trafficking and priming of T cell immunity in melanoma. Cancer Cell 30, 324–336.

Roche, P.A., Furuta, K., 2015. The ins and outs of MHC class II-mediated antigen processing and presentation. Nat. Rev. Immunol. 15, 203–216.

Rock, K.L., Reits, E., Neefjes, J., 2016. Present yourself! by MHC class I and MHC class II molecules. Trends Immunol. 37, 724–737.

Rodrigues, P.F., Alberti-Servera, L., Eremin, A., Grajales-Reyes, G.E., Ivanek, R., Tussiwand, R., 2018. Distinct progenitor lineages contribute to the heterogeneity of plasmacytoid dendritic cells. Nat. Immunol. 19, 711–722.

Rodriguez, A., Vigorito, E., Clare, S., Warren, M.V., Couttet, P., Soond, D.R., Van Dongen, S., Grocock, R.J., Das, P.P., Miska, E.A., Vetrie, D., Okkenhaug, K., Enright, A.J., Dougan, G., Turner, M., Bradley, A., 2007. Requirement of bic/microRNA-155 for normal immune function. Science 316, 608–611.

Romani, N., Clausen, B.E., Stoitzner, P., 2010. Langerhans cells and more: langerin-expressing dendritic cell subsets in the skin. Immunol. Rev. 234, 120–141.

Rothbart, S.B., Strahl, B.D., 2014. Interpreting the language of histone and DNA modifications. Biochim. Biophys. Acta 1839, 627–643.

Rubtsov, Y.P., Rasmussen, J.P., Chi, E.Y., Fontenot, J., Castelli, L., Ye, X., Treuting, P., Siewe, L., Roers, A., Henderson Jr., W.R., Muller, W., Rudensky, A.Y., 2008. Regulatory T cell-derived interleukin-10 limits inflammation at environmental interfaces. Immunity 28, 546–558.

Ruedl, C., Koebel, P., Bachmann, M., Hess, M., Karjalainen, K., 2000. Anatomical origin of dendritic cells determines their life span in peripheral lymph nodes. J. Immunol. 165, 4910–4916.

Sadaka, C., Marloie-Provost, M.-A., Soumelis, V., Benaroch, P., 2009. Developmental regulation of MHC II expression and transport in human plasmacytoid-derived dendritic cells. Blood 113, 2127–2135.

Sage, A.P., Murphy, D., Maffia, P., Masters, L.M., Sabir, S.R., Baker, L.L., Cambrook, H., Finigan, A.J., Ait-Oufella, H., Grassia, G., Harrison, J.E., Ludewig, B., Reith, W., Hansson, G.K., Reizis, B., Hugues, S., Mallat, Z., 2014. MHC class II-restricted antigen presentation by plasmacytoid dendritic cells drives proatherogenic T cell immunity. Circulation 130, 1363–1373.

Salmon, H., Idoyaga, J., Rahman, A., Leboeuf, M., Remark, R., Jordan, S., Casanova-Acebes, M., Khudoynazarova, M., Agudo, J., Tung, N., Chakarov, S., Rivera, C., Hogstad, B., Bosenberg, M., Hashimoto, D., Gnjatic, S., Bhardwaj, N., Palucka, A.K., Brown, B.D., Brody, J., Ginhoux, F., Merad, M., 2016. Expansion and activation of CD103(+) dendritic cell progenitors at the tumor site enhances tumor responses to therapeutic PD-L1 and BRAF inhibition. Immunity 44, 924–938.

Salvermoser, J., Van Blijswijk, J., Papaioannou, N.E., Rambichler, S., Pasztoi, M., Pakalniškytė, D., Rogers, N.C., Keppler, S.J., Straub, T., Reis, E., Sousa, C., Schraml, B.U., 2018. Clec9a-mediated ablation of conventional dendritic cells suggests a lymphoid path to generating dendritic cells in vivo. Front. Immunol. 9, 699.

Sancho, D., MourÃO-SÁ, D., Joffre, O.P., Schulz, O., Rogers, N.C., Pennington, D.J., Carlyle, J.R., Reis, E., Sousa, C., 2008. Tumor therapy in mice via antigen targeting to a novel, DC-restricted C-type lectin. J. Clin. Invest. 118, 2098–2110.

Sapoznikov, A., Fischer, J.A., Zaft, T., Krauthgamer, R., Dzionek, A., Jung, S., 2007. Organ-dependent in vivo priming of naive CD4+, but not CD8+, T cells by plasmacytoid dendritic cells. J. Exp. Med. 204, 1923–1933.

Saric, T., Chang, S.C., Hattori, A., York, I.A., Markant, S., Rock, K.L., Tsujimoto, M., Goldberg, A.L., 2002. An IFN-gamma-induced aminopeptidase in the ER, ERAP1, trims precursors to MHC class I-presented peptides. Nat. Immunol. 3, 1169–1176.

Sasaki, I., Hoshino, K., Sugiyama, T., Yamazaki, C., Yano, T., Iizuka, A., Hemmi, H., Tanaka, T., Saito, M., Sugiyama, M., Fukuda, Y., Ohta, T., Sato, K., Ainai, A., Suzuki, T., Hasegawa, H., Toyama-Sorimachi, N., Kohara, H., Nagasawa, T., Kaisho, T., 2012. Spi-B is critical for plasmacytoid dendritic cell function and development. Blood 120, 4733–4743.

Sathe, P., Pooley, J., Vremec, D., Mintern, J., Jin, J.-O., Wu, L., Kwak, J.-Y., Villadangos, J.A., Shortman, K., 2011. The acquisition of antigen cross-presentation function by newly formed dendritic cells. J. Immunol. 186, 5184.

Sathe, P., Vremec, D., Wu, L., Corcoran, L., Shortman, K., 2013. Convergent differentiation: myeloid and lymphoid pathways to murine plasmacytoid dendritic cells. Blood 121, 11–19.

Sathe, P., Metcalf, D., Vremec, D., Naik, S.H., Langdon, W.Y., Huntington, N.D., Wu, L., Shortman, K., 2014. Lymphoid tissue and plasmacytoid dendritic cells and macrophages do not share a common macrophage-dendritic cell-restricted progenitor. Immunity 41, 104–115.

Satpathy, A.T., Murphy, K.M., Kc, W., 2011. Transcription factor networks in dendritic cell development. Semin. Immunol. 23, 388–397.
Satpathy, A.T., Kc, W., Albring, J.C., Edelson, B.T., Kretzer, N.M., Bhattacharya, D., Murphy, T.L., Murphy, K.M., 2012. Zbtb46 expression distinguishes classical dendritic cells and their committed progenitors from other immune lineages. J. Exp. Med. 209, 1135–1152.
Satpathy, A.T., Briseno, C.G., Lee, J.S., Ng, D., Manieri, N.A., Kc, W., Wu, X., Thomas, S.R., Lee, W.L., Turkoz, M., McDonald, K.G., Meredith, M.M., Song, C., Guidos, C.J., Newberry, R.D., Ouyang, W., Murphy, T.L., Stappenbeck, T.S., Gommerman, J.L., Nussenzweig, M.C., Colonna, M., Kopan, R., Murphy, K.M., 2013. Notch2-dependent classical dendritic cells orchestrate intestinal immunity to attaching-and-effacing bacterial pathogens. Nat. Immunol. 14, 937–948.
Sawai, C.M., Sisirak, V., Ghosh, H.S., Hou, E.Z., Ceribelli, M., Staudt, L.M., Reizis, B., 2013. Transcription factor Runx2 controls the development and migration of plasmacytoid dendritic cells. J. Exp. Med. 210, 2151–2159.
Saxena, M., Bhardwaj, N., 2018. Re-emergence of dendritic cell vaccines for cancer treatment. Trends Cancer 4, 119–137.
Schaft, N., Dorrie, J., Thumann, P., Beck, V.E., Muller, I., Schultz, E.S., Kampgen, E., Dieckmann, D., Schuler, G., 2005. Generation of an optimized polyvalent monocyte-derived dendritic cell vaccine by transfecting defined RNAs after rather than before maturation. J. Immunol. 174, 3087–3097.
Schaum, N., Karkanias, J., Neff, N.F., May, A.P., Quake, S.R., Wyss-Coray, T., Darmanis, S., Batson, J., Botvinnik, O., Chen, M.B., Chen, S., Green, F., Jones, R.C., Maynard, A., Penland, L., Pisco, A.O., Sit, R.V., Stanley, G.M., Webber, J.T., Zanini, F., Baghel, A.S., Bakerman, I., Bansal, I., Berdnik, D., Bilen, B., Brownfield, D., Cain, C., Chen, M.B., Chen, S., Cho, M., Cirolia, G., Conley, S.D., Darmanis, S., Demers, A., Demir, K., De Morree, A., Divita, T., Du Bois, H., Dulgeroff, L.B.T., Ebadi, H., Espinoza, F.H., Fish, M., Gan, Q., George, B.M., Gillich, A., Green, F., Genetiano, G., Gu, X., Gulati, G.S., Hang, Y., Hosseinzadeh, S., Huang, A., Iram, T., Isobe, T., Ives, F., Jones, R.C., Kao, K.S., Karnam, G., Kershner, A.M., Kiss, B.M., Kong, W., Kumar, M.E., Lam, J.Y., Lee, D.P., Lee, S.E., Li, G., Li, Q., Liu, L., Lo, A., Lu, W.-J., Manjunath, A., May, A.P., May, K.L., May, O.L., Maynard, A., McKay, M., Metzger, R.J., Mignardi, M., Min, D., Nabhan, A.N., Neff, N.F., Ng, K.M., Noh, J., Patkar, R., Peng, W.C., Penland, L., Puccinelli, R., Rulifson, E.J., Schaum, N., Sikandar, S.S., Sinha, R., Sit, R.V., Szade, K., Tan, W., Tato, C., Tellez, K., Travaglini, K.J., Tropini, C., Waldburger, L., Van Weele, L.J., et al., 2018. Single-cell transcriptomics of 20 mouse organs creates a Tabula Muris. Nature 562, 367–372.
Schiavoni, G., Mattei, F., Borghi, P., Sestili, P., Venditti, M., Morse III, H.C., Belardelli, F., Gabriele, L., 2004. ICSBP is critically involved in the normal development and trafficking of Langerhans cells and dermal dendritic cells. Blood 103, 2221–2228.
Schlitzer, A., Ginhoux, F., 2014. Organization of the mouse and human DC network. Curr. Opin. Immunol. 26, 90–99.
Schlitzer, A., Loschko, J., Mair, K., Vogelmann, R., Henkel, L., Einwachter, H., Schiemann, M., Niess, J.H., Reindl, W., Krug, A., 2011. Identification of CCR9- murine plasmacytoid DC precursors with plasticity to differentiate into conventional DCs. Blood 117, 6562–6570.
Schlitzer, A., Heiseke, A.F., Einwachter, H., Reindl, W., Schiemann, M., Manta, C.P., See, P., Niess, J.H., Suter, T., Ginhoux, F., Krug, A.B., 2012. Tissue-specific differentiation of a circulating CCR9- pDC-like common dendritic cell precursor. Blood 119, 6063–6071.

Schlitzer, A., McGovern, N., Teo, P., Zelante, T., Atarashi, K., Low, D., Ho, A.W., See, P., Shin, A., Wasan, P.S., Hoeffel, G., Malleret, B., Heiseke, A., Chew, S., Jardine, L., Purvis, H.A., Hilkens, C.M., Tam, J., Poidinger, M., Stanley, E.R., Krug, A.B., Renia, L., Sivasankar, B., Ng, L.G., Collin, M., Ricciardi-Castagnoli, P., Honda, K., Haniffa, M., Ginhoux, F., 2013. IRF4 transcription factor-dependent CD11b+ dendritic cells in human and mouse control mucosal IL-17 cytokine responses. Immunity 38, 970–983.

Schlitzer, A., McGovern, N., Ginhoux, F., 2015a. Dendritic cells and monocyte-derived cells: two complementary and integrated functional systems. Semin. Cell Dev. Biol. 41, 9–22.

Schlitzer, A., Sivakamasundari, V., Chen, J., Sumatoh, H.R.B., Schreuder, J., Lum, J., Malleret, B., Zhang, S., Larbi, A., Zolezzi, F., Renia, L., Poidinger, M., Naik, S., Newell, E.W., Robson, P., Ginhoux, F., 2015b. Identification of cDC1- and cDC2-committed DC progenitors reveals early lineage priming at the common DC progenitor stage in the bone marrow. Nat. Immunol. 16, 718.

Schmid, D., Munz, C., 2007. Innate and adaptive immunity through autophagy. Immunity 27, 11–21.

Schonheit, J., Kuhl, C., Gebhardt, M.L., Klett, F.F., Riemke, P., Scheller, M., Huang, G., Naumann, R., Leutz, A., Stocking, C., Priller, J., Andrade-Navarro, M.A., Rosenbauer, F., 2013. PU.1 level-directed chromatin structure remodeling at the Irf8 gene drives dendritic cell commitment. Cell Rep. 3, 1617–1628.

Schotte, R., Nagasawa, M., Weijer, K., Spits, H., Blom, B., 2004. The ETS transcription factor Spi-B is required for human plasmacytoid dendritic cell development. J. Exp. Med. 200, 1503–1509.

Schraml, B.U., Reis e Sousa, C., 2015. Defining dendritic cells. Curr. Opin. Immunol. 32, 13–20.

Schreiber, H.A., Loschko, J., Karssemeijer, R.A., Escolano, A., Meredith, M.M., Mucida, D., Guermonprez, P., Nussenzweig, M.C., 2013. Intestinal monocytes and macrophages are required for T cell polarization in response to *Citrobacter rodentium*. J. Exp. Med. 210, 2025–2039.

Schuler, G., 2003. Use of monocyte-derived dendritic cells in cancer immunotherapy. Cancer Immun. Arch. 3, 6.

Schulz, O., Jaensson, E., Persson, E.K., Liu, X., Worbs, T., Agace, W.W., Pabst, O., 2009. Intestinal CD103+, but not CX3CR1+, antigen sampling cells migrate in lymph and serve classical dendritic cell functions. J. Exp. Med. 206, 3101–3114.

Schulz, C., Gomez Perdiguero, E., Chorro, L., Szabo-Rogers, H., Cagnard, N., Kierdorf, K., Prinz, M., Wu, B., Jacobsen, S.E., Pollard, J.W., Frampton, J., Liu, K.J., Geissmann, F., 2012. A lineage of myeloid cells independent of Myb and hematopoietic stem cells. Science 336, 86–90.

Schwarz, D.S., Hutvagner, G., Du, T., Xu, Z., Aronin, N., Zamore, P.D., 2003. Asymmetry in the assembly of the RNAi enzyme complex. Cell 115, 199–208.

Schweisguth, F., 2004. Regulation of Notch signaling activity. Curr. Biol. 14, R129–R138.

Scott, C.L., Soen, B., Martens, L., Skrypek, N., Saelens, W., Taminau, J., Blancke, G., Van Isterdael, G., Huylebroeck, D., Haigh, J., Saeys, Y., Guilliams, M., Lambrecht, B.N., Berx, G., 2016. The transcription factor Zeb2 regulates development of conventional and plasmacytoid DCs by repressing Id2. J. Exp. Med. 213, 897–911.

See, P., Dutertre, C.A., Chen, J., Gunther, P., McGovern, N., Irac, S.E., Gunawan, M., Beyer, M., Handler, K., Duan, K., Sumatoh, H.R.B., Ruffin, N., Jouve, M., Gea-Mallorqui, E., Hennekam, R.C.M., Lim, T., Yip, C.C., Wen, M., Malleret, B., Low, I., Shadan, N.B., Fen, C.F.S., Tay, A., Lum, J., Zolezzi, F., Larbi, A., Poidinger, M., Chan, J.K.Y., Chen, Q., Renia, L., Haniffa, M., Benaroch, P., Schlitzer, A., Schultze, J.L., Newell, E.W., Ginhoux, F., 2017. Mapping the human DC lineage through the integration of high-dimensional techniques. Science 356. eaag3009.

Segura, E., Amigorena, S., 2013. Inflammatory dendritic cells in mice and humans. Trends Immunol. 34, 440–445.
Segura, E., Villadangos, J.A., 2011. A modular and combinatorial view of the antigen cross-presentation pathway in dendritic cells. Traffic 12, 1677–1685.
Segura, E., Albiston, A.L., Wicks, I.P., Chai, S.Y., Villadangos, J.A., 2009. Different cross-presentation pathways in steady-state and inflammatory dendritic cells. Proc. Natl. Acad. Sci. U. S. A. 106 pnas.0910295106.
Segura, E., Valladeau-Guilemond, J., Donnadieu, M.H., Sastre-Garau, X., Soumelis, V., Amigorena, S., 2012. Characterization of resident and migratory dendritic cells in human lymph nodes. J. Exp. Med. 209, 653–660.
Segura, E., Durand, M., Amigorena, S., 2013a. Similar antigen cross-presentation capacity and phagocytic functions in all freshly isolated human lymphoid organ–resident dendritic cells. J. Exp. Med. 210, 1035.
Segura, E., Touzot, M., Bohineust, A., Cappuccio, A., Chiocchia, G., Hosmalin, A., Dalod, M., Soumelis, V., Amigorena, S., 2013b. Human inflammatory dendritic cells induce Th17 cell differentiation. Immunity 38, 336–348.
Seillet, C., Jackson, J.T., Markey, K.A., Brady, H.J., Hill, G.R., Macdonald, K.P., Nutt, S.L., Belz, G.T., 2013. CD8alpha+ DCs can be induced in the absence of transcription factors Id2, Nfil3, and Batf3. Blood 121, 1574–1583.
Semmrich, M., Plantinga, M., Svensson-Frej, M., Uronen-Hansson, H., Gustafsson, T., Mowat, A.M., Yrlid, U., Lambrecht, B.N., Agace, W.W., 2012. Directed antigen targeting in vivo identifies a role for CD103+ dendritic cells in both tolerogenic and immunogenic T-cell responses. Mucosal Immunol. 5, 150–160.
Sere, K., Baek, J.H., Ober-Blobaum, J., Muller-Newen, G., Tacke, F., Yokota, Y., Zenke, M., Hieronymus, T., 2012. Two distinct types of Langerhans cells populate the skin during steady state and inflammation. Immunity 37, 905–916.
Serra, P., Santamaria, P., 2019. Antigen-specific therapeutic approaches for autoimmunity. Nat. Biotechnol. 37, 238–251.
Serwold, T., Gonzalez, F., Kim, J., Jacob, R., Shastri, N., 2002. ERAAP customizes peptides for MHC class I molecules in the endoplasmic reticulum. Nature 419, 480–483.
Seth, S., Oberdorfer, L., Hyde, R., Hoff, K., Thies, V., Worbs, T., Schmitz, S., Forster, R., 2011. CCR7 essentially contributes to the homing of plasmacytoid dendritic cells to lymph nodes under steady-state as well as inflammatory conditions. J. Immunol. 186, 3364–3372.
Shakhar, G., Lindquist, R.L., Skokos, D., Dudziak, D., Huang, J.H., Nussenzweig, M.C., Dustin, M.L., 2005. Stable T cell-dendritic cell interactions precede the development of both tolerance and immunity in vivo. Nat. Immunol. 6, 707–714.
Shin, C., Han, J.A., Koh, H., Choi, B., Cho, Y., Jeong, H., Ra, J.S., Sung, P.S., Shin, E.C., Ryu, S., Do, Y., 2015. CD8alpha(-) dendritic cells induce antigen-specific T follicular helper cells generating efficient humoral immune responses. Cell Rep. 11, 1929–1940.
Shortman, K., Naik, S.H., 2007. Steady-state and inflammatory dendritic-cell development. Nat. Rev. Immunol. 7, 19–30.
Sichien, D., Scott, C.L., Martens, L., Vanderkerken, M., Van Gassen, S., Plantinga, M., Joeris, T., De Prijck, S., Vanhoutte, L., Vanheerswynghels, M., Van Isterdael, G., Toussaint, W., Madeira, F.B., Vergote, K., Agace, W.W., Clausen, B.E., Hammad, H., Dalod, M., Saeys, Y., Lambrecht, B.N., Guilliams, M., 2016. IRF8 transcription factor controls survival and function of terminally differentiated conventional and plasmacytoid dendritic cells, respectively. Immunity 45, 626–640.
Siegal, F.P., Kadowaki, N., Shodell, M., Fitzgerald-Bocarsly, P.A., Shah, K., Ho, S., Antonenko, S., Liu, Y.J., 1999. The nature of the principal type 1 interferon-producing cells in human blood. Science 284, 1835–1837.

Smyth, L.A., Boardman, D.A., Tung, S.L., Lechler, R., Lombardi, G., 2015. MicroRNAs affect dendritic cell function and phenotype. Immunology 144, 197–205.
Sontag, S., Forster, M., Qin, J., Wanek, P., Mitzka, S., Schuler, H.M., Koschmieder, S., Rose-John, S., Sere, K., Zenke, M., 2017. Modelling IRF8 deficient human hematopoiesis and dendritic cell development with engineered iPS cells. Stem Cells 35, 898–908.
Spits, H., Couwenberg, F., Bakker, A.Q., Weijer, K., Uittenbogaart, C.H., 2000. Id2 and Id3 inhibit development of CD34(+) stem cells into predendritic cell (pre-DC)2 but not into pre-DC1. Evidence for a lymphoid origin of pre-DC2. J. Exp. Med. 192, 1775–1784.
Spranger, S., Dai, D., Horton, B., Gajewski, T.F., 2017. Tumor-residing Batf3 dendritic cells are required for effector T cell trafficking and adoptive T cell therapy. Cancer Cell 31, 711–723.e4.
Steinman, R.M., Banchereau, J., 2007. Taking dendritic cells into medicine. Nature 449, 419–426.
Steinman, R.M., Cohn, Z.A., 1973. Identification Of A Novel Cell Type In Peripheral Lymphoid Organs Of Mice. J. Exp. Med. 137, 1142.
Steinman, R.M., Nussenzweig, M.C., 2002. Avoiding horror autotoxicus: the importance of dendritic cells in peripheral T cell tolerance. Proc. Natl. Acad. Sci. 99, 351.
Steinman, R.M., Witmer, M.D., 1978. Lymphoid dendritic cells are potent stimulators of the primary mixed leukocyte reaction in mice. Proc. Natl. Acad. Sci. U. S. A. 75, 5132–5136.
Stoitzner, P., Holzmann, S., McLellan, A.D., Ivarsson, L., Stossel, H., Kapp, M., Kammerer, U., Douillard, P., Kampgen, E., Koch, F., Saeland, S., Romani, N., 2003. Visualization and characterization of migratory Langerhans cells in murine skin and lymph nodes by antibodies against Langerin/CD207. J. Invest. Dermatol. 120, 266–274.
Stoitzner, P., Tripp, C.H., Eberhart, A., Price, K.M., Jung, J.Y., Bursch, L., Ronchese, F., Romani, N., 2006. Langerhans cells cross-present antigen derived from skin. Proc. Natl. Acad. Sci. 103, 7783.
Stoitzner, P., Schaffenrath, S., Tripp, C.H., Reider, D., Komenda, K., Del Frari, B., Djedovic, G., Ebner, S., Romani, N., 2014. Human skin dendritic cells can be targeted in situ by intradermal injection of antibodies against lectin receptors. Exp. Dermatol. 23, 909–915.
Strandt, H., Pinheiro, D.F., Kaplan, D.H., Wirth, D., Gratz, I.K., Hammerl, P., Thalhamer, J., Stoecklinger, A., 2017. Neoantigen expression in steady-state langerhans cells induces CTL tolerance. J. Immunol. 199, ji1602098.
Su, Y.L., Swiderski, P., Marcucci, G., Kortylewski, M., 2019. Targeted delivery of miRNA antagonists to myeloid cells in vitro and in vivo. Methods Mol. Biol. 1974, 141–150.
Sun, C.M., Hall, J.A., Blank, R.B., Bouladoux, N., Oukka, M., Mora, J.R., Belkaid, Y., 2007. Small intestine lamina propria dendritic cells promote de novo generation of Foxp3 T reg cells via retinoic acid. J. Exp. Med. 204, 1775–1785.
Sun, Y., Varambally, S., Maher, C.A., Cao, Q., Chockley, P., Toubai, T., Malter, C., Nieves, E., Tawara, I., Wang, Y., Ward, P.A., Chinnaiyan, A., Reddy, P., 2011. Targeting of microRNA-142-3p in dendritic cells regulates endotoxin-induced mortality. Blood 117, 6172–6183.
Suzuki, Y., Orellana, M.A., Schreiber, R.D., Remington, J.S., 1988. Interferon-gamma: the major mediator of resistance against Toxoplasma gondii. Science 240, 516–518.
Suzuki, S., Honma, K., Matsuyama, T., Suzuki, K., Toriyama, K., Akitoyo, I., Yamamoto, K., Suematsu, T., Nakamura, M., Yui, K., Kumatori, A., 2004. Critical roles of interferon regulatory factor 4 in CD11b- dendritic cell development. Proc. Natl. Acad. Sci. 101, 8981.

Swain, S.L., Weinberg, A.D., English, M., Huston, G., 1990. IL-4 directs the development of Th2-like helper effectors. J. Immunol. 145, 3796–3806.
Swiecki, M., Colonna, M., 2015. The multifaceted biology of plasmacytoid dendritic cells. Nat. Rev. Immunol. 15, 471–485.
Szabo, S.J., Kim, S.T., Costa, G.L., Zhang, X., Fathman, C.G., Glimcher, L.H., 2000. A novel transcription factor, T-bet, directs Th1 lineage commitment. Cell 100, 655–669.
Tacken, P.J., De Vries, I.J., Torensma, R., Figdor, C.G., 2007. Dendritic-cell immunotherapy: from ex vivo loading to in vivo targeting. Nat. Rev. Immunol. 7, 790–802.
Tamura, T., Tailor, P., Yamaoka, K., Kong, H.J., Tsujimura, H., O'shea, J.J., Singh, H., Ozato, K., 2005. IFN regulatory factor-4 and -8 govern dendritic cell subset development and their functional diversity. J. Immunol. 174, 2573–2581.
Tamura, T., Yanai, H., Savitsky, D., Taniguchi, T., 2008. The IRF family transcription factors in immunity and oncogenesis. Annu. Rev. Immunol. 26, 535–584.
Tamura, T., Kurotaki, D., Koizumi, S., 2015. Regulation of myelopoiesis by the transcription factor IRF8. Int. J. Hematol. 101, 342–351.
Tavano, B., Galao, R.P., Graham, D.R., Neil, S.J., Aquino, V.N., Fuchs, D., Boasso, A., 2013. Ig-like transcript 7, but not bone marrow stromal cell antigen 2 (also known as HM1.24, tetherin, or CD317), modulates plasmacytoid dendritic cell function in primary human blood leukocytes. J. Immunol. 190, 2622–2630.
Tel, J., Schreibelt, G., Sittig, S.P., Mathan, T.S., Buschow, S.I., Cruz, L.J., Lambeck, A.J., Figdor, C.G., De Vries, I.J., 2013. Human plasmacytoid dendritic cells efficiently cross-present exogenous Ags to CD8+ T cells despite lower Ag uptake than myeloid dendritic cell subsets. Blood 121, 459–467.
Tian, Y., Meng, L., Zhang, Y., 2017. Epigenetic regulation of dendritic cell development and function. Cancer J. 23, 302–307.
Tserel, L., Runnel, T., Kisand, K., Pihlap, M., Bakhoff, L., Kolde, R., Peterson, H., Vilo, J., Peterson, P., Rebane, A., 2011. MicroRNA expression profiles of human blood monocyte-derived dendritic cells and macrophages reveal miR-511 as putative positive regulator of Toll-like receptor 4. J. Biol. Chem. 286, 26487–26495.
Tsukada, J., Yoshida, Y., Kominato, Y., Auron, P.E., 2011. The CCAAT/enhancer (C/EBP) family of basic-leucine zipper (bZIP) transcription factors is a multifaceted highly-regulated system for gene regulation. Cytokine 54, 6–19.
Turner, M.L., Schnorfeil, F.M., Brocker, T., 2011. MicroRNAs regulate dendritic cell differentiation and function. J. Immunol. 187, 3911–3917.
Tussiwand, R., Onai, N., Mazzucchelli, L., Manz, M.G., 2005. Inhibition of natural type I IFN-producing and dendritic cell development by a small molecule receptor tyrosine kinase inhibitor with Flt3 affinity. J. Immunol. 175, 3674–3680.
Tussiwand, R., Lee, W.L., Murphy, T.L., Mashayekhi, M., Kc, W., Albring, J.C., Satpathy, A.T., Rotondo, J.A., Edelson, B.T., Kretzer, N.M., Wu, X., Weiss, L.A., Glasmacher, E., Li, P., Liao, W., Behnke, M., Lam, S.S., Aurthur, C.T., Leonard, W.J., Singh, H., Stallings, C.L., Sibley, L.D., Schreiber, R.D., Murphy, K.M., 2012. Compensatory dendritic cell development mediated by BATF-IRF interactions. Nature 490, 502–507.
Tussiwand, R., Everts, B., Grajales-Reyes, G.E., Kretzer, N.M., Iwata, A., Bagaitkar, J., Wu, X., Wong, R., Anderson, D.A., Murphy, T.L., Pearce, E.J., Murphy, K.M., 2015. Klf4 expression in conventional dendritic cells is required for T helper 2 cell responses. Immunity 42, 916–928.
Urban Jr., J.F., Noben-Trauth, N., Donaldson, D.D., Madden, K.B., Morris, S.C., Collins, M., Finkelman, F.D., 1998. IL-13, IL-4Ralpha, and Stat6 are required for the expulsion of the gastrointestinal nematode parasite Nippostrongylus brasiliensis. Immunity 8, 255–264.

Valladeau, J., Ravel, O., Dezutter-Dambuyant, C., Moore, K., Kleijmeer, M., Liu, Y., Duvert-Frances, V., Vincent, C., Schmitt, D., Davoust, J., Caux, C., Lebecque, S., Saeland, S., 2000. Langerin, a novel C-type lectin specific to Langerhans cells, is an endocytic receptor that induces the formation of Birbeck granules. Immunity 12, 71–81.
Van Blijswijk, J., Schraml, B.U., Reis, E., Sousa, C., 2013. Advantages and limitations of mouse models to deplete dendritic cells. Eur. J. Immunol. 43, 22–26.
Van de Laar, L., Van Den Bosch, A., Wierenga, A.T., Janssen, H.L., Coffer, P.J., Woltman, A.M., 2011. Tight control of STAT5 activity determines human CD34-derived interstitial dendritic cell and langerhans cell development. J. Immunol. 186, 7016–7024.
Van Den Berg, W.B., Miossec, P., 2009. IL-17 as a future therapeutic target for rheumatoid arthritis. Nat. Rev. Rheumatol. 5, 549.
Van Dinther, D., Stolk, D.A., Van De Ven, R., Van Kooyk, Y., De Gruijl, T.D., Den Haan, J.M.M., 2017. Targeting C-type lectin receptors: a high-carbohydrate diet for dendritic cells to improve cancer vaccines. J. Leukoc. Biol. 102, 1017–1034.
Van Endert, P., 2011. Post-proteasomal and proteasome-independent generation of MHC class I ligands. Cell. Mol. Life Sci. 68 (9), 1553–1567.
Van Endert, P., 2016. Intracellular recycling and cross-presentation by MHC class I molecules. Immunol. Rev. 272, 80–96.
Vander Lugt, B., Khan, A.A., Hackney, J.A., Agrawal, S., Lesch, J., Zhou, M., Lee, W.P., Park, S., Xu, M., Devoss, J., Spooner, C.J., Chalouni, C., Delamarre, L., Mellman, I., Singh, H., 2014. Transcriptional programming of dendritic cells for enhanced MHC class II antigen presentation. Nat. Immunol. 15, 161–167.
Velásquez-Lopera, M.M., Correa, L.A., GarcíA, L.F., 2008. Human spleen contains different subsets of dendritic cells and regulatory T lymphocytes. Clin. Exp. Immunol. 154, 107–114.
Veldhoen, M., Hocking, R.J., Atkins, C.J., Locksley, R.M., Stockinger, B., 2006. TGFbeta in the context of an inflammatory cytokine milieu supports de novo differentiation of IL-17-producing T cells. Immunity 24, 179–189.
Vijay, K., 2018. Toll-like receptors in immunity and inflammatory diseases: past, present, and future. Int. Immunopharmacol. 59, 391–412.
Villadangos, J.A., Schnorrer, P., 2007. Intrinsic and cooperative antigen-presenting functions of dendritic-cell subsets in vivo. Nat. Rev. Immunol. 7, 543–555.
Villadangos, J.A., Young, L., 2008. Antigen-presentation properties of plasmacytoid dendritic cells. Immunity 29, 352–361.
Villadangos, J.A., Riese, R.J., Peters, C., Chapman, H.A., Ploegh, H.L., 1997. Degradation of mouse invariant chain: roles of cathepsins S and D and the influence of major histocompatibility complex polymorphism. J. Exp. Med. 186, 549–560.
Villadangos, J.A., Cardoso, M., Steptoe, R.J., Van Berkel, D., Pooley, J., Carbone, F.R., Shortman, K., 2001. MHC class II expression is regulated in dendritic cells independently of invariant chain degradation. Immunity 14, 739–749.
Villani, A.C., Satija, R., Reynolds, G., Sarkizova, S., Shekhar, K., Fletcher, J., Griesbeck, M., Butler, A., Zheng, S., Lazo, S., Jardine, L., Dixon, D., Stephenson, E., Nilsson, E., Grundberg, I., McDonald, D., Filby, A., Li, W., De Jager, P.L., Rozenblatt-Rosen, O., Lane, A.A., Haniffa, M., Regev, A., Hacohen, N., 2017. Single-cell RNA-seq reveals new types of human blood dendritic cells, monocytes, and progenitors. Science 356.
Vitali, C., Mingozzi, F., Broggi, A., Barresi, S., Zolezzi, F., Bayry, J., Raimondi, G., Zanoni, I., Granucci, F., 2012. Migratory, and not lymphoid-resident, dendritic cells maintain peripheral self-tolerance and prevent autoimmunity via induction of iTreg cells. Blood 120, 1237–1245.

Vremec, D., Lieschke, G.J., Dunn, A.R., Robb, L., Metcalf, D., Shortman, K., 1997. The influence of granulocyte/macrophage colony-stimulating factor on dendritic cell levels in mouse lymphoid organs. Eur. J. Immunol. 27, 40–44.

Vyas, J.M., Van Der Veen, A.G., Ploegh, H.L., 2008. The known unknowns of antigen processing and presentation. Nat. Rev. Immunol. 8, 607–618.

Wakim, L.M., Waithman, J., Van Rooijen, N., Heath, W.R., Carbone, F.R., 2008. Dendritic cell-induced memory T cell activation in nonlymphoid tissues. Science 319, 198–202.

Walter, W., Scheuer, C., Lingnau, K., Reichert, T.E., Schmitt, E., Loos, M., Maeurer, M.J., 2000. H2-M, a facilitator of MHC class II peptide loading, and its negative modulator H2-O are differentially expressed in response to proinflammatory cytokines. Immunogenetics 51, 794–804.

Wang, K., Zhang, S., Weber, J., Baxter, D., Galas, D.J., 2010. Export of microRNAs and microRNA-protective protein by mammalian cells. Nucleic Acids Res. 38, 7248–7259.

Wang, Y., Szretter, K.J., Vermi, W., Gilfillan, S., Rossini, C., Cella, M., Barrow, A.D., Diamond, M.S., Colonna, M., 2012. IL-34 is a tissue-restricted ligand of CSF1R required for the development of Langerhans cells and microglia. Nat. Immunol. 13, 753.

Wang, Y., Sun, H.Y., Kumar, S., Puerta, M.D.M., Jo, H., Rezvan, A., 2019. ZBTB46 is a shear-sensitive transcription factor inhibiting endothelial cell proliferation via gene expression regulation of cell cycle proteins. Lab. Investig. 99, 305–318.

Waskow, C., Liu, K., Darrasse-Jeze, G., Guermonprez, P., Ginhoux, F., Merad, M., Shengelia, T., Yao, K., Nussenzweig, M., 2008. The receptor tyrosine kinase Flt3 is required for dendritic cell development in peripheral lymphoid tissues. Nat. Immunol. 9, 676–683.

Watchmaker, P.B., Lahl, K., Lee, M., Baumjohann, D., Morton, J., Kim, S.J., Zeng, R., Dent, A., Ansel, K.M., Diamond, B., Hadeiba, H., Butcher, E.C., 2013. Comparative transcriptional and functional profiling defines conserved programs of intestinal DC differentiation in humans and mice. Nat. Immunol. 15, 98.

Weissman, I.L., Shizuru, J.A., 2008. The origins of the identification and isolation of hematopoietic stem cells, and their capability to induce donor-specific transplantation tolerance and treat autoimmune diseases. Blood 112, 3543.

Whartenby, K.A., Calabresi, P.A., Mccadden, E., Nguyen, B., Kardian, D., Wang, T., Mosse, C., Pardoll, D.M., Small, D., 2005. Inhibition of FLT3 signaling targets DCs to ameliorate autoimmune disease. Proc. Natl. Acad. Sci. U. S. A. 102, 16741–16746.

Wightman, B., Ha, I., Ruvkun, G., 1993. Posttranscriptional regulation of the heterochronic gene lin-14 by lin-4 mediates temporal pattern formation in C. elegans. Cell 75, 855–862.

Williams, J.W., Tjota, M.Y., Clay, B.S., Vander Lugt, B., Bandukwala, H.S., Hrusch, C.L., Decker, D.C., Blaine, K.M., Fixsen, B.R., Singh, H., Sciammas, R., Sperling, A.I., 2013. Transcription factor IRF4 drives dendritic cells to promote Th2 differentiation. Nat. Commun. 4, 2990.

Wilson, D.C., Matthews, S., Yap, G.S., 2008. IL-12 signaling drives CD8+ T cell IFN-gamma production and differentiation of KLRG1+ effector subpopulations during Toxoplasma gondii Infection. J. Immunol. 180, 5935–5945.

Wilson, N.K., Kent, D.G., Buettner, F., Shehata, M., Macaulay, I.C., Calero-Nieto, F.J., Sanchez Castillo, M., Oedekoven, C.A., Diamanti, E., Schulte, R., Ponting, C.P., Voet, T., Caldas, C., Stingl, J., Green, A.R., Theis, F.J., Gottgens, B., 2015. Combined single-cell functional and gene expression analysis resolves heterogeneity within stem cell populations. Cell Stem Cell 16, 712–724.

Wollenberg, A., Mommaas, M., Oppel, T., Schottdorf, E.M., Gunther, S., Moderer, M., 2002. Expression and function of the mannose receptor CD206 on epidermal dendritic cells in inflammatory skin diseases. J. Invest. Dermatol. 118, 327–334.

Wu, L., D'amico, A., Winkel, K.D., Suter, M., Lo, D., Shortman, K., 1998. RelB is essential for the development of myeloid-related CD8alpha- dendritic cells but not of lymphoid-related CD8alpha+ dendritic cells. Immunity 9, 839–847.

Wu, X., Satpathy, A.T., Kc, W., Liu, P., Murphy, T.L., Murphy, K.M., 2013. Bcl11a controls Flt3 expression in early hematopoietic progenitors and is required for pDC development in vivo. PLoS One 8, e64800.

Wu, X., Gao, H., Bleday, R., Zhu, Z., 2014. Homeobox transcription factor VentX regulates differentiation and maturation of human dendritic cells. J. Biol. Chem. 289, 14633–14643.

Wu, S.J., Niknafs, Y.S., Kim, S.H., Oravecz-Wilson, K., Zajac, C., Toubai, T., Sun, Y., Prasad, J., Peltier, D., Fujiwara, H., Hedig, I., Mathewson, N.D., Khoriaty, R., Ginsburg, D., Reddy, P., 2017. A critical analysis of the role of SNARE protein SEC22B in antigen cross-presentation. Cell Rep. 19, 2645–2656.

Wu, J., Wu, H., An, J., Ballantyne, C.M., Cyster, J.G., 2018. Critical role of integrin CD11c in splenic dendritic cell capture of missing-self CD47 cells to induce adaptive immunity. Proc. Natl. Acad. Sci. 115, 6786.

Xiao, C., Srinivasan, L., Calado, D.P., Patterson, H.C., Zhang, B., Wang, J., Henderson, J.M., Kutok, J.L., Rajewsky, K., 2008. Lymphoproliferative disease and autoimmunity in mice with increased miR-17-92 expression in lymphocytes. Nat. Immunol. 9, 405–414.

Xing, Y., Hogquist, K.A., 2012. T-cell tolerance: central and peripheral. Cold Spring Harb. Perspect. Biol. 4, a006957. https://doi.org/10.1101/cshperspect.a006957.

Xing, Z., Gauldie, J., Cox, G., Baumann, H., Jordana, M., Lei, X.F., Achong, M.K., 1998. IL-6 is an antiinflammatory cytokine required for controlling local or systemic acute inflammatory responses. J. Clin. Invest. 101, 311–320.

Xu, Y., Zhan, Y., Lew, A.M., Naik, S.H., Kershaw, M.H., 2007. Differential development of murine dendritic cells by GM-CSF versus Flt3 ligand Has implications for inflammation and trafficking. J. Immunol. 179, 7577.

Xue, X., Feng, T., Yao, S., Wolf, K.J., Liu, C.G., Liu, X., Elson, C.O., Cong, Y., 2011. Microbiota downregulates dendritic cell expression of miR-10a, which targets IL-12/IL-23p40. J. Immunol. 187, 5879–5886.

Yadav, T., Quivy, J.P., Almouzni, G., 2018. Chromatin plasticity: a versatile landscape that underlies cell fate and identity. Science 361, 1332–1336.

Yamazaki, S., Dudziak, D., Heidkamp, G.F., Fiorese, C., Bonito, A.J., Inaba, K., Nussenzweig, M.C., Steinman, R.M., 2008. CD8+ CD205+ splenic dendritic cells are specialized to induce Foxp3+ regulatory T cells. J. Immunol. 181, 6923–6933.

Yamazaki, C., Miyamoto, R., Hoshino, K., Fukuda, Y., Sasaki, I., Saito, M., Ishiguchi, H., Yano, T., Sugiyama, T., Hemmi, H., Tanaka, T., Hamada, E., Hirashima, T., Amakawa, R., Fukuhara, S., Nomura, S., Ito, T., Kaisho, T., 2010. Conservation of a chemokine system, XCR1 and its ligand, XCL1, between human and mice. Biochem. Biophys. Res. Commun. 397, 756–761.

Yao, C., Zurawski, S.M., Jarrett, E.S., Chicoine, B., Crabtree, J., Peterson, E.J., Zurawski, G., Kaplan, D.H., Igyarto, B.Z., 2015. Skin dendritic cells induce follicular helper T cells and protective humoral immune responses. J. Allergy Clin. Immunol. 136, 1387–1397.e1-7.

Yasmin, N., Konradi, S., Eisenwort, G., Schichl, Y.M., Seyerl, M., Bauer, T., Stockl, J., Strobl, H., 2013. beta-Catenin promotes the differentiation of epidermal Langerhans dendritic cells. J. Invest. Dermatol. 133, 1250–1259.

Yasumi, T., Katamura, K., Okafuji, I., Yoshioka, T., Meguro, T.-A., Nishikomori, R., Kusunoki, T., Heike, T., Nakahata, T., 2005. Limited ability of antigen-specific Th1 responses to inhibit Th2 cell development in vivo. J. Immunol. 174, 1325.

Yewdell, J.W., Anton, L.C., Bennink, J.R., 1996. Defective ribosomal products (DRiPs): a major source of antigenic peptides for MHC class I molecules? J. Immunol. 157, 1823–1826.

Yewdell, J.W., Reits, E., Neefjes, J., 2003. Making sense of mass destruction: quantitating MHC class I antigen presentation. Nat. Rev. Immunol. 3, 952–961.

Yoshida, H., Lareau, C.A., Ramirez, R.N., Rose, S.A., Maier, B., Wroblewska, A., Desland, F., Chudnovskiy, A., Mortha, A., Dominguez, C., Tellier, J., Kim, E., Dwyer, D., Shinton, S., Nabekura, T., Qi, Y., Yu, B., Robinette, M., Kim, K.W., Wagers, A., Rhoads, A., Nutt, S.L., Brown, B.D., Mostafavi, S., Buenrostro, J.D., Benoist, C., 2019. The cis-regulatory atlas of the mouse immune system. Cell 176, 897–912.e20.

Zaba, L.C., Fuentes-Duculan, J., Steinman, R.M., Krueger, J.G., Lowes, M.A., 2007. Normal human dermis contains distinct populations of CD11c+BDCA-1+ dendritic cells and CD163+FXIIIA+ macrophages. J. Clin. Invest. 117, 2517–2525.

Zaba, L.C., Krueger, J.G., Lowes, M.A., 2009. Resident and "inflammatory" dendritic cells in human skin. J. Invest. Dermatol. 129, 302–308.

Zaric, M., Lyubomska, O., Poux, C., Hanna, M.L., Mccrudden, M.T., Malissen, B., Ingram, R.J., Power, U.F., Scott, C.J., Donnelly, R.F., Kissenpfennig, A., 2015. Dissolving microneedle delivery of nanoparticle-encapsulated antigen elicits efficient cross-priming and Th1 immune responses by murine Langerhans cells. J. Invest. Dermatol. 135, 425–434.

Zehner, M., Marschall, A.L., Bos, E., Schloetel, J.G., Kreer, C., Fehrenschild, D., Limmer, A., Ossendorp, F., Lang, T., Koster, A.J., Dubel, S., Burgdorf, S., 2015. The translocon protein Sec61 mediates antigen transport from endosomes in the cytosol for cross-presentation to CD8(+) T cells. Immunity 42, 850–863.

Zelenay, S., Keller, A.M., Whitney, P.G., Schraml, B.U., Deddouche, S., Rogers, N.C., Schulz, O., Sancho, D., Reis, E., Sousa, C., 2012. The dendritic cell receptor DNGR-1 controls endocytic handling of necrotic cell antigens to favor cross-priming of CTLs in virus-infected mice. J. Clin. Invest. 122, 1615–1627.

Zeng, R., Bscheider, M., Lahl, K., Lee, M., Butcher, E.C., 2016. Generation and transcriptional programming of intestinal dendritic cells: essential role of retinoic acid. Mucosal Immunol. 9, 183–193.

Zhan, Y., Wu, L., 2012. Functional regulation of monocyte-derived dendritic cells by microRNAs. Protein Cell 3, 497–507.

Zhang, J., Raper, A., Sugita, N., Hingorani, R., Salio, M., Palmowski, M.J., Cerundolo, V., Crocker, P.R., 2006. Characterization of Siglec-H as a novel endocytic receptor expressed on murine plasmacytoid dendritic cell precursors. Blood 107, 3600–3608.

Zhang, T.T., Liu, D., Calabro, S., Eisenbarth, S.C., Cattoretti, G., Haberman, A.M., 2014. Dynamic expression of BCL6 in murine conventional dendritic cells during in vivo development and activation. PLoS One 9, e101208.

Zhang, X., Gu, J., Yu, F.S., Zhou, L., Mi, Q.S., 2016. Tgf-β1-induced transcription factor networks in Langerhans cell development and maintenance. Allergy 71, 758–764.

Zheng, Y., Danilenko, D.M., Valdez, P., Kasman, I., Eastham-Anderson, J., Wu, J., Ouyang, W., 2007. Interleukin-22, a T(H)17 cytokine, mediates IL-23-induced dermal inflammation and acanthosis. Nature 445, 648–651.

Zhou, L., Ivanov, I.I., Spolski, R., Min, R., Shenderov, K., Egawa, T., Levy, D.E., Leonard, W.J., Littman, D.R., 2007. IL-6 programs T(H)-17 cell differentiation by promoting sequential engagement of the IL-21 and IL-23 pathways. Nat. Immunol. 8, 967–974.

Zhou, H., Huang, X., Cui, H., Luo, X., Tang, Y., Chen, S., Wu, L., Shen, N., 2010. miR-155 and its star-form partner miR-155* cooperatively regulate type I interferon production by human plasmacytoid dendritic cells. Blood 116, 5885–5894.

Zhu, J., Min, B., Hu-Li, J., Watson, C.J., Grinberg, A., Wang, Q., Killeen, N., Urban, J.F., Guo, L., Paul, W.E., 2004. Conditional deletion of Gata3 shows its essential function in TH1-TH2 responses. Nat. Immunol. 5, 1157–1165.

Zhu, J., Yamane, H., Cote-Sierra, J., Guo, L., Paul, W.E., 2006. GATA-3 promotes Th2 responses through three different mechanisms: induction of Th2 cytokine production, selective growth of Th2 cells and inhibition of Th1 cell-specific factors. Cell Res. 16, 3–10.

Zhu, J., Wang, F.L., Wang, H.B., Dong, N., Zhu, X.M., Wu, Y., Wang, Y.T., Yao, Y.M., 2017. TNF-alpha mRNA is negatively regulated by microRNA-181a-5p in maturation of dendritic cells induced by high mobility group box-1 protein. Sci. Rep. 7 12239.

Zigmond, E., Varol, C., Farache, J., Elmaliah, E., Satpathy, A.T., Friedlander, G., Mack, M., Shpigel, N., Boneca, I.G., Murphy, K.M., Shakhar, G., Halpern, Z., Jung, S., 2012. Ly6C hi monocytes in the inflamed colon give rise to proinflammatory effector cells and migratory antigen-presenting cells. Immunity 37, 1076–1090.

CHAPTER THREE

The impact of endoplasmic reticulum stress responses in dendritic cell immunobiology

Camilla Salvagno[a,b], Juan R. Cubillos-Ruiz[a,b,c,*]
[a]Department of Obstetrics and Gynecology, Weill Cornell Medicine, New York, NY, United States
[b]Sandra and Edward Meyer Cancer Center, Weill Cornell Medicine, New York, NY, United States
[c]Weill Cornell Graduate School of Medical Sciences, Cornell University, New York, NY, United States
*Corresponding author: e-mail address: jur2016@med.cornell.edu

Contents

1. A brief introduction to dendritic cell biology 154
2. ER stress responses in DCs 156
   2.1 The unfolded protein response (UPR) 156
   2.2 The UPR in DC homeostasis 160
   2.3 The UPR in DCs activated via pattern recognition receptors 162
   2.4 The UPR in DCs under pathological conditions 164
3. Concluding remarks 169
Acknowledgments 170
Competing interest 170
References 170

Abstract

Dendritic cells (DCs) are critical for bridging innate and adaptive immunity. They do so by presenting antigens to T cells, and by expressing diverse molecules that further promote T cell activation, differentiation and memory formation. During this process, intracellular and extracellular factors can perturb the protein-folding capacity of endoplasmic reticulum (ER) and induce a cellular state of "ER stress," which is controlled and resolved by the unfolded protein response (UPR). Interestingly, various studies have shown that DCs can activate UPR-related pathways even in the absence of global ER stress, and that this process can modulate their normal activity. In other settings, such as cancer, adverse microenvironmental conditions have been demonstrated to evoke severe ER stress and persistent activation of the UPR in tumor-infiltrating DCs. This process disrupts their metabolism and local antigen-presenting capacity, hence impeding the initiation and maintenance of anti-cancer immunity. Here, we review recent findings on how canonical and non-canonical UPR activation impacts DC immunobiology at the steady-state, upon activation via pattern recognition receptors, and under diverse pathological conditions. We also discuss the potential therapeutic implications that targeting the UPR in DCs may have in the context of cancer and in other pathologies such as graft-versus-host disease.

International Review of Cell and Molecular Biology, Volume 349
ISSN 1937-6448
https://doi.org/10.1016/bs.ircmb.2019.08.004

During their lifetime, cells experience a variety of physiological and pathophysiological conditions that perturb their protein-folding homeostasis. The endoplasmic reticulum (ER) is the organelle where transmembrane and secreted proteins undergo maturation via N-linked glycosylation, folding and disulfide bond formation. Since cells produce approximately 4×10^6 proteins every minute (Princiotta et al., 2003), monitoring and ensuring their correct folding and modification is critical. Several conditions such as accumulation of reactive oxygen species (ROS), hypoxia, acidosis, or nutrient deprivation, might interfere with the protein-folding capacity of the ER. The consequent accumulation of misfolded or unfolded proteins in the ER induces a cellular state of "ER stress" that can result in both a shortage of secreted proteins and cellular toxicity. Cells have evolved a quality control checkpoint in the ER to cope with the accumulation of misfolded or unfolded proteins. The sensing of and adaptation to ER stress is governed by three ER-resident proteins, which activate an adaptive signaling pathway called the unfolded protein response (UPR). The UPR is particularly important for secretory and immune cells that rely on efficient protein synthesis and secretion for their main functions. In this review, we focus on how ER stress response pathways influence the immunobiology of dendritic cells (DCs) under normal physiological conditions and in the setting of various pathologies.

1. A brief introduction to dendritic cell biology

Since the discovery of DCs in the early 1970s (Steinman and Cohn, 1973, 1974), accumulating evidence has demonstrated the unique and powerful role that these myeloid cells have in activating the adaptive immune system. DCs are short-lived cells whose main function is to sample the tissue environment for foreign antigens. Upon pathogen encounter, DCs play a central role in orchestrating $CD8^+$ and $CD4^+$ T cell priming in the lymph node by presenting antigens to T cells in the context of MHC-I and MHC-II, respectively, and creating a favorable ecosystem for T cell activation, expansion and differentiation (Borst et al., 2018). In general, DCs are defined by the expression of CD11c and MHC-II, but various DC subtypes exist based on the combination of surface markers expressed and the key transcription factors involved in their development. In steady-state conditions, plasmacytoid DCs (pDCs) represent a small fraction of DCs that are potent in producing type-I interferon (IFN) upon recognition of danger signals (Reizis, 2019). The most common types of DCs are conventional DCs (cDCs), which are subdivided into cDC1 and cDC2 types. The term cDC1 refer to both lymphoid resident DCs expressing the CD8αα homodimer, and non-lymphoid tissue-resident DCs expressing the integrin CD103. These cells potently cross-present antigens to $CD8^+$ T cells, both during infection and in the tumor context (Bedoui et al., 2009; Kim and Braciale, 2009; Roberts et al., 2016; Salmon et al., 2016). In addition, cDC1s express the C-type lectin Clec9a and the chemokine receptor XCR1 (Murphy et al., 2016; Schraml et al., 2013). cDC2, instead, are characterized by the expression of the integrin CD11b and the glycoprotein SIRPα and preferentially stimulate $CD4^+$ T cell responses (Calabro et al., 2016; Gao et al., 2013; Krishnaswamy et al., 2017; Murphy et al., 2016; Pooley et al., 2001). Moreover, another classification distinguishes cDCs as either lymph node-resident or migratory, where the latter are tissue-resident DCs able to migrate to the draining lymph node upon antigen encounter due to the expression of the chemokine receptor 7 (CCR7) (Calabro et al., 2016; Roberts et al., 2016). Various transcription factors controlling the development of specific DC subtypes have been identified and characterized. The basic leucine zipper transcription factor ATF-like 3 (Batf3), interferon regulatory factor (Irf) 8 and Id2 are critical for both $CD8^+$ and $CD103^+$ cDC1s, while Irf4, Notch 2 and Zeb2 are key for the development of $CD11b^+$ cDC2s (Broz et al., 2014; Hacker et al., 2003; Hildner et al., 2008; Lewis et al., 2011; Scott et al., 2016; Suzuki et al., 2004). In addition, pDC generation has been demonstrated to rely on the expression of Irf8 and Zeb2, as well as helix-loop-helix E protein E2-2 (Cisse et al., 2008; Scott et al., 2016; Tsujimura et al., 2003).

In normal conditions, DCs develop in the bone marrow from the common myeloid progenitors (CMPs) following the granulocyte-macrophage progenitors (GMP), macrophage-dendritic cell progenitors (MDP) and common dendritic cell progenitors (CDPs), which lack monocyte-macrophage differentiation potential (Murphy et al., 2016). CDPs then give rise to pDCs and pre-conventional dendritic cells (pre-cDCs) (Murphy et al., 2016). The latter population can migrate to lymphoid and non-lymphoid tissues where they differentiate into cDC1 or cDC2 subtypes (Liu et al., 2009; Murphy et al., 2016). Yet, in pathologic conditions such as cancer, DCs may also originate from differentiated monocytes, and are therefore named monocyte-derived DCs (MoDCs) or inflammatory DCs (Segura and Amigorena, 2013). MoDCs are usually induced in inflamed tissues and can behave as cDCs (Cheong et al., 2010; Naik et al., 2006). Indeed, they express the DC-lineage specific transcription factor Zbtb46 similarly to cDCs (Satpathy et al., 2012; Zigmond et al., 2012). Because of their promiscuous expression of some surface markers that are also shared with monocytes, macrophages and cDCs, such as CD11b, Ly6C and CD11c, it is commonly difficult to distinguish MoDCs in inflamed tissues and in the tumor. However, it has been proposed that the marker FcεRI allows for the discrimination of MoDC from other DC subtypes (Segura and Amigorena, 2013). For additional details on the diverse DC subtypes and their precise origin and function, we refer the reader to specialized reviews published elsewhere (Eisenbarth, 2019; Merad et al., 2013; Segura and Amigorena, 2013). Throughout the text, we will outline the type of DCs investigated in each study based on the markers used for their identification and the current nomenclature. However, for those studies in which DCs are incompletely identified or characterized, the general term of "DCs" will be used.

2. ER stress responses in DCs

2.1 The unfolded protein response (UPR)

The term UPR has its origin in studies conducted in the late 1980s showing that the accumulation of misfolded proteins in the ER induced the expression of ER chaperones, which are critical for the correct folding of proteins (Kozutsumi et al., 1988). In the following years, studies in the yeast *Saccharomyces cerevisiae* provided key insights into the molecular mediators of the UPR (Cox et al., 1993; Mori, 2015; Mori et al., 1993). These studies paved the way to establish that the mammalian UPR is activated by three ER-resident transmembrane sensors: inositol-requiring enzyme 1α (IRE1α, encoded by *ERN1*), PKR-like ER kinase (PERK, encoded by *EIF2AK3*), and activating transcription factor 6 (ATF6) (Fig. 1). Under steady-state conditions, binding-immunoglobulin protein, BiP, also called glucose-regulated protein 78 (GRP78), binds to and inactivates these three sensors (Bertolotti et al., 2000; Shen et al., 2002). When misfolded or unfolded proteins accumulate in the ER, BiP dissociates from the sensors, allowing their activation to trigger the UPR (Bertolotti et al., 2000; Shen et al., 2002). This leads to a series of downstream signaling events that enable the recovery of the ER from the accumulation of misfolded proteins. These events include: attenuation of protein synthesis initiation by activation of the PERK pathway (Harding et al., 2000); induction of the ER-associated protein degradation (ERAD) machinery, a process that removes malfolded proteins via their ubiquitination and proteasomal degradation (Hwang and Qi, 2018);

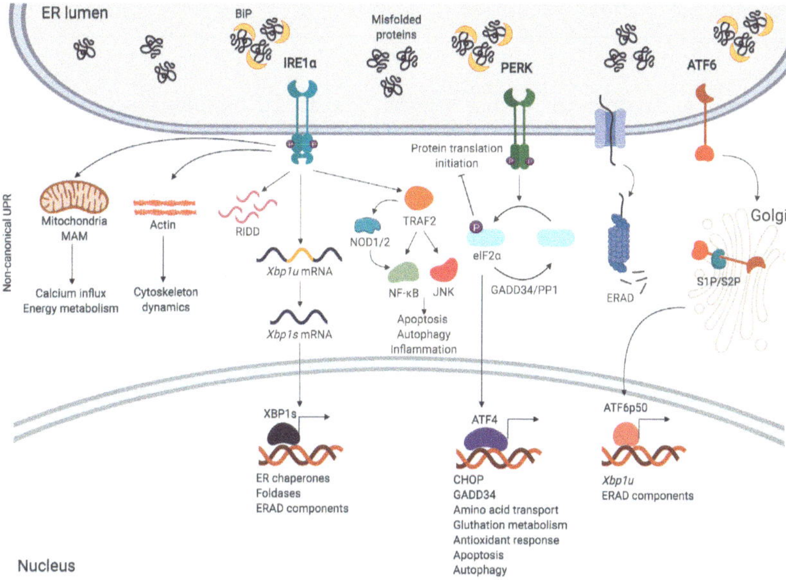

Fig. 1 The UPR in mammalian cells. The UPR comprises three signaling pathways mediated by the ER transmembrane proteins IRE1α, PERK and ATF6. Under steady-state conditions, these three sensors are inactivated by the binding of BiP. The dissociation of BiP upon accumulation of misfolded proteins in the ER lumen triggers the UPR: IRE1α dimerizes, autophosphorylates and activates its RNase domain, which removes 26 nucleotides from the *Xbp1u* mRNA to generate a spliced version (*Xbp1s*) that encodes the functionally active transcription factor XBP1 (XBP1s). XBP1s triggers the expression of chaperones and foldases to promote the correct folding of proteins. IRE1α hyperactivation can result in RIDD, which triggers global degradation of mRNAs. IRE1α can also activate JNK and NF-κB to induce autophagy and apoptosis. In addition, TRAF2 recruits NOD1/2 to mediate NF-kB-dependent ER stress-induced inflammation. Besides its canonical role as an ER stress sensor, IRE1α regulates actin cytoskeleton remodeling by binding to filamin A, and controls mitochondrial bioenergetics by acting as a scaffold in MAMs, which favors calcium transfer between the ER and the mitochondrion. The second arm of the UPR comprises PERK, which upon activation phosphorylates eIF2α to reduce protein translation initiation. This process leads to the selective translation of ATF4, which induces the expression of genes including the pro-apoptotic CHOP and GADD34, which together with PP1 dephosphorylates eIF2α to restore protein translation. In addition, ATF4 elicits the expression of proteins involved in amino acid transport, glutathione metabolism, antioxidant responses, apoptosis and autophagy. The third ER stress sensor is ATF6, which translocates into the Golgi apparatus upon ER stress to undergo cleavage by proteases S1P and S2P. This process releases its N-terminal segment that functions as a transcription factor. ATF6 induce the expression of the unspliced form of *Xbp1* mRNA. Both the IRE1α and ATF6 axis drive the expression of ERAD components, a process that removes unfolded proteins via proteasomal degradation. Abbreviations: ATF, activating transcription factor; BiP, binding-immunoglobulin protein; CHOP, C/EBP homologous protein; eIF2α, eukaryotic translation initiation factor 2α; ER, endoplasmic reticulum; ERAD, ER-associated protein degradation; GADD34, growth arrest and DNA damage-inducible gene 34; IRE1α, inositol-requiring enzyme 1α; JNK, c-Jun N-terminal kinase; MAM, mitochondria-associated membranes; NF-kB, nuclear factor κ light chain enhancer of activated B cells; NOD1/2, nucleotide-binding oligomerization domain; PERK, PKR-like ER kinase; PP1, protein phosphatase 1; RIDD, regulated IRE1α dependent decay of mRNA; S1P and S2P, site 1 and site 2 proteases; TRAF2, tumor necrosis factor receptor-associated factor 2; UPR, unfolded protein response; *Xbp1s*, spliced *Xbp1*; *Xbp1u*, unspliced *Xbp1*. *Figure created with BioRender.com.*

autophagy, a process in which pieces of the ER containing malfolded proteins are delivered to the lysosomes for degradation (Bernales et al., 2006; Yorimitsu et al., 2006); and production of foldases and chaperones, such as BiP, to promote protein folding and the secretion of properly-folded proteins (Kozutsumi et al., 1988; Lee et al., 2003). If none of these events are able to recover ER homeostasis, cells will undergo apoptosis (Ron and Walter, 2007).

The most evolutionally conserved arm of the UPR is controlled by the dual enzyme IRE1α, which is an ER-resident transmembrane protein that contains both a kinase and RNase domain. When BiP detaches, IRE1α dimerizes, autophosphorylates and induces a conformational change that activates its RNase domain. Consequently, the RNase domain of IRE1α removes 26 nucleotides from the cytosolic *Xbp1* mRNA, resulting in the generation of a spliced form, *Xbp1s*, that encodes the functionally active transcription factor XBP1 (XBP1s) (Calfon et al., 2002). As a result, XBP1s induces the expression of several proteins, including chaperones, foldases and components of the ERAD system (Lee et al., 2003) (Fig. 1). Furthermore, in some settings, IRE1α can also alleviate protein translation by triggering global mRNA degradation through regulated IRE1α dependent decay of mRNA (RIDD) (Hollien and Weissman, 2006). Indeed, IRE1α RNase activity cleaves and degrades several other mRNA substrates that bear a consensus sequence (CUGCAG) and a stem-loop structure similar to the one recognized in the *Xbp1* mRNA (Oikawa et al., 2010). In addition, IRE1α can activate the c-Jun N-terminal kinase (JNK) and nuclear transcription factor-kB (NF-kB) by binding to the adaptor protein tumor necrosis factor (TNF) receptor-associated factor (TRAF)2 resulting in autophagy and apoptosis (Fig. 1) (Haberzettl and Hill, 2013; Kaneko et al., 2003; Urano et al., 2000). By binding to nucleotide-binding oligomerization domain (NOD) 1 and 2, TRAF2 also mediates ER stress-induced inflammation (Keestra-Gounder et al., 2016). Interestingly, recent findings have reported non-canonical (UPR-unrelated) roles for IRE1α that are independent of its RNase activity (Carreras-Sureda et al., 2019; Urra et al., 2018). IRE1α was first shown to regulate cytoskeletal remodeling and cell migration in several animal species by directly binding to filamin A (Urra et al., 2018). The same group then reported that IRE1α further operates as a scaffold in mitochondria-associated membranes (MAMs), which is the site of physical interaction between the ER and mitochondria, hence allowing calcium uptake by mitochondria and favoring energy production (Carreras-Sureda et al., 2019) (Fig. 1).

During ER stress and subsequent BiP dissociation, PERK dimerizes and undergoes autophosphorylation, similar to IRE1α. Activated PERK phosphorylates eukaryotic translation initiation factor 2α (eIF2α), resulting in reduced rates of protein translation initiation (Harding et al., 1999, 2000). Nevertheless, active PERK can also stimulate the selective translation of the activating transcription factor 4 (ATF4), which induces the expression of genes involved in amino acid transport, glutathione metabolism and resistance to oxidative stress (Harding et al., 2003). In addition, ATF4 leads to the expression of C/EBP homologous protein (CHOP) (Oyadomari and Mori, 2004) that can promote autophagy and apoptosis (B'Chir et al., 2013). ATF4 also leads to the expression of growth arrest and DNA damage-inducible gene 34 (GADD34), which in concert with protein phosphatase 1 (PP1) dephosphorylates eIF2α to restore protein translation (Fig. 1) (Connor et al., 2001).

The third arm of the UPR is governed by ATF6. During conditions of ER stress, ATF6 translocates to the Golgi apparatus where it is processed by site-1 and -2 proteases (S1P and S2P), in order to release the N-terminal segment, ATF6p50, which operates as a transcription factor (Haze et al., 1999; Shen et al., 2002). Consequently, ATF6p50 triggers the expression of ER chaperones and the unspliced *Xbp1* transcript, a process that cooperates with the induction of ERAD components (Fig. 1) (Lee et al., 2002; Yamamoto et al., 2007).

It is clear that the UPR is able to control cell survival or cell death, but the precise mechanisms involved in this process remain largely unexplored. One explanation might lie on the duration of the activity of individual UPR arms (Lin et al., 2007). In human embryonic kidney 293 cells chronically stimulated with tunicamycin or thapsigargin, which induce ER stress by blocking N-linked glycosylation or inhibiting the ER calcium transport, respectively, all three branches of the UPR were activated. However, while the response of IRE1α and ATF6 decreased after few hours, PERK activity and pro-apoptotic *Chop* expression persisted longer (Lin et al., 2007). Enforced IRE1α expression in these cells enhanced their viability, suggesting that changes in UPR signaling affect the balance between pro-survival and pro-apoptotic cues (Lin et al., 2007). However, this mechanism might be cell-, tissue-, or context-dependent, as distinct cell types have specific functional and metabolic requirements. For example, in response to pathogens, immune cells increase the production and the secretion of proteins like chemokines and cytokines by maintaining the proper folding of the proteins while preventing ER stress-induced apoptosis. Indeed, in murine splenic macrophages and other cells, Toll-like receptor 4 (TLR4) engagement reduced *Chop* expression and apoptosis (Woo et al., 2009), suggesting that the type of stimulus and the tissue context are key factors in determining such life-death decisions.

2.2 The UPR in DC homeostasis

Various studies reported that some DC populations demonstrate constitutive (ER stress-independent) activation of the IRE1α arm of the UPR, which has been suggested to modulate multiple DC functions (Fig. 2, left). The first evidence that the UPR is important for DC biology derived from studies using XBP1/RAG2-deficient chimeric mice, in which animals bear a disrupted XBP1 mainly in lymphoid organs (Iwakoshi et al., 2007). Mice lacking XBP1 in this system displayed a pronounced decrease in splenic DCs, with pDCs being reduced the most (Iwakoshi et al., 2007). These results are in line with the highest levels of *Xbp1* splicing in pDC compared to cDCs and other immune cells in wild-type mice under steady-state conditions (Iwakoshi et al., 2007). However, by using the ER stress-activated indicator (ERAI) mouse model, where cells undergoing IRE1α activation express a variant of GFP (Iwawaki et al., 2004), it was shown that cDCs constitute the population with the highest IRE1α activity in the spleen (Osorio et al., 2014). In contrast to previous work using chimeric mice, splenic cDC1s in the ERAI system demonstrated higher IRE1α activity than cDC2s and pDCs (Osorio et al., 2014), likely reflecting differences in mouse strain background, age and/or microbiota that might differentially activate the UPR in specific cell subtypes. In addition, conditional deletion of XBP1 in CD11c-expressing cells did not result in decreased number of splenic cDC1s, cDC2s or pDCs (Osorio et al., 2014), in contrast to previous reports using XBP1/RAG2-deficient chimeric mice (Iwakoshi et al., 2007). These differences likely reflect the stage of DC development at which loss of XBP1 occurs. Indeed, it is possible that in chimeric mice early loss of XBP1 might affect DC precursors and thus impair DC development, whereas loss of XBP1 in committed CD11c-expressing DCs does not influence their survival, at least in the spleen. Interestingly, a deeper characterization of DCs in several organs of mice lacking XBP1 in CD11c-expressing cells showed that loss of XBP1 induced apoptosis in cDC1s residing in the lung, while this effect was not observed in intestinal or splenic cDC1 populations (Tavernier et al., 2017). By using mice lacking both XBP1 and IRE1α, the authors reported that the survival of intestinal cDC1s was mediated

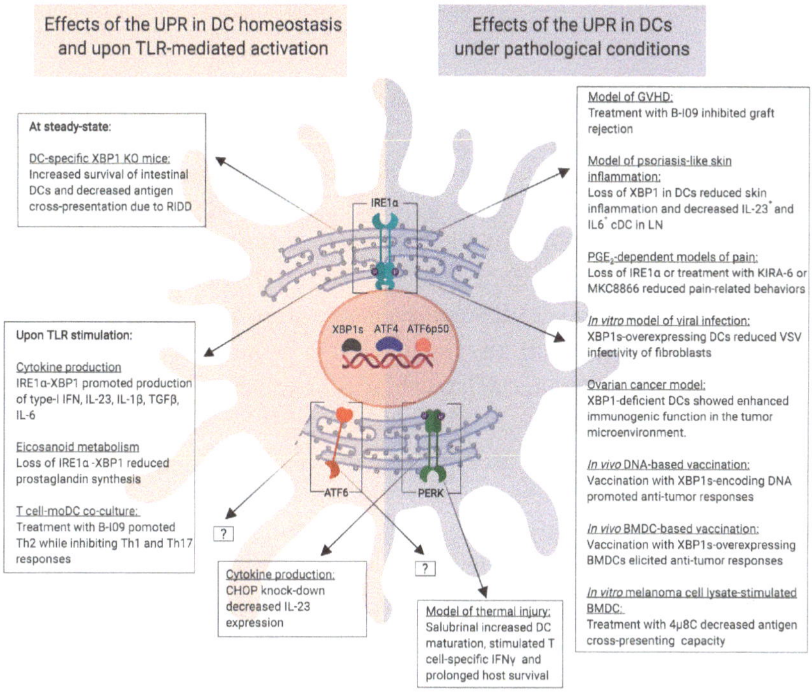

Fig. 2 DC-intrinsic roles of the UPR in homeostatic and pathological conditions. The left side of the schematic depicts the roles of ER stress sensors in DCs under steady-state conditions and after TLR-mediated activation: At the steady-state, RIDD, mainly induced by XBP1 deletion, promotes the survival of intestinal DCs. Upon TLR activation, IRE1α plays a role in inducing cytokine production. This control in cytokine production indirectly influences $CD4^+$ T cell differentiation in co-culture systems by favoring Th1 and Th17 responses. Besides IRE1α, PERK can also regulate cytokine production, in particular IL-23. Recently, the IRE1α-XBP1 arm of the UPR was shown to control eicosanoid metabolism and prostaglandin synthesis. The right side of the schematic, in contrast, shows the effects of DC UPR signaling in several disease models. The IRE1α arm of the UPR mediates detrimental effects in a model of GVHD as it promotes graft rejection. It also exacerbates inflammation in a model of psoriasis-like skin and induces pain-related behaviors in PGE_2-dependent models of pain. It has, however, positive outcomes in viral infection, as it protects cells from VSV infection *in vitro*. In ovarian cancer models, the sustained IRE1α-XBP1 signaling provokes DC dysfunction at tumor sites, which is characterized by abnormal lipid accumulation and reduced antigen-presenting capacity that ultimately impedes the development of anti-tumor T cell responses. Instead, in DNA- and BMDC-based vaccination models where DCs are not persistently exposed to the harsh conditions of the tumor microenvironment, XBP1 expression appears to be necessary for eliciting anti-tumor responses. In an *in vivo* model for thermal injury, PERK decreased the expression of co-stimulatory molecules and hampered the production of IFNγ by T cells in co-culture experiments. The role of ATF6 in DC remains largely unexplored and deserves further investigation. B-I09, 4μ8C and MKC8866 are inhibitors of the RNase domain of IRE1α; KIRA6 is an inhibitor of the kinase domain of IRE1α; Salubrinal is an inhibitor of eIF2α. Abbreviations: ATF, activating transcription factor; BMDC, bone marrow-derived dendritic cell; cDC, conventional dendritic cell; CHOP, C/EBP homologous protein; DC, dendritic cell; GVHD, graft-versus-host disease; IRE1α, inositol-requiring enzyme 1α; LN, lymph node; PEG_2, prostaglandin E2; PERK, PKR-like ER kinase; RIDD, regulated IRE1α dependent decay of mRNA; type I IFN, type I interferon; UPR, unfolded protein response; VSV, vesicular stomatitis virus. *Figure created with BioRender.com.*

by increased IRE1α-dependent RIDD activity (Fig. 2), although the mechanism behind the protection against apoptosis was not determined (Tavernier et al., 2017). Of note, whether RIDD activity promotes survival of splenic cDC1s remains unknown. While these data suggest some roles for IRE1α in controlling cDC1 biology, the phenotypes reported appear to be a consequence of XBP1 ablation, which induces an abnormal feedback mechanism that provokes severe IRE1α hyperactivation and subsequent RIDD (Lee et al., 2008). Thus, it remains unclear whether DCs can experience endogenous RIDD that modulates their function in the setting of XBP1 sufficiency (Subramanian and Tabas, 2014).

Because of its location in the ER, the UPR can regulate processes involved in antigen processing and presentation, one of the main features of DCs (Osorio et al., 2018). In general, cytosolic antigens are presented to T cells via the MHC-I machinery, while presentation of extracellular antigens preferentially relies on MHC-II molecules. DCs, especially cDC1s, can also present extracellular antigens that are phagocytosed or endocytosed via the cross-presentation pathway (Embgenbroich and Burgdorf, 2018). A study using conditional knockout mice lacking XBP1 in CD11c-expressing cells showed that RIDD, triggered in response to XBP1 ablation, limited the ability of $CD8\alpha^+$ cDC1s to cross-present dead cell-derived antigens to $CD8^+$ T cells (Osorio et al., 2014). These defects were associated to reduced expression of genes involved in antigen uptake and processing specifically in XBP1-deficient cDC1s undergoing RIDD, compared with their wild-type counterparts (Osorio et al., 2014). Interestingly, IRE1α-XBP1 signaling in epithelial cells has been shown to regulate microRNA-346 expression, which inhibits the peptide transporter 1 (TAP1) that is important for the transport of peptides into the ER and correct MHC-I-dependent antigen presentation (Bartoszewski et al., 2011). Although XBP1s was first discovered to regulate MHC-II expression in B cells (Liou et al., 1990), its precise role in governing antigen presentation by diverse DC subsets remains unclear. It is possible that the UPR plays different functions in determining the ability of DCs to present antigens onto MHC-I or MHC-II molecules depending on the type of stimulus that DCs receive, the type of antigen, or the specific DC subtype in question.

2.3 The UPR in DCs activated via pattern recognition receptors

Similar to other myeloid cells like macrophages (Martinon et al., 2010), the UPR is important for DC cytokine production in response to engagement of pattern recognition receptors (PRR) (Fig. 2). For example, loss of XBP1 in murine pDCs results in a poorly-developed ER and reduced type-I IFN production upon stimulation with a TLR9 agonist (Iwakoshi et al., 2007). The reduction in type-I IFN production was also observed after treatment with 4μ8C, an inhibitor of the RNase domain of IRE1α, in TLR7-stimulated human PBMCs (Beisel et al., 2017) and in *Chlamydia trachomatis*-infected human MoDCs (Webster et al., 2016). Although interesting, these data were obtained using high concentrations of 4μ8C (60 and 30 μM, respectively), which has recently been shown to also act as a potent ROS scavenger (Chan et al., 2018). In light of the potential off-target effect of 4μ8C, additional experiments using different IRE1α inhibitors or the genetic ablation of IRE1α could be useful to confirm these observations. Furthermore, *in vitro* treatment of human pDCs with the proteasome inhibitor bortezomib, which has been shown to induce ER stress-mediated apoptosis in multiple myeloma cells (Obeng et al., 2006), suppressed *Xbp1* splicing, induced cell apoptosis and reduced type-I IFN production after TLR9 and TLR7 stimulation (Hirai et al., 2011). In particular, bortezomib blocked IFNα production by inhibiting the trafficking of TLR9 from the ER to the endolysosomes and this halted the nuclear translocation of two important transcription factors for type-I IFN production, IRF-7 and NF-κB (Hirai et al., 2011). In addition, IL-23 expression, a cytokine that is important for the development of IL-17-producing $CD4^+$ T cells (Iwakura and Ishigame, 2006), relies on the UPR after PRR activation. In particular, initial studies using *in vitro*-differentiated human MoDCs have shown that in the presence of the hexokinase inhibitor 2-deoxy-D-glucose (2-DG), production of IL-23 by MoDCs in response to bacterial lipopolysaccharide (LPS) depends on both PERK and IRE1α, while response to the fungal β-glucan zymosan in the same setting mostly depends on IRE1α (Fig. 2) (Goodall et al., 2010; Marquez et al., 2017). Several predicted binding sites for CHOP and XBP1s are found in the promoter regions of the gene encoding IL-23, and their binding are enhanced by various stimuli. Accordingly, chromatin immunoprecipitation analyses followed by real-time PCR in *in vitro*-differentiated human MoDCs stimulated with both LPS and the ER stress-inducing agent tunicamycin showed enrichment of CHOP binding to the *IL23* promoter, while XBP1s bound the *IL23* promoter in human MoDC stimulated with both zymosan and 2-DG (Goodall et al., 2010, Marquez et al., 2017). Interestingly, in murine bone marrow-derived DCs (BMDCs) both CHOP and XBP1s were found to bind the *IL23* promoter upon treatment with a TLR7/8 ligand and/or the fatty acid palmitic acid, which induced the UPR in these cells (Goodall et al., 2010; Mogilenko et al., 2019). These findings indicate that engagement of different PRRs can trigger distinct arms of the UPR for rapid and potent cytokine induction in DCs. Besides IL-23, *in vitro*-differentiated and LPS-stimulated human MoDCs treated with B-I09, a potent and selective inhibitor of the RNase domain of IRE1α, decreased their production of IL-1β and TGFβ (Betts et al., 2018). In addition, it was shown that *in vitro* co-culture of B-I09-pretreated MoDCs with $CD4^+$ T cells under continuous B-I09 exposure hampered their differentiation into Th1 and Th17, while increasing Th2 responses (Fig. 2) (Betts et al., 2018). Of note, IRE1α-inhibited MoDCs were also found to exhibit reduced capacity to generate regulatory T cells *in vitro* (Betts et al., 2018), but the direct effect of B-I09 on T cell differentiation was not assessed in this study. These data suggest that IRE1α activation plays an important role in DCs by influencing their cytokine expression profile, which, in turn, can shape diverse T cell functions.

2.4 The UPR in DCs under pathological conditions

In light of their ability to produce pro-inflammatory cytokines and stimulate T cell activation, DCs have been implicated in certain diseases, such as graft-versus-host disease (GVHD), wherein donor-derived T cells in the graft recognize the host as foreign. Since IRE1α inhibition in DC-T cell co-cultures altered Th1 and Th17 responses that characterize GVHD, B-I09 was further tested in an established disease model based on human skin xenografts implanted into NOD-*scid* $IL2r\gamma^{null}$ (NSG) mice (Betts et al., 2018). After transplantation, mice were injected with human PBMCs allogenic to the skin in order to drive GVHD. Interestingly, treatment with B-I09 reduced both the skin graft rejection and the GVHD against murine tissues, while decreasing the levels of human Th1 and Th17 cells in the mouse spleen, suggesting that the IRE1α arm of the UPR is a potential therapeutic target against GVHD (Fig. 2) (Betts et al., 2018). Future studies are needed to define whether the effects of IRE1α pharmacological inhibition in these models of GVHD solely depend on DC targeting or whether other immune cells could implicated in the process.

Another recent study reported that DCs adapt their metabolism to a fatty acid-enriched environment after PRR activation, resulting in metabolic stress, UPR overactivation and ultimately, in an inflammatory response mediated by IL-23 and IL-6 (Mogilenko et al., 2019). Specifically, in a model of psoriasis-like skin inflammation induced by treatment with

TLR7/8 ligands in high fat diet-fed mice, decreased numbers of IL-23$^+$ and IL-6$^+$ DCs in the lymph node and reduced skin inflammation were found in mice selectively lacking XBP1 in CD11c$^+$ cells, compared with their wild-type counterparts (Fig. 2) (Mogilenko et al., 2019). These effects on IL-23$^+$ DCs were anticipated since previous studies had reported that the IRE1α-XBP1 axis was critical for optimal IL-23 production by these innate immune cells, as discussed above. Yet, these data suggest that targeting the UPR in IL-23-dependent conditions, such as psoriasis, inflammatory bowel disease or rheumatoid arthritis (Tang et al., 2012), might be useful for the treatment of these autoimmune disorders.

We recently uncovered that activation of the IRE1α-XBP1 pathway under ER stress conditions or following stimulation with agonists for plasma membrane-bound PRRs drives eicosanoid metabolism in various myeloid cell types (Chopra et al., 2019), including murine BMDCs and *in vitro*-differentiated human MoDCs (Fig. 2). Mechanistically, we determined that upon activation by IRE1α the transcription factor XBP1s directly induces the expression of genes encoding enzymes necessary for inducible prostaglandin biosynthesis, namely, prostaglandin-endoperoxide synthase 2 (*Ptgs2*/Cox-2) and prostaglandin E synthase (*Ptges*/mPGES-1). IRE1α-XBP1 signaling was therefore required for optimal production of prostaglandin E2 (PGE_2), which is a bioactive lipid mediator involved in multiple physiological processes, including the development of pain. Accordingly, mice devoid of IRE1α-XBP1 in leukocytes or treated with pharmacological IRE1α inhibitors targeting its kinase or RNase domain demonstrated reduced pain-related behaviors in PGE_2-dependent models of pain (Chopra et al., 2019).

Blocking DC cytokine production by inhibiting the UPR, however, might not always be the best strategy. An example is represented by anti-viral responses, when certain cytokines are necessary for the elimination of the pathogen. It was reported that a murine DC line stimulated with polyI:C (a synthetic analogue of double stranded RNA that activates TLR3), in the presence of the ER stress inducer tunicamycin upregulated the production of IFNβ, an anti-viral molecule belonging to the type-I IFN family, and this effect was mediated by XBP1 (Hu et al., 2011). However, whether physiological ER stress-inducing stimuli synergize with PRR activation was not assessed. Enforced overexpression of XBP1s in this DC line stimulated type-I IFN production and protected fibroblasts from infection with a vesicular stomatitis virus *in vitro* (Fig. 2), indicating that activation of the IRE1α-XBP1 arm of UPR in this setting is beneficial (Hu et al., 2011).

Additional studies are necessary to confirm the impact of DC-specific XBP1 signaling in the control of viral replication *in vivo*.

In a model of thermal injury where DCs are thought to be immunosuppressive (Shen et al., 2012), splenic $CD11c^+$ DCs displaying activated PERK and marked upregulation of BiP and XBP1s were defective in stimulating $CD4^+$ T cell proliferation. In addition, these DCs induced the Th2-related cytokine IL-4 while decreasing IFNγ in co-cultured $CD4^+$ T cells (Zhu et al., 2017). The modest silencing of XBP1s in isolated splenic DCs, did not rescue the phenotype. However, treatment of injured mice with salubrinal, an inhibitor of eIF2α (Boyce et al., 2005), decreased CHOP protein levels in splenic DCs while increasing their expression of CD80, CD86, and MHC-II molecules. Co-culture of salubrinal-treated DCs with $CD4^+$ T cells thus augmented the production of IFNγ by T cells (Fig. 2). Importantly, salubrinal treatment, but not silencing of *Xbp1*, resulted in increased survival rate of thermal injured mice (Zhu et al., 2017). These data suggest that in certain circumstances, the PERK-CHOP axis of the UPR might be a relevant target to enhance Th1 responses and improve host survival.

Beyond the role of DC-intrinsic IRE1α-XBP1 signaling in inflammatory conditions, it is now appreciated that the UPR is also critical for the function of DCs in malignancy. Earlier studies showed that murine BMDCs cultured in conditioned medium obtained from ER-stressed cancer cell lines, upregulated UPR-related genes such as *Xbp1s* and *Chop* as well as pro-inflammatory cytokines and immunosuppressive molecules such as arginase I (Mahadevan et al., 2012). This transmissible ER stress (TERS) from cancer cell lines to BMDCs also resulted in decreased ability to cross-present antigens to $CD8^+$ T cells (Mahadevan et al., 2012). In line with this, an accelerated tumor growth was observed when B16·F10 melanoma cells were subcutaneously injected together with TERS-imprinted BMDCs, suggesting that ER stress responses in DCs can interfere with T cell-driven immunity and promote tumor growth (Mahadevan et al., 2012). However, the precise ER stress sensor mediating these effects remained unidentified.

Our group subsequently determined that tumor-associated DCs (tDCs), defined as $CD11c^+MHC\text{-}II^+CD11b^+$ cells expressing high levels of both *Zbtb46* and *Clec9a*, in various models of ovarian cancer exhibited aberrant activation of the IRE1α-XBP1 arm of the UPR, which inhibited their capacity to present antigens to anti-tumor T cells (Fig. 2) (Cubillos-Ruiz et al., 2015). Mechanistically, ROS accumulation in tDCs generated lipid peroxidation byproducts, such as the reactive aldehyde 4-hydroxynonenal (4-HNE), that formed stable adducts with ER-resident chaperones, hence inducing protein misfolding and ER stress. The consequent activation of the IRE1α-XBP1 axis induced a transcriptional program promoting lipogenesis and abnormal lipid droplet accumulation that impaired the antigen-presenting capacity of the tDC. Polyethylenimine (PEI)-based nanoparticles encapsulating siRNA were previously shown to be specifically engulfed by phagocytic tDCs (Cubillos-Ruiz et al., 2009, 2012) and thus were used to knock down *Xbp1s* in murine tDCs via intraperitoneal injection. This resulted in increased T cell activation and a consequent decrease in the metastatic burden in the peritoneal cavity of ovarian cancer-bearing mice (Cubillos-Ruiz et al., 2015). The therapeutic silencing of either IRE1α or XBP1s translated in an increased survival benefit of mice inoculated with ID8-*Defb29*/*Vegf-A* cancer cell line. Experiments performed in immunodeficient mice demonstrated that this therapeutic activity relied on the adaptive immune system (Cubillos-Ruiz et al., 2015). Therefore, controlling this UPR pathway in DCs specifically in the tumor microenvironment may represent a new form of immunotherapy that could be used to enhance the efficacy of current T cell-centric approaches such as checkpoint blockade, which unfortunately elicit limited effects in ovarian cancer patients (Matulonis et al., 2019). Importantly, although our study identified tDCs as $CD11b^{+}$ DCs with attributes compatible with the cDC2 subtype, a subsequent report using the same ovarian cancer model suggested that these tDCs might correspond to inflammatory MoDCs, based on their capacity to express the marker FcεRI (Tesone et al., 2016).

In the setting of DNA-based vaccination in mice, Tian et al. showed that XBP1s overexpression in DCs could be important for eliciting anti-tumor responses (Tian et al., 2012). In this study, mice were vaccinated using a gene gun approach with a plasmid encoding both the tumor antigen and XBP1s. Mice inoculated subcutaneously with cancer cells bearing the same antigen present in the vaccine either before (prophylactic) or after (therapeutic) immunization showed superior tumor control and prolonged survival (Fig. 2). Importantly, these results were dependent on $CD8^{+}$ T cells as antibody-mediated elimination of these cells abrogated the anti-tumor effects of vaccination (Tian et al., 2012). Later work by the same group then showed that these effects of DNA vaccination in a therapeutic setting were abolished in Batf3 knockout mice, which lack cDC1s (Zhang et al., 2015). Not only cDC1s were important for the therapeutic benefit of this type of DNA vaccination, but also pDCs, as shown in intervention studies where tumor-bearing mice were treated with a pDC-depleting antibody against PDAC1 (Zhang et al., 2015). In addition, similar anti-tumor results were observed in a DC-based vaccination setting where murine BMDCs modified to overexpress both the tumor antigen and XBP1s were infused intraperitoneally into mice bearing orthotopic HER2-expressing 4T1.2 breast tumors or glioma GL26 subcutaneous tumors (Fig. 2) (Zhang et al., 2015). *In vitro* studies showed that BMDCs transfected with this construct and co-cultured with antigen-specific $CD8^+$ T cells, stimulated IFNγ production by T cells in an XBP1-dependent manner (Zhang et al., 2015). More recently, Medel et al. reported that *in vitro*-differentiated BMDCs incubated with melanoma cell lysates activated IRE1α and induced *Xbp1* splicing, but not RIDD (Medel et al., 2018) (Fig. 2). Reduced antigen cross-presenting capacity was observed when BMDCs were treated with the IRE1α RNAse inhibitor 4μ8C, suggesting that IRE1α activation plays a role in this process (Medel et al., 2018). Yet, BMDCs were exposed to high concentrations (~50 μM) of 4μ8C, which is a compound that can also function as a direct ROS scavenger (Chan et al., 2018). Since low levels of ROS are critical for the normal cross-presenting capacity of DCs by increasing the pH of phagosomes and preventing antigen degradation (Savina et al., 2006), using additional IRE1α inhibitory compounds, such as B-I09, MKC8866 or KIRA6/8, could help confirm the specificity of the defects observed. Nonetheless, IRE1α-deficient BMDCs stimulated *ex vivo* with melanoma cell lysates and the antigen ovalbumin demonstrated impaired capacity to induce the proliferation of adoptively transferred ovalbumin-specific $CD8^+$ T cells in naïve mice (Medel et al., 2018). The potential therapeutic effects of vaccination using *ex vivo*-pulsed BMDCs lacking IRE1α, or other ER stress sensors, in cancer-bearing hosts remains to be determined.

At first glance, these reports may appear contradictory to the detrimental effects that persistent ER stress responses and IRE1α-XBP1 overactivation evoke in ovarian cancer-associated DCs or in BMDCs in the TERS context. However, it is conceivable that skin DCs targeted via DNA-based vaccines, or BMDCs adoptively transferred into cancer-free mice, do not experience the same level of ER stress that DCs persistently face in the tumor microenvironment. Furthermore, the adverse conditions in the tumor milieu that trigger ER stress-driven immunometabolic changes in DCs may not be fully recapitulated in BMDCs *in vitro*. Hence, these findings underscore the importance of understanding the precise pathophysiological context wherein DCs experience ER stress and/or UPR activation, and highlight that interfering with this pathway might represent a therapeutic option for certain disorders, although additional research is needed.

3. Concluding remarks

While the UPR was discovered almost four decades ago, its role as a key regulator of immune cell functions has just begun to be appreciated and understood. Interestingly, emerging evidence indicates that not only accumulation of misfolded proteins activates the UPR in DCs. Indeed, other stimuli, such as specific TLR ligands, can induce selective activation of UPR-related factors in the absence of drastic ER stress. Recent discoveries have uncovered the remarkable therapeutic potential that targeting the UPR in immune DCs may have in the setting of various diseases. Several inhibitors of IRE1α have been developed that target either its RNAse domain or its kinase domain (Almanza et al., 2019; Jiang et al., 2015). Although these compounds have shown specificity and activity in cell lines and in some pre-clinical models of disease, further studies are necessary to fully understand the effects of these molecules in a cell type- and pathophysiological context-specific manner. Combination of these agents with other therapeutic approaches should also be explored. In particular, since DCs link innate and adaptive immunity, these drugs may enhance the efficacy of immunotherapeutic drugs targeting effector T cells, i.e., immune checkpoint inhibitors for the treatment of cancer. Furthermore, given the potential therapeutic use of these inhibitors, future studies will investigate the timing of treatment to achieve the best results. Moreover, the majority of the studies present in the literature focuses on the IRE1α-XBP1 arm of the UPR. However, a further understanding of the contribution of the PERK and ATF6 arms of the UPR in DC function is needed. Interestingly, expression of PERK-induced *Chop* in ovarian cancer-associated DCs isolated from patients with late stage disease negatively correlated with intratumoral T cell infiltration (Cubillos-Ruiz et al., 2015), suggesting that this arm of the UPR may also play key immunosuppressive roles in DCs residing in the tumor microenvironment. Likewise, future studies examining the potential interplay among the three arms of the UPR, both in homeostasis and in pathological conditions, would be of immense interest.

Finally, recent studies have also reported non-canonical roles for IRE1α in regulating mitochondrial functions and cell migration (Carreras-Sureda et al., 2019; Urra et al., 2018). Although these effects were discovered in non-immune cells, it is interesting to speculate that the ER stress sensors may also be involved in DC bioenergetics and chemotaxis or recruitment. Additional studies should address these exciting possibilities, and elucidate the implications that targeting UPR factors in these myeloid cells may have in diverse pathological settings.

Acknowledgments

Our research has been supported by the Cancer Research Institute, the Ovarian Cancer Research Fund Alliance, the Ovarian Cancer Academy Early-Career Investigator Award W81XWH-16-1-0438 of the Department of Defense, the Stand Up to Cancer Innovative Research Grant SU2C-AACR-IRG-03-16, and the Pershing Square Sohn Cancer Research Alliance. We thank Tito Sandoval Medina for assisting with illustrations.

Competing interest

J.R.C.-R. is co-founder of and scientific advisor for Quentis Therapeutics, Inc.

References

Almanza, A., Carlesso, A., Chintha, C., Creedican, S., Doultsinos, D., Leuzzi, B., Luis, A., McCarthy, N., Montibeller, L., More, S., Papaioannou, A., Puschel, F., Sassano, M.L., Skoko, J., Agostinis, P., de Belleroche, J., Eriksson, L.A., Fulda, S., Gorman, A.M., Healy, S., Kozlov, A., Munoz-Pinedo, C., Rehm, M., Chevet, E., Samali, A., 2019. Endoplasmic reticulum stress signalling—from basic mechanisms to clinical applications. FEBS J. 286, 241–278.

Bartoszewski, R., Brewer, J.W., Rab, A., Crossman, D.K., Bartoszewska, S., Kapoor, N., Fuller, C., Collawn, J.F., Bebok, Z., 2011. The unfolded protein response (UPR)-activated transcription factor X-box-binding protein 1 (XBP1) induces microRNA-346 expression that targets the human antigen peptide transporter 1 (TAP1) mRNA and governs immune regulatory genes. J. Biol. Chem. 286, 41862–41870.

B'Chir, W., Maurin, A.C., Carraro, V., Averous, J., Jousse, C., Muranishi, Y., Parry, L., Stepien, G., Fafournoux, P., Bruhat, A., 2013. The eIF2alpha/ATF4 pathway is essential for stress-induced autophagy gene expression. Nucleic Acids Res. 41, 7683–7699.

Bedoui, S., Whitney, P.G., Waithman, J., Eidsmo, L., Wakim, L., Caminschi, I., Allan, R.S., Wojtasiak, M., Shortman, K., Carbone, F.R., Brooks, A.G., Heath, W.R., 2009. Cross-presentation of viral and self antigens by skin-derived CD103+ dendritic cells. Nat. Immunol. 10, 488–495.

Beisel, C., Ziegler, S., Martrus Zapater, G., Chapel, A., Griesbeck, M., Hildebrandt, H., Lohse, A.W., Altfeld, M., 2017. TLR7-mediated activation of XBP1 correlates with the IFNalpha production in humans. Cytokine 94, 55–58.

Bernales, S., McDonald, K.L., Walter, P., 2006. Autophagy counterbalances endoplasmic reticulum expansion during the unfolded protein response. PLoS Biol. 4, e423.

Bertolotti, A., Zhang, Y., Hendershot, L.M., Harding, H.P., Ron, D., 2000. Dynamic interaction of BiP and ER stress transducers in the unfolded-protein response. Nat. Cell Biol. 2, 326–332.

Betts, B.C., Locke, F.L., Sagatys, E.M., Pidala, J., Walton, K., Menges, M., Reff, J., Saha, A., Djeu, J.Y., Kiluk, J.V., Lee, M.C., Kim, J., Kang, C.W., Tang, C.H., Frieling, J., Lynch, C.C., List, A., Rodriguez, P.C., Blazar, B.R., Conejo-Garcia, J.R., Del Valle, J.R., Hu, C.C., Anasetti, C., 2018. Inhibition of human dendritic cell ER stress response reduces T cell alloreactivity yet spares donor anti-tumor immunity. Front. Immunol. 9, 2887.

Borst, J., Ahrends, T., Babala, N., Melief, C.J.M., Kastenmuller, W., 2018. CD4(+) T cell help in cancer immunology and immunotherapy. Nat. Rev. Immunol. 18, 635–647.

Boyce, M., Bryant, K.F., Jousse, C., Long, K., Harding, H.P., Scheuner, D., Kaufman, R.J., Ma, D., Coen, D.M., Ron, D., Yuan, J., 2005. A selective inhibitor of eIF2alpha dephosphorylation protects cells from ER stress. Science 307, 935–939.
Broz, M.L., Binnewies, M., Boldajipour, B., Nelson, A.E., Pollack, J.L., Erle, D.J., Barczak, A., Rosenblum, M.D., Daud, A., Barber, D.L., Amigorena, S., Van't Veer, L.J., Sperling, A.I., Wolf, D.M., Krummel, M.F., 2014. Dissecting the tumor myeloid compartment reveals rare activating antigen-presenting cells critical for T cell immunity. Cancer Cell 26, 638–652.
Calabro, S., Liu, D., Gallman, A., Nascimento, M.S., Yu, Z., Zhang, T.T., Chen, P., Zhang, B., Xu, L., Gowthaman, U., Krishnaswamy, J.K., Haberman, A.M., Williams, A., Eisenbarth, S.C., 2016. Differential intrasplenic migration of dendritic cell subsets tailors adaptive immunity. Cell Rep. 16, 2472–2485.
Calfon, M., Zeng, H., Urano, F., Till, J.H., Hubbard, S.R., Harding, H.P., Clark, S.G., Ron, D., 2002. IRE1 couples endoplasmic reticulum load to secretory capacity by processing the XBP-1 mRNA. Nature 415, 92–96.
Carreras-Sureda, A., Jana, F., Urra, H., Durand, S., Mortenson, D.E., Sagredo, A., Bustos, G., Hazari, Y., Ramos-Fernandez, E., Sassano, M.L., Pihan, P., van Vliet, A.R., Gonzalez-Quiroz, M., Torres, A.K., Tapia-Rojas, C., Kerkhofs, M., Vicente, R., Kaufman, R.J., Inestrosa, N.C., Gonzalez-Billault, C., Wiseman, R.L., Agostinis, P., Bultynck, G., Court, F.A., Kroemer, G., Cardenas, J.C., Hetz, C., 2019. Non-canonical function of IRE1alpha determines mitochondria-associated endoplasmic reticulum composition to control calcium transfer and bioenergetics. Nat. Cell Biol. 21, 755–767.
Chan, S.M.H., Lowe, M.P., Bernard, A., Miller, A.A., Herbert, T.P., 2018. The inositol-requiring enzyme 1 (IRE1alpha) RNAse inhibitor, 4micro8C, is also a potent cellular antioxidant. Biochem. J. 475, 923–929.
Cheong, C., Matos, I., Choi, J.H., Dandamudi, D.B., Shrestha, E., Longhi, M.P., Jeffrey, K.L., Anthony, R.M., Kluger, C., Nchinda, G., Koh, H., Rodriguez, A., Idoyaga, J., Pack, M., Velinzon, K., Park, C.G., Steinman, R.M., 2010. Microbial stimulation fully differentiates monocytes to DC-SIGN/CD209(+) dendritic cells for immune T cell areas. Cell 143, 416–429.
Chopra, S., Giovanelli, P., Alvarado-Vazquez, P.A., Alonso, S., Song, M., Sandoval, T.A., Chae, C.S., Tan, C., Fonseca, M.M., Gutierrez, S., Jimenez, L., Subbaramaiah, K., Iwawaki, T., Kingsley, P.J., Marnett, L.J., Kossenkov, A.V., Crespo, M.S., Dannenberg, A.J., Glimcher, L.H., Romero-Sandoval, E.A., Cubillos-Ruiz, J.R., 2019. IRE1alpha-XBP1 signaling in leukocytes controls prostaglandin biosynthesis and pain. Science 365, eaau6499.
Cisse, B., Caton, M.L., Lehner, M., Maeda, T., Scheu, S., Locksley, R., Holmberg, D., Zweier, C., den Hollander, N.S., Kant, S.G., Holter, W., Rauch, A., Zhuang, Y., Reizis, B., 2008. Transcription factor E2-2 is an essential and specific regulator of plasmacytoid dendritic cell development. Cell 135, 37–48.
Connor, J.H., Weiser, D.C., Li, S., Hallenbeck, J.M., Shenolikar, S., 2001. Growth arrest and DNA damage-inducible protein GADD34 assembles a novel signaling complex containing protein phosphatase 1 and inhibitor 1. Mol. Cell. Biol. 21, 6841–6850.
Cox, J.S., Shamu, C.E., Walter, P., 1993. Transcriptional induction of genes encoding endoplasmic reticulum resident proteins requires a transmembrane protein kinase. Cell 73, 1197–1206.
Cubillos-Ruiz, J.R., Engle, X., Scarlett, U.K., Martinez, D., Barber, A., Elgueta, R., Wang, L., Nesbeth, Y., Durant, Y., Gewirtz, A.T., Sentman, C.L., Kedl, R., Conejo-Garcia, J.R., 2009. Polyethylenimine-based siRNA nanocomplexes reprogram tumor-associated dendritic cells via TLR5 to elicit therapeutic antitumor immunity. J. Clin. Invest. 119, 2231–2244.
Cubillos-Ruiz, J.R., Baird, J.R., Tesone, A.J., Rutkowski, M.R., Scarlett, U.K., Camposeco-Jacobs, A.L., Anadon-Arnillas, J., Harwood, N.M., Korc, M.,

Fiering, S.N., Sempere, L.F., Conejo-Garcia, J.R., 2012. Reprogramming tumor-associated dendritic cells in vivo using miRNA mimetics triggers protective immunity against ovarian cancer. Cancer Res. 72, 1683–1693.
Cubillos-Ruiz, J.R., Silberman, P.C., Rutkowski, M.R., Chopra, S., Perales-Puchalt, A., Song, M., Zhang, S., Bettigole, S.E., Gupta, D., Holcomb, K., Ellenson, L.H., Caputo, T., Lee, A.H., Conejo-Garcia, J.R., Glimcher, L.H., 2015. ER stress sensor XBP1 controls anti-tumor immunity by disrupting dendritic cell homeostasis. Cell 161, 1527–1538.
Eisenbarth, S.C., 2019. Dendritic cell subsets in T cell programming: location dictates function. Nat. Rev. Immunol. 19, 89–103.
Embgenbroich, M., Burgdorf, S., 2018. Current concepts of antigen cross-presentation. Front. Immunol. 9, 1643.
Gao, Y., Nish, S.A., Jiang, R., Hou, L., Licona-Limon, P., Weinstein, J.S., Zhao, H., Medzhitov, R., 2013. Control of T helper 2 responses by transcription factor IRF4-dependent dendritic cells. Immunity 39, 722–732.
Goodall, J.C., Wu, C., Zhang, Y., McNeill, L., Ellis, L., Saudek, V., Gaston, J.S., 2010. Endoplasmic reticulum stress-induced transcription factor, CHOP, is crucial for dendritic cell IL-23 expression. Proc. Natl. Acad. Sci. U. S. A. 107, 17698–17703.
Haberzettl, P., Hill, B.G., 2013. Oxidized lipids activate autophagy in a JNK-dependent manner by stimulating the endoplasmic reticulum stress response. Redox Biol. 1, 56–64.
Hacker, C., Kirsch, R.D., Ju, X.S., Hieronymus, T., Gust, T.C., Kuhl, C., Jorgas, T., Kurz, S.M., Rose-John, S., Yokota, Y., Zenke, M., 2003. Transcriptional profiling identifies Id2 function in dendritic cell development. Nat. Immunol. 4, 380–386.
Harding, H.P., Zhang, Y., Ron, D., 1999. Protein translation and folding are coupled by an endoplasmic-reticulum-resident kinase. Nature 397, 271–274.
Harding, H.P., Zhang, Y., Bertolotti, A., Zeng, H., Ron, D., 2000. Perk is essential for translational regulation and cell survival during the unfolded protein response. Mol. Cell 5, 897–904.
Harding, H.P., Zhang, Y., Zeng, H., Novoa, I., Lu, P.D., Calfon, M., Sadri, N., Yun, C., Popko, B., Paules, R., Stojdl, D.F., Bell, J.C., Hettmann, T., Leiden, J.M., Ron, D., 2003. An integrated stress response regulates amino acid metabolism and resistance to oxidative stress. Mol. Cell 11, 619–633.
Haze, K., Yoshida, H., Yanagi, H., Yura, T., Mori, K., 1999. Mammalian transcription factor ATF6 is synthesized as a transmembrane protein and activated by proteolysis in response to endoplasmic reticulum stress. Mol. Biol. Cell 10, 3787–3799.
Hildner, K., Edelson, B.T., Purtha, W.E., Diamond, M., Matsushita, H., Kohyama, M., Calderon, B., Schraml, B.U., Unanue, E.R., Diamond, M.S., Schreiber, R.D., Murphy, T.L., Murphy, K.M., 2008. Batf3 deficiency reveals a critical role for CD8alpha+ dendritic cells in cytotoxic T cell immunity. Science 322, 1097–1100.
Hirai, M., Kadowaki, N., Kitawaki, T., Fujita, H., Takaori-Kondo, A., Fukui, R., Miyake, K., Maeda, T., Kamihira, S., Miyachi, Y., Uchiyama, T., 2011. Bortezomib suppresses function and survival of plasmacytoid dendritic cells by targeting intracellular trafficking of Toll-like receptors and endoplasmic reticulum homeostasis. Blood 117, 500–509.
Hollien, J., Weissman, J.S., 2006. Decay of endoplasmic reticulum-localized mRNAs during the unfolded protein response. Science 313, 104–107.
Hu, F., Yu, X., Wang, H., Zuo, D., Guo, C., Yi, H., Tirosh, B., Subjeck, J.R., Qiu, X., Wang, X.Y., 2011. ER stress and its regulator X-box-binding protein-1 enhance polyIC-induced innate immune response in dendritic cells. Eur. J. Immunol. 41, 1086–1097.
Hwang, J., Qi, L., 2018. Quality control in the endoplasmic reticulum: crosstalk between ERAD and UPR pathways. Trends Biochem. Sci. 43, 593–605.

Iwakoshi, N.N., Pypaert, M., Glimcher, L.H., 2007. The transcription factor XBP-1 is essential for the development and survival of dendritic cells. J. Exp. Med. 204, 2267–2275.
Iwakura, Y., Ishigame, H., 2006. The IL-23/IL-17 axis in inflammation. J. Clin. Invest. 116, 1218–1222.
Iwawaki, T., Akai, R., Kohno, K., Miura, M., 2004. A transgenic mouse model for monitoring endoplasmic reticulum stress. Nat. Med. 10, 98–102.
Jiang, D., Niwa, M., Koong, A.C., 2015. Targeting the IRE1alpha-XBP1 branch of the unfolded protein response in human diseases. Semin. Cancer Biol. 33, 48–56.
Kaneko, M., Niinuma, Y., Nomura, Y., 2003. Activation signal of nuclear factor-kappa B in response to endoplasmic reticulum stress is transduced via IRE1 and tumor necrosis factor receptor-associated factor 2. Biol. Pharm. Bull. 26, 931–935.
Keestra-Gounder, A.M., Byndloss, M.X., Seyffert, N., Young, B.M., Chavez-Arroyo, A., Tsai, A.Y., Cevallos, S.A., Winter, M.G., Pham, O.H., Tiffany, C.R., de Jong, M.F., Kerrinnes, T., Ravindran, R., Luciw, P.A., McSorley, S.J., Baumler, A.J., Tsolis, R.M., 2016. NOD1 and NOD2 signalling links ER stress with inflammation. Nature 532, 394–397.
Kim, T.S., Braciale, T.J., 2009. Respiratory dendritic cell subsets differ in their capacity to support the induction of virus-specific cytotoxic CD8+ T cell responses. PLoS One 4, e4204.
Kozutsumi, Y., Segal, M., Normington, K., Gething, M.J., Sambrook, J., 1988. The presence of malfolded proteins in the endoplasmic reticulum signals the induction of glucose-regulated proteins. Nature 332, 462–464.
Krishnaswamy, J.K., Gowthaman, U., Zhang, B., Mattsson, J., Szeponik, L., Liu, D., Wu, R., White, T., Calabro, S., Xu, L., Collet, M.A., Yurieva, M., Alsen, S., Fogelstrand, P., Walter, A., Heath, W.R., Mueller, S.N., Yrlid, U., Williams, A., Eisenbarth, S.C., 2017. Migratory CD11b(+) conventional dendritic cells induce T follicular helper cell-dependent antibody responses. Sci. Immunol. 2, eaam9169.
Lee, K., Tirasophon, W., Shen, X., Michalak, M., Prywes, R., Okada, T., Yoshida, H., Mori, K., Kaufman, R.J., 2002. IRE1-mediated unconventional mRNA splicing and S2P-mediated ATF6 cleavage merge to regulate XBP1 in signaling the unfolded protein response. Genes Dev. 16, 452–466.
Lee, A.H., Iwakoshi, N.N., Glimcher, L.H., 2003. XBP-1 regulates a subset of endoplasmic reticulum resident chaperone genes in the unfolded protein response. Mol. Cell. Biol. 23, 7448–7459.
Lee, A.H., Scapa, E.F., Cohen, D.E., Glimcher, L.H., 2008. Regulation of hepatic lipogenesis by the transcription factor XBP1. Science 320, 1492–1496.
Lewis, K.L., Caton, M.L., Bogunovic, M., Greter, M., Grajkowska, L.T., Ng, D., Klinakis, A., Charo, I.F., Jung, S., Gommerman, J.L., Ivanov, I.I., Liu, K., Merad, M., Reizis, B., 2011. Notch2 receptor signaling controls functional differentiation of dendritic cells in the spleen and intestine. Immunity 35, 780–791.
Lin, J.H., Li, H., Yasumura, D., Cohen, H.R., Zhang, C., Panning, B., Shokat, K.M., Lavail, M.M., Walter, P., 2007. IRE1 signaling affects cell fate during the unfolded protein response. Science 318, 944–949.
Liou, H.C., Boothby, M.R., Finn, P.W., Davidon, R., Nabavi, N., Zeleznik-Le, N.J., Ting, J.P., Glimcher, L.H., 1990. A new member of the leucine zipper class of proteins that binds to the HLA DR alpha promoter. Science 247, 1581–1584.
Liu, K., Victora, G.D., Schwickert, T.A., Guermonprez, P., Meredith, M.M., Yao, K., Chu, F.F., Randolph, G.J., Rudensky, A.Y., Nussenzweig, M., 2009. In vivo analysis of dendritic cell development and homeostasis. Science 324, 392–397.
Mahadevan, N.R., Anufreichik, V., Rodvold, J.J., Chiu, K.T., Sepulveda, H., Zanetti, M., 2012. Cell-extrinsic effects of tumor ER stress imprint myeloid dendritic cells and impair CD8(+) T cell priming. PLoS One 7, e51845.

Marquez, S., Fernandez, J.J., Teran-Cabanillas, E., Herrero, C., Alonso, S., Azogil, A., Montero, O., Iwawaki, T., Cubillos-Ruiz, J.R., Fernandez, N., Crespo, M.S., 2017. Endoplasmic reticulum stress sensor IRE1alpha enhances IL-23 expression by human dendritic cells. Front. Immunol. 8, 639.

Martinon, F., Chen, X., Lee, A.H., Glimcher, L.H., 2010. TLR activation of the transcription factor XBP1 regulates innate immune responses in macrophages. Nat. Immunol. 11, 411–418.

Matulonis, U.A., Shapira-Frommer, R., Santin, A.D., Lisyanskaya, A.S., Pignata, S., Vergote, I., Raspagliesi, F., Sonke, G.S., Birrer, M., Provencher, D.M., Sehouli, J., Colombo, N., Gonzalez-Martin, A., Oaknin, A., Ottevanger, P.B., Rudaitis, V., Katchar, K., Wu, H., Keefe, S., Ruman, J., Ledermann, J.A., 2019. Antitumor activity and safety of pembrolizumab in patients with advanced recurrent ovarian cancer: results from the phase 2 KEYNOTE-100 study. Ann. Oncol. 30, 1080–1087.

Medel, B., Costoya, C., Fernandez, D., Pereda, C., Lladser, A., Sauma, D., Pacheco, R., Iwawaki, T., Salazar-Onfray, F., Osorio, F., 2018. IRE1alpha activation in bone marrow-derived dendritic cells modulates innate recognition of melanoma cells and favors CD8(+) T cell priming. Front. Immunol. 9, 3050.

Merad, M., Sathe, P., Helft, J., Miller, J., Mortha, A., 2013. The dendritic cell lineage: ontogeny and function of dendritic cells and their subsets in the steady state and the inflamed setting. Annu. Rev. Immunol. 31, 563–604.

Mogilenko, D.A., Haas, J.T., L'Homme, L., Fleury, S., Quemener, S., Levavasseur, M., Becquart, C., Wartelle, J., Bogomolova, A., Pineau, L., Molendi-Coste, O., Lancel, S., Dehondt, H., Gheeraert, C., Melchior, A., Dewas, C., Nikitin, A., Pic, S., Rabhi, N., Annicotte, J.S., Oyadomari, S., Velasco-Hernandez, T., Cammenga, J., Foretz, M., Viollet, B., Vukovic, M., Villacreces, A., Kranc, K., Carmeliet, P., Marot, G., Boulter, A., Tavernier, S., Berod, L., Longhi, M.P., Paget, C., Janssens, S., Staumont-Salle, D., Aksoy, E., Staels, B., Dombrowicz, D., 2019. Metabolic and innate immune cues merge into a specific inflammatory response via the UPR. Cell 177, 1201–1216.e19.

Mori, K., 2015. The unfolded protein response: the dawn of a new field. Proc. Jpn. Acad. Ser. B Phys. Biol. Sci. 91, 469–480.

Mori, K., Ma, W., Gething, M.J., Sambrook, J., 1993. A transmembrane protein with a cdc2+/CDC28-related kinase activity is required for signaling from the ER to the nucleus. Cell 74, 743–756.

Murphy, T.L., Grajales-Reyes, G.E., Wu, X., Tussiwand, R., Briseno, C.G., Iwata, A., Kretzer, N.M., Durai, V., Murphy, K.M., 2016. Transcriptional control of dendritic cell development. Annu. Rev. Immunol. 34, 93–119.

Naik, S.H., Metcalf, D., van Nieuwenhuijze, A., Wicks, I., Wu, L., O'Keeffe, M., Shortman, K., 2006. Intrasplenic steady-state dendritic cell precursors that are distinct from monocytes. Nat. Immunol. 7, 663–671.

Obeng, E.A., Carlson, L.M., Gutman, D.M., Harrington Jr., W.J., Lee, K.P., Boise, L.H., 2006. Proteasome inhibitors induce a terminal unfolded protein response in multiple myeloma cells. Blood 107, 4907–4916.

Oikawa, D., Tokuda, M., Hosoda, A., Iwawaki, T., 2010. Identification of a consensus element recognized and cleaved by IRE1 alpha. Nucleic Acids Res. 38, 6265–6273.

Osorio, F., Tavernier, S.J., Hoffmann, E., Saeys, Y., Martens, L., Vetters, J., Delrue, I., De Rycke, R., Parthoens, E., Pouliot, P., Iwawaki, T., Janssens, S., Lambrecht, B.N., 2014. The unfolded-protein-response sensor IRE-1alpha regulates the function of CD8alpha+ dendritic cells. Nat. Immunol. 15, 248–257.

Osorio, F., Lambrecht, B.N., Janssens, S., 2018. Antigen presentation unfolded: identifying convergence points between the UPR and antigen presentation pathways. Curr. Opin. Immunol. 52, 100–107.

Oyadomari, S., Mori, M., 2004. Roles of CHOP/GADD153 in endoplasmic reticulum stress. Cell Death Differ. 11, 381–389.
Pooley, J.L., Heath, W.R., Shortman, K., 2001. Cutting edge: intravenous soluble antigen is presented to CD4 T cells by CD8- dendritic cells, but cross-presented to CD8 T cells by CD8+ dendritic cells. J. Immunol. 166, 5327–5330.
Princiotta, M.F., Finzi, D., Qian, S.B., Gibbs, J., Schuchmann, S., Buttgereit, F., Bennink, J.R., Yewdell, J.W., 2003. Quantitating protein synthesis, degradation, and endogenous antigen processing. Immunity 18, 343–354.
Reizis, B., 2019. Plasmacytoid dendritic cells: development, regulation, and function. Immunity 50, 37–50.
Roberts, E.W., Broz, M.L., Binnewies, M., Headley, M.B., Nelson, A.E., Wolf, D.M., Kaisho, T., Bogunovic, D., Bhardwaj, N., Krummel, M.F., 2016. Critical role for CD103(+)/CD141(+) dendritic cells bearing CCR7 for tumor antigen trafficking and priming of T cell immunity in melanoma. Cancer Cell 30, 324–336.
Ron, D., Walter, P., 2007. Signal integration in the endoplasmic reticulum unfolded protein response. Nat. Rev. Mol. Cell Biol. 8, 519–529.
Salmon, H., Idoyaga, J., Rahman, A., Leboeuf, M., Remark, R., Jordan, S., Casanova-Acebes, M., Khudoynazarova, M., Agudo, J., Tung, N., Chakarov, S., Rivera, C., Hogstad, B., Bosenberg, M., Hashimoto, D., Gnjatic, S., Bhardwaj, N., Palucka, A.K., Brown, B.D., Brody, J., Ginhoux, F., Merad, M., 2016. Expansion and activation of CD103(+) dendritic cell progenitors at the tumor site enhances tumor responses to therapeutic PD-L1 and BRAF inhibition. Immunity 44, 924–938.
Satpathy, A.T., Kc, W., Albring, J.C., Edelson, B.T., Kretzer, N.M., Bhattacharya, D., Murphy, T.L., Murphy, K.M., 2012. Zbtb46 expression distinguishes classical dendritic cells and their committed progenitors from other immune lineages. J. Exp. Med. 209, 1135–1152.
Savina, A., Jancic, C., Hugues, S., Guermonprez, P., Vargas, P., Moura, I.C., Lennon-Dumenil, A.M., Seabra, M.C., Raposo, G., Amigorena, S., 2006. NOX2 controls phagosomal pH to regulate antigen processing during crosspresentation by dendritic cells. Cell 126, 205–218.
Schraml, B.U., van Blijswijk, J., Zelenay, S., Whitney, P.G., Filby, A., Acton, S.E., Rogers, N.C., Moncaut, N., Carvajal, J.J., Reis e Sousa, C., 2013. Genetic tracing via DNGR-1 expression history defines dendritic cells as a hematopoietic lineage. Cell 154, 843–858.
Scott, C.L., Soen, B., Martens, L., Skrypek, N., Saelens, W., Taminau, J., Blancke, G., Van Isterdael, G., Huylebroeck, D., Haigh, J., Saeys, Y., Guilliams, M., Lambrecht, B.N., Berx, G., 2016. The transcription factor Zeb2 regulates development of conventional and plasmacytoid DCs by repressing Id2. J. Exp. Med. 213, 897–911.
Segura, E., Amigorena, S., 2013. Inflammatory dendritic cells in mice and humans. Trends Immunol. 34, 440–445.
Shen, J., Chen, X., Hendershot, L., Prywes, R., 2002. ER stress regulation of ATF6 localization by dissociation of BiP/GRP78 binding and unmasking of Golgi localization signals. Dev. Cell 3, 99–111.
Shen, H., de Almeida, P.E., Kang, K.H., Yao, P., Chan, C.W., 2012. Burn injury triggered dysfunction in dendritic cell response to TLR9 activation and resulted in skewed T cell functions. PLoS One 7, e50238.
Steinman, R.M., Cohn, Z.A., 1973. Identification of a novel cell type in peripheral lymphoid organs of mice. I. Morphology, quantitation, tissue distribution. J. Exp. Med. 137, 1142–1162.
Steinman, R.M., Cohn, Z.A., 1974. Identification of a novel cell type in peripheral lymphoid organs of mice. II. Functional properties in vitro. J. Exp. Med. 139, 380–397.

Subramanian, M., Tabas, I., 2014. A new RIDDle in DC-mediated cross-presentation. Nat. Immunol. 15, 213–215.

Suzuki, S., Honma, K., Matsuyama, T., Suzuki, K., Toriyama, K., Akitoyo, I., Yamamoto, K., Suematsu, T., Nakamura, M., Yui, K., Kumatori, A., 2004. Critical roles of interferon regulatory factor 4 in CD11bhighCD8alpha- dendritic cell development. Proc. Natl. Acad. Sci. U. S. A. 101, 8981–8986.

Tang, C., Chen, S., Qian, H., Huang, W., 2012. Interleukin-23: as a drug target for autoimmune inflammatory diseases. Immunology 135, 112–124.

Tavernier, S.J., Osorio, F., Vandersarren, L., Vetters, J., Vanlangenakker, N., Van Isterdael, G., Vergote, K., De Rycke, R., Parthoens, E., van de Laar, L., Iwawaki, T., Del Valle, J.R., Hu, C.C., Lambrecht, B.N., Janssens, S., 2017. Regulated IRE1-dependent mRNA decay sets the threshold for dendritic cell survival. Nat. Cell Biol. 19, 698–710.

Tesone, A.J., Rutkowski, M.R., Brencicova, E., Svoronos, N., Perales-Puchalt, A., Stephen, T.L., Allegrezza, M.J., Payne, K.K., Nguyen, J.M., Wickramasinghe, J., Tchou, J., Borowsky, M.E., Rabinovich, G.A., Kossenkov, A.V., Conejo-Garcia, J.R., 2016. Satb1 overexpression drives tumor-promoting activities in cancer-associated dendritic cells. Cell Rep. 14, 1774–1786.

Tian, S., Liu, Z., Donahue, C., Falo Jr., L.D., You, Z., 2012. Genetic targeting of the active transcription factor XBP1s to dendritic cells potentiates vaccine-induced prophylactic and therapeutic antitumor immunity. Mol. Ther. 20, 432–442.

Tsujimura, H., Tamura, T., Ozato, K., 2003. Cutting edge: IFN consensus sequence binding protein/IFN regulatory factor 8 drives the development of type I IFN-producing plasmacytoid dendritic cells. J. Immunol. 170, 1131–1135.

Urano, F., Wang, X., Bertolotti, A., Zhang, Y., Chung, P., Harding, H.P., Ron, D., 2000. Coupling of stress in the ER to activation of JNK protein kinases by transmembrane protein kinase IRE1. Science 287, 664–666.

Urra, H., Henriquez, D.R., Canovas, J., Villarroel-Campos, D., Carreras-Sureda, A., Pulgar, E., Molina, E., Hazari, Y.M., Limia, C.M., Alvarez-Rojas, S., Figueroa, R., Vidal, R.L., Rodriguez, D.A., Rivera, C.A., Court, F.A., Couve, A., Qi, L., Chevet, E., Akai, R., Iwawaki, T., Concha, M.L., Glavic, A., Gonzalez-Billault, C., Hetz, C., 2018. IRE1alpha governs cytoskeleton remodelling and cell migration through a direct interaction with filamin A. Nat. Cell Biol. 20, 942–953.

Webster, S.J., Ellis, L., O'Brien, L.M., Tyrrell, B., Fitzmaurice, T.J., Elder, M.J., Clare, S., Chee, R., Gaston, J.S., Goodall, J.C., 2016. IRE1alpha mediates PKR activation in response to *Chlamydia trachomatis* infection. Microbes Infect. 18, 472–483.

Woo, C.W., Cui, D., Arellano, J., Dorweiler, B., Harding, H., Fitzgerald, K.A., Ron, D., Tabas, I., 2009. Adaptive suppression of the ATF4-CHOP branch of the unfolded protein response by toll-like receptor signalling. Nat. Cell Biol. 11, 1473–1480.

Yamamoto, K., Sato, T., Matsui, T., Sato, M., Okada, T., Yoshida, H., Harada, A., Mori, K., 2007. Transcriptional induction of mammalian ER quality control proteins is mediated by single or combined action of ATF6alpha and XBP1. Dev. Cell 13, 365–376.

Yorimitsu, T., Nair, U., Yang, Z., Klionsky, D.J., 2006. Endoplasmic reticulum stress triggers autophagy. J. Biol. Chem. 281, 30299–30304.

Zhang, Y., Chen, G., Liu, Z., Tian, S., Zhang, J., Carey, C.D., Murphy, K.M., Storkus, W.J., Falo Jr., L.D., You, Z., 2015. Genetic vaccines to potentiate the effective CD103+ dendritic cell-mediated cross-priming of antitumor immunity. J. Immunol. 194, 5937–5947.

Zhu, X.M., Dong, N., Wang, Y.B., Zhang, Q.H., Yu, Y., Yao, Y.M., Liang, H.P., 2017. The involvement of endoplasmic reticulum stress response in immune dysfunction of dendritic cells after severe thermal injury in mice. Oncotarget 8, 9035–9052.

Zigmond, E., Varol, C., Farache, J., Elmaliah, E., Satpathy, A.T., Friedlander, G., Mack, M., Shpigel, N., Boneca, I.G., Murphy, K.M., Shakhar, G., Halpern, Z., Jung, S., 2012. Ly6C hi monocytes in the inflamed colon give rise to proinflammatory effector cells and migratory antigen-presenting cells. Immunity 37, 1076–1090.

CHAPTER FOUR

The versatile plasmacytoid dendritic cell: Function, heterogeneity, and plasticity

Rebecca Leylek, Juliana Idoyaga*
Department of Microbiology and Immunology, and Immunology Program, Stanford University School of Medicine, Stanford, CA, United States
*Corresponding author: e-mail address: jidoyaga@stanford.edu

Contents

1. Introduction 178
2. Functional diversity of plasmacytoid dendritic cells 178
   2.1 pDC capacity for interferon production and its role during innate immune responses 178
   2.2 pDC capacity to present antigen and modulate adaptive T cell responses 181
   2.3 pDC capacity to modulate immune tolerance 183
3. Sources of pDC functional diversity: Developmentally encoded subsets versus environmentally mediated plasticity 185
   3.1 Evidence for pDC subpopulations 185
   3.2 Evidence for pDC plasticity 193
4. Concluding remarks 197
Acknowledgments 198
References 199

Abstract

Since their identification as the natural interferon-producing cell two decades ago, plasmacytoid dendritic cells (pDCs) have been attributed diverse functions in the immune response. Their most well characterized function is innate, i.e., their rapid and robust production of type-I interferon (IFN-I) in response to viruses. However, pDCs have also been implicated in antigen presentation, activation of adaptive immune responses and immunoregulation. The mechanisms by which pDCs enact these diverse functions are poorly understood. One central debate is whether these functions are carried out by different pDC subpopulations or by plasticity in the pDC compartment. This chapter summarizes the latest reports regarding pDC function, heterogeneity, cell conversion and environmentally influenced plasticity, as well as the role of pDCs in infection, autoimmunity and cancer.

International Review of Cell and Molecular Biology, Volume 349
ISSN 1937-6448
https://doi.org/10.1016/bs.ircmb.2019.10.002

1. Introduction

Plasmacytoid dendritic cells (pDCs) were first described in the 1950s as cells localized in human lymphoid organs harboring morphological features of plasma cells (Lennert and Remmele, 1958). These cells were later found to have high capacity to produce type-I interferons (α and β, abbreviated IFN-I) in response to viruses, followed by a capacity to differentiate into *bona fide* DCs (Cella et al., 1999, 2000; Grouard et al., 1997; Rissoan et al., 1999; Siegal et al., 1999). Soon after, the mouse equivalent of this human population was identified, indicating that pDCs are a conserved immune cell type (Asselin-Paturel et al., 2001; Björck, 2001; Nakano et al., 2001). In the last two decades, we have learned much about the complex biology of pDCs, but many aspects of their regulation remain enigmatic. In particular, it has become clear that pDCs are capable of wide-ranging functions, i.e., innate and adaptive, activating and regulatory, protective and pathogenic. An overarching question is whether pDC functions are mediated by distinct subpopulations of pDCs or if a homogeneous pDC population undergoes functional diversification after stimulation.

Here, we review recent advances in our understanding of pDC functional regulation and the latest evidence regarding the existence of functionally specialized pDC subpopulations and environmentally modulated plasticity. Understanding how pDCs regulate their distinct functions is an exciting area of research. As technical advances make it possible to address these questions, they will inform the rational design of future pDC-based therapeutic approaches.

2. Functional diversity of plasmacytoid dendritic cells

2.1 pDC capacity for interferon production and its role during innate immune responses

Although pDCs are involved in the immune response against various bacterial, fungal, and parasitic infections, they are most well studied for their role during viral infection (Raieli et al., 2019; reviewed in Alculumbre et al., 2018a). Plasmacytoid DCs primarily sense viruses via two endosomal toll-like receptors (TLRs), TLR7 and TLR9 (Edwards et al., 2003; Gilliet et al., 2008; Jarrossay et al., 2001). Upon stimulation, pDCs rapidly produce large amounts of IFN-I, but they do not appear to maintain IFN-I mRNA transcripts or protein stores (Ito et al., 2006). Rather, pDC capacity to promptly secrete IFN-I is mediated by constitutively high levels of the transcription factor interferon regulatory factor 7 (IRF7), which promotes *de novo* IFN-I transcription (Izaguirre et al., 2003; Kerkmann et al., 2003; Takauji et al., 2002). Triggering of TLR7 or TLR9 by viral nucleic acids activates the adaptor myeloid differentiation primary response gene 88 (MyD88) and nuclear translocation of IRF7, leading to the rapid and massive production of IFN-I by mouse and human pDCs (Asselin-Paturel et al., 2001; Dalod et al., 2002; Siegal et al., 1999; Uematsu and Akira, 2007). Similar to other DC subsets, TLR sensing in pDCs also triggers activation of the nuclear factor kappa B (NF-κB) pathway for production of other pro-inflammatory cytokines, such as tumor necrosis factor alpha (TNFα), and the expression of costimulatory molecules (Gilliet et al., 2008). In addition to the endosomal pathogen sensing pathway, it has been proposed that pDCs can also produce IFN-I following sensing of DNA and RNA viruses through cyclic GMP-AMP synthase and stimulator of interferon genes (cGAS-STING) and retinoic acid-inducible gene I-like receptor (RLR) cytoplasmic pathways (Bode et al., 2016; Bruni et al., 2015; Ishikawa et al., 2009; Li et al., 2013). IFN-I, in turn, plays a major role during viral infection by promoting the expression of interferon-stimulated genes (ISGs) that restrict viral entry and replication (McNab et al., 2015).

It is clear that pDCs produce large amounts of IFN-I in response to viruses *in vivo*; however, the overall importance of pDC-derived IFN-I for the antiviral immune response is controversial. Patients with Pitt-Hopkins syndrome, which have functionally impaired pDCs as a consequence of TCF4/E2-2 haploinsufficiency, do not appear to have increased susceptibility to viral infections (Amiel et al., 2007; Peippo and Ignatius, 2012; Zweier et al., 2007). In mice, pDCs are essential for rapid IFN-I production upon viral infection (Dalod et al., 2002; Krug et al., 2004; Swiecki et al., 2010). However, they do not appear to be absolutely necessary for the control of certain infections such as influenza virus, murine cytomegalovirus virus (MCMV), vesicular stomatitis virus (VSV), acute lymphocytic choriomeningitis virus (LCMV), or local herpes simplex virus (HSV) (Cervantes-Barragan et al., 2012; Swiecki et al., 2010, 2013; Wolf et al., 2009). Nevertheless, there are examples in which pDCs seem to have a critical, non-redundant role. A lack of pDCs during infection with respiratory syncytial virus (RSV) or mouse hepatitis virus (MHV) results in delayed viral clearance and more severe pathology (Cervantes-Barragan et al., 2007, 2012; Smit et al., 2006). Furthermore, pDC depletion reduces the survival during acute systemic HSV infection (Swiecki et al., 2013). Lastly, pDC-restricted IRF7 signaling is sufficient to control chikungunya and dengue virus infections (Webster et al., 2018).

The conflicting *in vivo* results presented above may be a consequence of multiple factors. First, it is possible that there are compensatory mechanisms that function in the absence of pDCs. For example, IFN-I production from other innate cells may be sufficient to clear some pathogens. In line with this, lack of the interferon alpha receptor (IFNAR) itself results in greater lethality than pDC depletion during infection with MHV or VSV (Cervantes-Barragan et al., 2007; Müller et al., 1994). Additionally, other innate mechanisms such as killing of infected cells by natural killer cells may compensate for the lack of pDCs (Swiecki et al., 2010). Second, pDCs may only be required to develop optimal immune responses against certain viruses and routes of infections. Third, there are major discrepancies in the pDC depletion mouse models used. Early methods depleted pDCs using the anti-BST2/CD317/PDCA1 antibody 120G8; however, this is not an ideal strategy because BST2 is upregulated in other cell types upon exposure to IFN-I or IFN-II (IFN-γ) (Blasius et al., 2006b). Similarly, diphtheria toxin inoculation of mice expressing the diphtheria toxin receptor under the SiglecH promoter (SiglecH-DTR mice) results in depletion of pDCs and some populations of splenic macrophages (Swiecki et al., 2014). Newer genetic depletion models such as the transgenic CLEC4C-DTR mouse and the CD11c-Cre $Tcf4^{fl/-}$ conditional knockouts are available and have been used to demonstrate a non-redundant role for pDCs during viral infections (Cervantes-Barragan et al., 2012; Cisse et al., 2008; Ghosh et al., 2010; Swiecki et al., 2010). Finally, the apparent dispensability for pDCs in many acute infection models could be an artifact of using viruses with human tropism in laboratory mouse hosts, which usually harbor mutations in several ISGs. Collectively, these discrepancies point to the need of (a) developing new models to study pDC function *in vivo*, including humanized mice (Douam et al., 2018); and (b) systematically and rigorously assessing pDC role during different types of viral infections. Ultimately, the IFN-I production capacity of pDCs is a promising therapeutic target for antiviral responses and warrants detailed study.

Although pDC-derived IFN-I may be advantageous during the early stages of an acute immune response, dysfunctional or dysregulated IFN-I production can cause pathology (Barrat and Su, 2019). Decreased IFN-I production by exhausted pDCs during chronic infection with hepatitis C virus (human) or LCMV (mouse) results in a limited antiviral response (Macal et al., 2018; Yonkers et al., 2007). Conversely, elevated levels of

IFN-I produced by activated pDCs for a prolonged period of time contributes to chronic HIV infection pathology by promoting CD4$^+$ T cell death (Li et al., 2014). A dysregulated IFN-I signature is also found in multiple autoimmune diseases, including systemic lupus erythematosus (SLE), psoriasis, and systemic sclerosis (reviewed in Crow et al., 2019). In these autoimmune diseases, pDCs are suggested to be a major cellular source of IFN-I production (Means et al., 2005; Panda et al., 2017). Mouse models have been useful to demonstrate a central role for pDCs in SLE (Davison and Jørgensen, 2015; Rowland et al., 2014; Sisirak et al., 2014); however, as with viral infections, mouse models have shown contradictory results regarding the role of pDCs in other autoimmune diseases such as psoriasis (Glitzner et al., 2014; Nestle et al., 2005; Wohn et al., 2013). The exact mechanisms that cause dysregulated production of IFN-I by pDCs in SLE and other autoimmune diseases remain unclear. Interestingly, a preliminary report suggests that patients receiving anti-TNFα therapy for various autoimmune diseases harbor pDCs that are trapped in a pathological state of unchecked IFN-I production, leading to the development of paradoxical psoriasis (Conrad et al., 2018). In agreement with these clinical observations, TNFα signaling has been shown to promote pDC maturation accompanied by a loss of pDC capacity to produce IFN-I (Alculumbre et al., 2018b; Palucka et al., 2005). These results suggest that IFN-I and TNFα may have antagonistic effects on pDCs, promoting or repressing IFN-I production, respectively. Further characterization of dysregulated IFN-I production by pDCs may unveil promising therapeutic strategies.

2.2 pDC capacity to present antigen and modulate adaptive T cell responses

Plasmacytoid dendritic cells' capacity to present antigen to T cells is associated with their activation status. Unlike classical dendritic cell (cDC) subsets, steady state pDCs have a round plasma cell-like morphology and express intermediate levels of MHC-II (HLA-DR). However, human pDCs can convert into cDC-like cells in response to culture with interleukin 3 (IL-3), IL-3 + CD40L, or viral stimulation (Grouard et al., 1997; Kadowaki et al., 2000). pDC cell conversion includes morphological changes, i.e., from plasmacytoid to dendritic morphology, as well as upregulation of MHC-II/HLA-DR and costimulatory molecules. These changes increase pDC capacity to present antigen to CD4$^+$ T cells (Cella et al., 2000; Rissoan et al., 1999; Tel et al., 2011, 2012; Villadangos and Young, 2008). Importantly, the capacity of pDCs to activate T cells has been recently brought into question by the identification of a rare human pDC-like population expressing the TAM (TYRO3-AXL-MERTK) family receptor tyrosine kinase AXL (See et al., 2017; Villani et al., 2017). As discussed below, steady state AXL$^+$ DCs have a relatively higher capacity to promote T cell proliferation compared to steady state pDCs; however, there is already evidence suggesting that activated pDCs can promote T cell proliferation independently of AXL$^+$ DCs. First, AXL$^+$ DCs represent a minor fraction (~5–10%) of cells that fall in the traditional pDC gate (CD123$^+$ HLA-DR$^+$), and these cells are unable to proliferate in culture (Villani et al., 2017). Consequently, it is difficult to consider that AXL$^+$ DCs would account for all of the changes observed in stimulated pDCs after only 24–48 h of culture (Cella et al., 2000). Second, we have recently shown that pDCs purified using a sorting strategy that removes AXL$^+$ DCs are still able to promote T cell proliferation following stimulation with IL-3 + CD40L (Alcántara-Hernández et al., 2017). Importantly, this was also observed during stimulation of pDCs with influenza virus (Alculumbre et al., 2018b). Altogether, these observations suggest that pDC capacity to differentiate into cDC-like cells and promote T cell proliferation cannot be solely attributed to a contamination with a minor population of AXL$^+$ DCs. Importantly, this capacity of pDCs to activate T cells has also been demonstrated for mouse pDCs, suggesting that this conserved pathway has an important role in the immune response (Asselin-Paturel et al., 2001; Nakano et al., 2001; Remer et al., 2007; Salio et al., 2004; Schlecht et al., 2004; Young et al., 2008).

Despite their capacity to promote T cell proliferation, it is still unclear how pDCs acquire antigen. pDCs seem to be efficient at presenting endogenous and virus-derived antigens via MHC-I and MHC-II; however, relative to cDCs, pDCs have lower capacity to internalize extracellular antigen via macropinocytosis and phagocytosis (Tel et al., 2012; Villadangos and Young, 2008). Despite the fact that mouse pDCs are not efficient at cross-presentation (Jaehn et al., 2008; Sapoznikov et al., 2007; Shinohara et al., 2006), human pDCs are reported to cross-present exogenous soluble antigen to activate CD8$^+$ T cells *in vitro* (Di Pucchio et al., 2008; Hoeffel et al., 2007; Segura et al., 2013). Interestingly, this appears to be a general difference between mouse and human DC subsets. In mouse, cDC1 are the main cross-presenting cell while human cDC1, cDC2, and pDC appear to be equally capable of cross-presentation, at least *in vitro* (Segura et al., 2013).

There is evidence that pDCs could present virus-derived antigen or virions at the infection site *in vivo* (Di Pucchio et al., 2008; McGill et al., 2008). However, given the lack of reliable pDC depletion models, it is unclear what role pDC-mediated antigen presentation plays in the overall immune response. A great approach to dissect the antigen presentation capacity of different DC subsets *in vivo* is to deliver model or disease-associated antigens to these cells using subset-specific anti-receptor monoclonal antibodies (mAbs). Indeed, we have used this strategy to compare the capacity of lymphoid-resident and migratory cDC1 and cDC2 to present antigen and promote T cell activation and tolerance *in vivo* (Idoyaga et al., 2011, 2013). Ideally, receptors used for this purpose would not trigger responses upon mAb binding, which has been a main challenge in the case of antigen-targeting to pDCs. In fact, many pDC-specific surface targets for antibody-mediated antigen delivery have inhibitory functions that are triggered upon mAb engagement, confounding the interpretation of results regarding the inherent capacity of pDCs to modulate immune responses. For instance, SiglecH has been used for antigen-targeting to pDCs; however, this inhibitory receptor impairs IFN-I production upon mAb binding (Blasius et al., 2006a; Loschko et al., 2011a; Zhang et al., 2006). Similarly, mAb binding to the human pDC-specific C-type lectin receptor BDCA2 (CLEC4C/CD303) in transgenic mice results in inhibitory signaling that promotes T cell tolerance (Chappell et al., 2014). Conversely, BST2 does not have inhibitory function upon mAb binding, and antigen delivery to this protein promotes protective immunity (Loschko et al., 2011b; Sapoznikov et al., 2007). However, BST2 is upregulated in other DC subsets during inflammation, making it unsuitable to evaluate the sole contribution of pDC after anti-BST2-antigen delivery in the context of infection (Blasius et al., 2006b). Importantly, these effects do not seem to be mediated by binding through Fc receptors as these targeting mAbs usually carried mutations in the Fc region or were digested to $F(ab')_2$ fragment (Loschko et al., 2011b; Sapoznikov et al., 2007). Thus, there are technical limitations that preclude a clear comparison of the antigen presentation capacity of pDCs versus cDCs *in vivo* following anti-receptor mAb targeting. Identification of novel pDC-specific receptors in mice or development of transgenic mice that express human pDC receptors could be used to answer this important question of pDC function *in vivo*.

2.3 pDC capacity to modulate immune tolerance

Multiple reports have shown a role for pDCs in maintaining tolerance in mouse models. One mechanism through which pDCs promote tolerance is induction of $Foxp3^+$ regulatory T cells (Tregs). pDCs can promote antigen-specific Tregs in a cardiac allograft model and an atherosclerosis model (Ochando et al., 2006; Yun et al., 2016). Also, pDCs promote antigen-specific Tregs that ameliorate experimental autoimmune encephalitis (Duraes et al., 2015; Irla et al., 2010). Furthermore, pDCs are required for promoting IL-10 producing Tregs against the commensal microbial molecule polysaccharide A (Dasgupta et al., 2014). Similarly, human pDCs have also demonstrated the ability to promote IL-10 producing Tregs *in vitro* (Ito et al., 2007; Martín-Gayo et al., 2010). In addition to peripheral tolerance, pDCs are reported to transport antigen to the thymus to promote central tolerance through clonal deletion (Hadeiba et al., 2012). Lastly, pDCs can exert immunosuppressive effects via mechanisms that involve T cell anergy, deletion and the upregulation of programmed death ligand 1 (PD-L1) (Goubier et al., 2008; Kool et al., 2009; Loschko et al., 2011a). This demonstrates that pDCs have multiple mechanisms devoted to immune tolerance, and these may be engaged in a context-dependent manner.

As expected, these tolerance mechanisms are detrimental in the tumor microenvironment. Indeed, pDCs can fail to produce IFN-I and instead induce Tregs that result in tumor escape (Faget et al., 2013; Munn et al., 2004; Sisirak et al., 2012). Accordingly, pDC infiltration in tumors often correlates with a poor prognosis (Labidi-Galy et al., 2012; Treilleux et al., 2004). However, correlations between pDC infiltration and tumor prognosis are primarily drawn from immunohistochemistry analysis using only one pDC marker, e.g., CD123. This strategy may be insufficient to identify pDCs, and thus, it is important to reevaluate these correlations using approaches that rely on more than 1–3 markers to identify pDCs. Advances in high-dimensional techniques that allow measuring 40 or more protein markers at the single-cell level will enable a better resolution to unequivocally define pDCs in tumors. If pDCs are present in the tumor microenvironment and, indeed, correlated with a poor prognosis, it is possible that these harmful tolerance mechanisms can be reversed with proper stimulation (Drobits et al., 2012; Kalb et al., 2012; Vermi et al., 2011). Indeed, topical application of the TLR7 agonist Imiquimod is FDA-approved for treatment of skin cancers, and it has been reported to function by recruiting IFN-I-producing pDCs (Urosevic et al., 2005). Additionally, TLR9 agonists such as CpG have shown promise in mouse tumor models and are in various stages of clinical trials in humans (Krieg, 2008, 2012). More recently, these trials have combined TLR9 agonists with checkpoint blockade therapy (Adamus and Kortylewski, 2018). This combination strategy may be beneficial to overcome potential immune exhaustion due to prolonged IFN-I production in the tumor (Snell et al., 2017; Wang et al., 2016). Alternatively, tumor antigen-loaded pDCs have demonstrated potential to induce favorable immune responses in clinical trials (Tel et al., 2013). Understanding the role of pDCs in the tumor microenvironment and the factors that influence their functional regulation could inform novel cancer treatments that better harness pDC potential to activate anti-tumor immune responses.

3. Sources of pDC functional diversity: Developmentally encoded subsets versus environmentally mediated plasticity

Plasmacytoid DC diverse functions have generated an ongoing debate regarding mechanistic regulation that essentially distills down to nature versus nurture. Are diverse pDC functions distributed among developmentally encoded specialized subsets within the population (nature)? Or are these diverse functions environmentally encoded, i.e., functional heterogeneity arising from plasticity within the pDC pool (nurture)? These competing models are illustrated in Fig. 1. Recent publications describing a pDC-like population contaminating the traditional pDC gate have brought this question to the forefront of the field and raised doubts about the range of functions previously described for pDCs. Below, we review evidence for these two models.

3.1 Evidence for pDC subpopulations

The broad range of functions attributed to pDCs has endorsed the hypothesis of the existence of distinct, functionally specialized subsets within the pDC compartment (Fig. 1, Model 1). It has been challenging, however, to define pDC subpopulations with distinct functions, and to synthesize findings between human and mouse. We start this section by reviewing pDC development since the hypothesis of subpopulations presumes that pDC functional heterogeneity is developmentally encoded.

3.1.1 Developmental origins of pDCs

The developmental origin of pDCs has been a matter of debate for some time. Plasmacytoid dendritic cell development is strictly dependent on the E protein transcription factor 4 (TCF4), also known as E2-2 (Cisse et al., 2008; Nagasawa et al., 2008). TCF4 directly binds the promoters

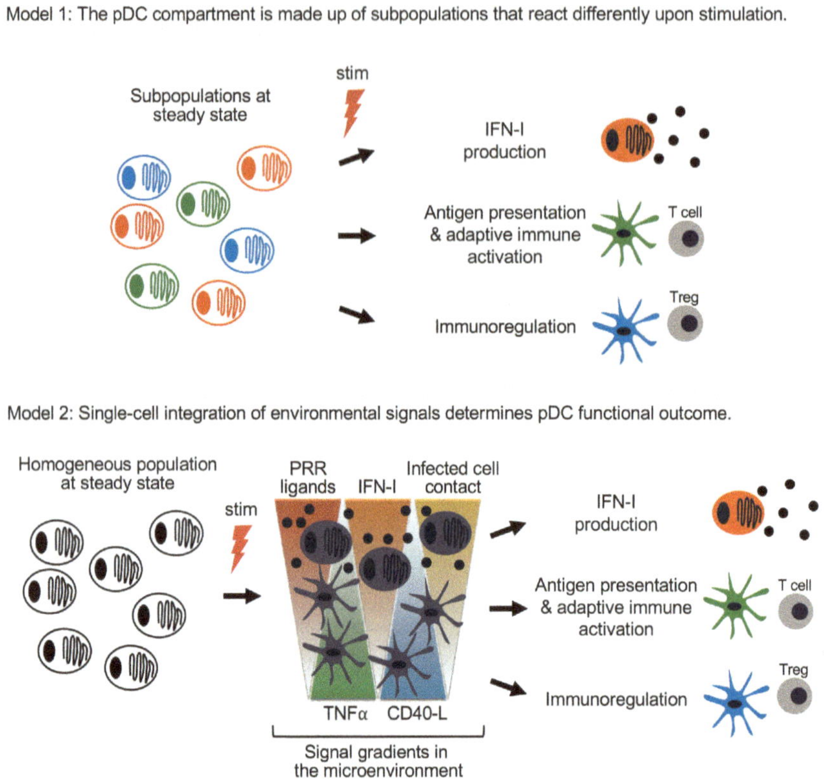

Fig. 1 Models of pDC functional diversity. Model 1: pDC functions are enacted by pre-determined subpopulations of pDCs. Upon activation, there is a division of labor among these functionally biased subpopulations. Model 2: neighboring pDCs receive slightly varied environmental cues at the single-cell level (e.g., dose or duration of TLR ligand stimulation, interaction with infected cells, feedback from neighboring pDCs, and concentration of cytokines) that lead to functional heterogeneity.

of multiple pDC-specific markers and transcription factors such as LILRA4 (ILT7) and SPIB (Ghosh et al., 2010). In addition to TCF4, pDC development is dependent on Runx2, SpiB, Bcl11a, Zeb2, IKZF1, and the TCF4 co-factor Mtg16 (Cytlak et al., 2018; Ghosh et al., 2014; Ippolito et al., 2014; Sawai et al., 2013; Schotte et al., 2004; Scott et al., 2016; Wu et al., 2013, 2016). The transcription factor Irf8 was initially thought to be required for pDC development; however, recent analyses showed that pDCs are in fact present in Irf8$^{-/-}$ mice, albeit with altered phenotype and function (Durai et al., 2019; Schiavoni et al., 2002; Sichien et al., 2016; Tsujimura et al., 2003).

One unsolved question regarding pDC development is whether pDCs are derived from a myeloid or lymphoid progenitor. Similar to cDCs, pDCs express the receptor Fms related tyrosine kinase 3 (FLT3) and demonstrate developmental dependence on the hematopoietin FLT3 ligand (FLT3L) (Durai et al., 2018). Indeed, FLT3L alone drives pDC differentiation from myeloid precursors, and pDCs are greatly expanded by administration of exogenous FLT3L in both humans and mice (Blom et al., 2000; Brawand et al., 2002; Breton et al., 2015; Gilliet et al., 2002; Pulendran et al., 2000). Furthermore, pDCs have been shown to arise from a common dendritic progenitor (CDP) shared with cDCs (Liu et al., 2009; Naik et al., 2007; Onai et al., 2013). While a population of $CD115^{-}$ CDPs appears preferentially primed toward pDCs, $CD115^{+}$ CDPs can also produce pDCs in addition to cDC1 and cDC2 (Bagadia et al., 2019; Onai et al., 2013). Indeed, similar to cDCs, at least ~80% of pDCs are labeled in Csf1r (CD115) lineage tracing mice that express YFP at the level of CDPs (Loschko et al., 2016). Moreover, genetic deletion of transcription factors or co-factors required for pDC development, e.g., Mtg16, Zeb2 or IKZF1, results in greater numbers of cDCs, particularly cDC1 (Cytlak et al., 2018; Ghosh et al., 2014; Scott et al., 2016). Finally, as elegantly shown recently, pDCs and cDC1 share regulatory features such as reliance on a specific +41-kb Irf8 enhancer (Bagadia et al., 2019; Durai et al., 2019). Altogether, these are evidence that pDCs can develop, to a certain extent, from a precursor that is shared with cDCs.

Evidence that has been used to support a lymphoid past for pDCs includes expression of the pre-T cell receptor alpha (PTCRA) in human pDCs and D-J rearrangements of the immunoglobulin heavy chain (IgH) locus in mouse pDCs (Bendriss-Vermare et al., 2001; Corcoran et al., 2003; Res et al., 1999). Also, pDCs have been shown to arise, to a certain extent, from the common lymphocyte progenitor (CLP) (Manz et al., 2001; Shigematsu et al., 2004). Recently, a pDC-biased $SiglecH^{+}$ $Ly6D^{+}$ $IL\text{-}7R^{+}$ progenitor was identified in the bone marrow of mice (Dress et al., 2019; Rodrigues et al., 2018). Nevertheless, the lymphoid origin of pDCs has to be analyzed carefully. It is clear that pDCs acquire lymphoid features during development, independently of their lymphoid or myeloid origin. Indeed, IgH D-J rearrangements are also observed in pDCs derived from a myeloid precursor, indicating that transcriptional programs can converge to produce the same phenotype regardless of developmental origin (Onai et al., 2013; Sathe et al., 2013; Shigematsu et al., 2004). This convergence may be merely attributable to expression of the transcription factor TCF4, as suggested (Reizis, 2019; Reizis et al., 2011). Thus, it is possible that several other lymphoid features currently used to indicate a lymphoid past, e.g., expression of IL-7R and CD2, could also arise as a consequence of TCF4 expression. Further lineage tracing studies using lymphoid markers that are not present in mature pDCs are needed to determine the contribution of lymphoid and myeloid precursors to mature peripheral pDCs.

While the exact path of pDC differentiation remains unclear, there is little evidence to support emergence of pDC subsets during development. Only one report suggests that a minor pDC-like population in the periphery, distinguished by expression of the cDC transcription factor Zbtb46, is the outcome of a myeloid path to pDC development (Rodrigues et al., 2018). However, this presumes that the majority of peripheral pDCs are exclusively lymphoid-derived. Alternatively, the capacity of pDCs to arise from both myeloid and lymphoid precursors could be interpreted as plasticity within the precursor compartment. Indeed, it is logical to hypothesize that the immune system has evolved redundant pathways that lead to DC differentiation given their essential role in maintaining homeostasis and responding to a wide array of immunogenic insults. In favor of developmental plasticity, transferred $TCF4^{high}$ bone marrow pDC-biased precursors can differentiate into cDCs in the small intestine (Onai et al., 2017). Furthermore, in a mouse model with genetic depletion of cDC precursors (pre-DCs) and their progeny, cells that were phenotypically and transcriptionally similar to cDC2 were still found in the spleen but carried IgH D-J rearrangements, suggesting that a lymphoid origin for cDC2 is possible under extenuating circumstances (Salvermoser et al., 2018). Similarly, *in vitro* models of human differentiation revealed that a multipotent early lymphoid progenitor can give rise to pDCs and cDC1 that are virtually indistinguishable from DC subsets derived from myeloid progenitors (Helft et al., 2017). Collectively, these results are indicative of a continuous rather than linear model of DC development (Brown et al., 2018; Haas et al., 2018). Accordingly, the developmental origin of pDCs may not have significant bearing on the final phenotype and function of these cells in the periphery. Instead, a core set of transcription factors activated in either the lymphoid or myeloid progenitor might be essential and sufficient to dictate pDC identity and function.

3.1.2 pDC diversity before the high-dimensional era

In humans, shortly after pDC identification as IFN-I-producing cells, one group noted a myeloid-like subpopulation that falls within the

HLA-DR^{+}CD123^{+} pDC gate (Comeau et al., 2002). These cells expressed CD2, CD5, and CD11c, and displayed antigen-presenting functions prior to *in vitro* maturation. Subsequently, CD2 alone was proposed as a marker to distinguish two subsets of pDCs in blood (Matsui et al., 2009). Although CD2^{high} pDCs have enhanced capacity to promote T cell stimulation when activated, both CD2^{high} and CD2^{low} subpopulations secrete IFN-I upon stimulation. Wilhelm et al. described a similar subpopulation among pDCs distinguished by SIGLEC1 expression (Wilhelm et al., 2016). SIGLEC1^{+} pDCs appear semi-mature, and do not produce IFN-I upon stimulation. Most recently, a subpopulation of human pDCs expressing CD2, CD5 and CD81 with reduced potential for IFN-I production was described (Zhang et al., 2017). This subpopulation of pDCs has a stronger capacity to induce T cell proliferation, regulatory T cell induction, and B cell antibody production than CD5^{-}, CD81^{-} pDCs. Notably, multiple reports identified the receptor tyrosine kinase AXL as a marker for this CD2^{+}, CD5^{+}, CD81^{+} subpopulation of pDCs (Bryant et al., 2016; Matsui et al., 2009; Zhang et al., 2017).

In mice, evidence of pDC subpopulations is scarcer. Two populations of pDCs could be distinguished in the bone marrow using RAG1-GFP mice (Pelayo et al., 2005). RAG1-GFP^{+} pDCs had reduced capacity for both IFN-I production and allogeneic T cell activation in comparison to RAG1-GFP^{-} pDCs. CCR9 was proposed to be a marker that distinguishes tolerogenic pDCs in the gut, which can contribute to central tolerance (Hadeiba et al., 2008, 2012). However, it has been shown that B220^{+}CCR9^{-} cells actually represent a precursor of pDCs and other DC subsets (Schlitzer et al., 2011; Segura et al., 2009). Björck et al. identified CD9 as a distinguishing marker for two pDC subpopulations in FLT3L-inoculated mice (Björck et al., 2011). Whereas CD9^{+} pDCs produce IFN-I upon stimulation, CD9^{-} pDCs do not produce IFN-I and appear specialized for regulatory T cell induction. Importantly, CD9^{+} pDCs do not express CCR9 and can become CD9^{-} after adoptive transfer, suggesting that these subpopulations likely represent distinct differentiation or maturation stages. Similarly, Sca-1 (Ly6a) was suggested to distinguish an earlier Sca-1^{-} pDC developmental stage from a later, functionally distinct Sca-1^{+} stage (Niederquell et al., 2013). This Sca1^{-} stage in the bone marrow appears to overlap descriptions of CCR9^{-} and Ly49Q^{-} pDC populations (Kamogawa-Schifter et al., 2005; Omatsu et al., 2005; Schlitzer et al., 2011). More recently, a CD5^{+}CD81^{+} mouse pDC subpopulation with decreased IFN-I capacity was identified in mouse spleen, purportedly a murine equivalent to the $CD5^+CD81^+$ human pDC subset described above (Zhang et al., 2017). A more comprehensive analysis is needed to determine if $CD5^+CD81^+$ pDCs are conserved between mouse and human.

3.1.3 pDC diversity during the high-dimensional era

Historically, one of the greatest challenges to the study of pDCs has been their low frequency (<1%) among immune cells in blood and tissues. Furthermore, previous attempts to discern subpopulations of pDCs (see above section) were hindered by approaches that rely on *a priori* gating of limited flow cytometry marker panels. Advances in high-dimensional single-cell transcriptomic and proteomic technologies have provided enhanced resolution, facilitating the use of unbiased analyses that have enabled insights into minor immune cell populations.

Recently, we and others have applied high-dimensional single-cell approaches to the study of human DCs, including pDCs. Single-cell RNA-Seq and mass cytometry (CyTOF) revealed a population within the human DC network that is pDC-like in phenotype (Alcántara-Hernández et al., 2017; See et al., 2017; Villani et al., 2017). These cells express similar or low levels of multiple pDC markers, including CD123, BDCA4, and BDCA2. Interestingly, these cells also express cDC markers such as CD33 and CD11c, resulting in an intermediate phenotype between that of pDCs and cDCs (Fig. 2). From their phenotype, it appears that this population overlaps with previously defined human pDC subpopulations (listed above) identified using flow cytometry. Indeed, this population expresses CD2, CD5, SIGLEC1, and AXL. However, not all CD2-expressing pDCs fall within this population, indicating that the originally described $CD2^{high}$ pDCs (Matsui et al., 2009) may have been a mixed population. Furthermore, AXL cannot be used as a single defining marker of this subpopulation given that some blood cDC2 also express AXL (Fig. 2; Alcántara-Hernández et al., 2017), as do tissue macrophages (Fujimori et al., 2015). Therefore, it is necessary to use a combination of pDC markers and AXL to distinguish these cells from other DC subsets.

The nature of these cells remains unclear. Villani et al. defined this subpopulation as a distinct subset of DCs, termed DC5 or $AXL^+SIGLEC6^+$ DCs (AS DCs). In contrast, See et al. suggested that they are a circulating cDC precursor (pre-DC), given their capacity to acquire some markers of cDCs in an *in vitro* differentiation assay. In our own analysis, we were struck by the apparent transitional relationship between these cells, pDCs, and cDC2.

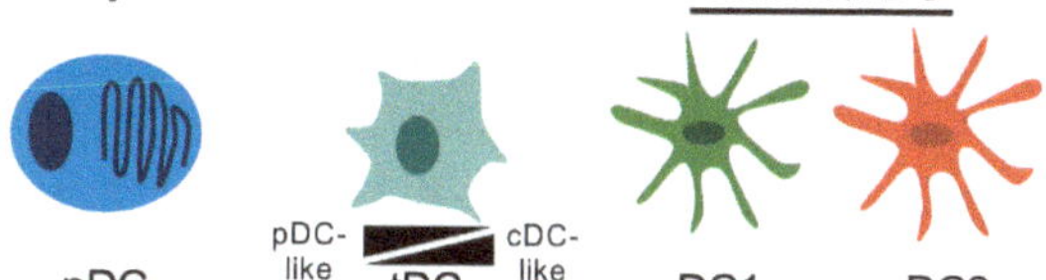

DC Populations in Human Blood		Plasmacytoid DC pDC	Transitional DC tDC	Classical DC cDC1	Classical DC cDC2
Surface Markers	BDCA2	+++	++	–	–
	BDCA4	+++	++	–	–
	CD123	+++	++	–	–
	AXL	–	+++	–	5-10% +
	SIGLEC1	–	++	–	–
	SIGLEC6	–	++	–	–
	BDCA1	–	–	–	+++
	BDCA3	+	++	+++	+
Functions	IFN-I	+++	–	–	–
	Ag presentation	+ (++ after stim)	++	+++	+++
	Cross-presentation (soluble antigen)	++	?	++	++
	IL-12p40	–	+	++	++
	IL-12p70	–	–/+	++	++
Transcription Factors	TCF4	+++	++	–	–
	ID2	–	+	+++	++
	IRF4	++	++	–	++
	IRF8	+++	+++	+++	–

Fig. 2 Summary of tDC features in relation to pDCs and other human DC subsets. Human transitional DCs (tDCs, also known as AXL^{+} DCs, ASDC or pre-DCs) have features of both pDCs and cDCs. Summary of surface marker expression (from protein expression data), functional capacities determined by *ex vivo* assays, and transcription factors measured by RNA expression (ID2, IRF4, IRF8) or protein expression (TCF4).

Indeed, in any high-dimensional analysis at the protein level and in all the tissues analyzed, these cells appear as a continuum of heterogeneous phenotypes ranging from pDC-like to cDC-like, i.e., "connecting" pDCs to cDC2 (Alcántara-Hernández et al., 2017). This phenomenon was also observed by single-cell RNA-Seq analysis (Villani et al., 2017). Notably, Villani et al. reported limited proliferation of these cells *in vitro*, as well as a limited capacity for conversion into cDC1 and a preference for conversion into cDC2.

These observations suggest that while these cells may have the ability to differentiate into cDCs, it remains unclear whether they are the equivalent to the traditional bone marrow DC precursor (Liu et al., 2009). Given their unique phenotype and transcriptome, as well as their still unclear developmental origin, we have named these cells transitional DCs, or tDCs, and will refer to them as such for the rest of this chapter.

Several features of tDCs associate them more closely with the pDC lineage. Most strikingly, tDCs express the pDC lineage-defining transcription factor TCF4, which is likely responsible for establishing their pDC-like phenotype as discussed above (Cisse et al., 2008). Indeed, TCF4 expression levels vary across the population correlating with how pDC-like or cDC-like the cells' phenotype appears. Second, similar to pDCs, tDCs are barely detected in healthy human skin, indicating that their tissue distribution aligns with that of pDCs more than cDCs (Alcántara-Hernández et al., 2017). Despite these similarities, tDCs demonstrate several functional differences from pDCs (Fig. 2). For one, tDCs do not display great capacity for IFN-I production after TLR stimulation. Furthermore, unlike freshly-isolated pDCs, tDCs are able to initiate an allogeneic T cell reaction comparable to that of the cDC subsets (Alcántara-Hernández et al., 2017). Interestingly, tDCs have been suggested to be more similar to cDCs morphologically (See et al., 2017; Villani et al., 2017). However, given the heterogeneity of this population, care should be taken when analyzing their morphology. To definitely establish tDC morphological features, it may be necessary to perform a detailed analysis on a larger number of cells occupying different parts of their phenotypic spectrum. Indeed, we have observed that tDCs that are more similar to pDCs in phenotype harbor a plasmacytoid morphology (unpublished results). Finally, another functional difference among tDCs and pDCs is their proposed capacity to produce the Th1-priming cytokine IL-12. Indeed, it has been suggested that the previously detected secretion of IL-12 by pDCs could be attributed to contamination with tDCs. However, it is important to observe that most studies that reported IL-12 production by pDCs evaluated expression of the IL-12p40 subunit and not the active cytokine IL-12p70 (Ito et al., 2006; Matsui et al., 2009). Similarly, tDCs are reported to produce IL-12p40, but minimal to non-detectable levels of IL-12p70 (See et al., 2017; Villani et al., 2017; Zhang et al., 2017). Studies that use updated sorting strategies and compare production of IL-12p70 between DC subsets after stimulation are necessary to clarify the capacity of human tDCs and pDCs to secrete this cytokine. Importantly, it is still unclear if tDCs have a unique function that distinguishes them from pDCs and cDCs.

Given their shared transcriptome with pDCs, tDCs may be specialized to detect viral infection and cooperate with pDCs for the activation of T cell responses. Indeed, the marker AXL itself is the main entry receptor for Zika virus (Sun et al., 2017). Further characterization is needed to determine functional differences between tDCs, pDCs and cDCs.

To date, tDCs have only been described in humans. It remains to be determined whether this population is conserved in mice, and if the mouse homolog shares the human tDC phenotype including expression of CD2, CD5, Siglec1 and Axl (SIGLEC6 does not have a murine homolog). Of note, the combination of pDC and cDC features observed in tDCs is reminiscent of the noncanonical DCs described in mice (Bar-On et al., 2010; Lau et al., 2016), a DC population with features of both pDCs and cDC2, suggesting that these cells may be the homologous population. To determine whether noncanonical DCs are related to human tDCs, we have recently performed an integrated high-dimensional cross-species analysis of the DC compartment in mouse and human. We found significant overlap between the noncanonical DCs and human tDCs; however, we observed that noncanonical DCs only represent a fraction (~33%) of a larger population that we termed mouse tDCs (unpublished results). Our analysis is the first one to align mouse and human tDCs in terms of phenotype, transcriptome, and function. Identification of mouse tDCs is critical to characterize their developmental origin and function, which will ultimately determine whether these cells in fact represent a peripheral DC precursor, a subpopulation of pDCs or another DC subset.

In conclusion, the identification of tDCs with high-dimensional approaches has unified multiple descriptions of human pDC subpopulations, but further study is needed to determine their function and identity. Overall, there is still not sufficient consensus to conclude that mouse and human pDC functionality is regulated by subpopulations.

3.2 Evidence for pDC plasticity

An alternative hypothesis to the presence of developmentally encoded subsets is that the capacity of pDCs to enact multiple functions is a consequence of environmentally regulated functional plasticity. For this discussion, we define plasticity as the ability of a differentiated cell (in this case, pDCs) to alter their phenotype and function in response to environmental stimuli, even taking on features of other cell types. This process is also referred to as cell state conversion or transdifferentiation (Merrell and Stanger, 2016).

Plasticity is widely documented in the immune system, for example, among T cell subsets (Hirahara et al., 2013). It has also been described among myeloid cells including macrophages, mast cells, and neutrophils (Galli et al., 2011). In support of environmental rather than developmental regulation of pDCs, single-cell transcriptomic analyses of purified human pDCs (i.e., without tDC contamination) suggest that they are relatively homogeneous during homeostasis (Villani et al., 2017; Wimmers et al., 2018). Accordingly, functional diversity of pDCs could be driven by slight differences in signal integration at the single-cell level (Fig. 1, Model 2). For example, pDCs localized in the same microenvironment may be exposed to slight variations in the intensity of multiple signals, such as cytokines or TLR agonists. Understanding the factors that influence this functional plasticity could provide therapeutic targets to selectively engage the different functions of pDCs.

3.2.1 Functional diversity of pDCs emerges after stimulation

One of the most striking characteristics of pDC plasticity is their capacity to undergo cell state conversion and become cDC-like *in vitro*. Importantly, we and others recently showed that pDCs purified without tDCs maintain the capacity to become cDC-like and induce T cell proliferation after being cultured in the presence of IL-3 + CD40L or virus (Alcántara-Hernández et al., 2017; Alculumbre et al., 2018b). Therefore, the antigen presentation capacity of pDCs cannot be attributed solely to tDC contamination, suggesting that pDCs are truly plastic. It remains to be confirmed that this is true *in vivo*, but some insight can be drawn from mouse models. During MCMV infection, pDCs adopt a cDC-like transcriptional program, indicative of cell state conversion (Manh et al., 2013). Additionally, in the context of LCMV infection, bone marrow pDCs can differentiate into cDCs upon adoptive transfer (Liou et al., 2008). Nevertheless, it is important to reevaluate these *in vivo* experiments with updated strategies to remove any mouse tDCs.

If pDC functions are determined environmentally rather than by predisposed subpopulations, then a homogeneous population of pDCs must regulate the acquisition of distinct functions. This can either occur (1) through homogeneous activation, with a uniform population acquiring different functions sequentially throughout the course of a response or (2) via diversification, with individual pDCs acquiring distinct functions upon stimulation. There is little evidence to support the first hypothesis. In particular, only a fraction of pDCs in the population produce IFN-I in response to stimulation as shown by reporter mouse models and during single-cell stimulation in fluidics systems (Bauer et al., 2016; Kumagai et al., 2009; Wimmers et al., 2018). Moreover, pDCs can convert to cDC-like cells in the absence of IFN-I production, for example upon stimulation with IL-3 or CpG-B (Guiducci et al., 2006; Kadowaki et al., 2000). Thus, it is more likely that pDCs administer their varied functions by diversifying functionally upon stimulation (Fig. 1, Model 2). Indeed, this diversification strategy was observed in human pDCs activated with influenza virus (Alculumbre et al., 2018b). Alculumbre et al. report that pDCs diversify into three stable populations when activated with a single stimulus. This diversification was defined by the durable and differential expression of CD80 and PD-L1, and distinct functional features such as IFN-I production or T cell activation. Importantly, the presence or absence of tDCs did not influence these results, indicating that diversification is not a consequence of this confounding subpopulation.

The mechanisms that promote this diversity remain undetermined. We hypothesize that slight variations in the signals received and integrated at the single-cell level are actively responsible for enacting distinct functional programs in pDCs (Fig. 1, Model 2). This proposed model is drawn from single-cell studies in which homogeneous cells in seemingly identical stimulation conditions *in vitro* show variation in signal processing, leading to phenotypic and functional diversity (Fischer et al., 2019; Gutschow et al., 2019; Kellogg et al., 2017). In disease settings, the plastic process of parallel diversification may be advantageous, allowing pDCs to fill the necessary roles demanded by each unique pathogen. Furthermore, it offers a way to control cytokine responses, diverting pDCs with IFN-I potential into alternative functional maturation programs. Future studies will need to develop strategies to address the relative importance of these diversified functional states during infection. Lastly, the development of new, specific pDC-tracking mouse models may be necessary to interrogate to what extent this occurs *in vivo*.

3.2.2 Environmental determinants of pDC function

How are multiple signals present in the tissue microenvironment integrated to determine pDC function? Dissecting these signals begins with cataloging the effect of different microenvironmental factors on pDCs in controlled experiments. It is clear that individual TLR ligands trigger IFN-I production, but engagement of more than one pattern recognition receptor (PRR) pathway in pDCs may yield a different outcome. For example, signaling through cytosolic PRRs is reported to limit the TLR-mediated IFN-I response in pDCs (Yu et al., 2016).

Additionally, different pathogens and doses can have dramatic effects on pDC functional responses (Thomas et al., 2014). Beyond PRR ligands, there are many other environmental signals that influence pDC function at the single-cell level. Autocrine or paracrine IFN-I promotes a positive feedback loop that maintains IFN-I production (Kerkmann et al., 2003; Kim et al., 2014; Kumagai et al., 2009; Wimmers et al., 2018). In addition to IFN-I itself, cell-to-cell contact with surrounding infected cells also promotes IFN-I production by pDCs (Assil et al., 2019; Saitoh et al., 2017; Tomasello et al., 2018). In opposition to signals that promote IFN-I production, other environmental signals are reported to inhibit IFN-I by pDCs. Factors in the mucosal environment and tumors, such as IL-10, prostaglandin E_2 (PGE_2) and transforming growth factor β (TGFβ), suppress pDC IFN-I production (Bekeredjian-Ding et al., 2009; Contractor et al., 2007). Additionally, hormones such as estradiol, progesterone, and testosterone affect IFN-I production by pDCs, which may contribute to the reported sex differences in pDC function (Hughes et al., 2008; Seillet et al., 2012; Wang et al., 2012). These signals would not be uniformly present in the tissue microenvironment, thus creating a gradient that regulates pDC function *in situ* at the single-cell level.

There is less information regarding environmental signals that modulate pDC capacity to convert into cDC-like cells. CD40L was the first stimulation reported to have this effect (Grouard et al., 1997); however, it has since been shown that other stimuli, such as influenza virus, can also promote pDC conversion into cDC-like cells (Alculumbre et al., 2018b; Kadowaki et al., 2000). Similarly, TNFα induces signaling that drives pDC maturation away from IFN-I production, toward a cDC-like antigen-presenting cell (Alculumbre et al., 2018b; Palucka et al., 2005). The transcriptional mechanisms downstream of the signals that promote pDC-to-cDC conversion also remain poorly defined; however, a transcription factor that may play a role is TCF4. Conditional deletion of TCF4 from mature peripheral pDCs results in their loss of pDC features and a spontaneous conversion to cDC-like cells, indicating that altering expression of this single transcription factor can change pDC cell state (Ghosh et al., 2010). Interestingly, tDCs have similarly low expression of TCF4, accompanied by limited IFN-I production potential and increased antigen presentation capacity. The transitional nature of tDCs has led us to hypothesize that these cells may represent the *in vivo* state of a pDC conversion into cDC-like cells that occurs under homeostatic conditions. Accordingly, tDCs could be a "maturation state" of pDCs that parallels migratory cDCs, i.e., a homeostatic maturation program of tissue DCs that involves upregulation of costimulatory markers and MHC-II for the induction of immunity or tolerance (Idoyaga et al., 2013; Worbs et al., 2017). In this way, transcriptional regulation of cell state conversion may maintain homeostasis by curtailing pDC potential to produce IFN-I, and perhaps promoting tolerance.

Beyond signal processing and transcriptional regulation, epigenetic regulation via chromatin remodeling is likely involved in maintaining stable cell states after activation. If cells proliferate, these epigenetic modifications could be passed down to progeny cells to maintain a form of innate memory, as recently reported in cDCs (Hole et al., 2019). In homeostasis, pDCs appear to undergo limited proliferation (Liu et al., 2007); however, it was recently shown that pDCs do proliferate during chronic viral infection, thereby maintaining a pool of exhausted pDCs (Macal et al., 2018). Due to technical limitations requiring large cell numbers, epigenetic regulation has not yet been well studied in primary DCs. With recent advances in chromatin accessibility assays, future studies could address alterations in chromatin accessibility in homeostatic and stimulated pDCs in mouse and human.

Collectively, this evidence demonstrates that as plastic sensor cells, pDCs are uniquely sensitive to their microenvironment. Therefore, conclusions drawn from naïve mice held in specific pathogen-free environments may translate poorly for the diversity of environmental influences in human biology. Additionally, pDC responses differ depending on the pathogen, and may be particularly poorly represented by murine studies using viruses that have evolved with tropism for human hosts. Consequently, we emphasize the importance of translating findings in mice to human pDC biology.

4. Concluding remarks

pDCs have been attributed diverse functions in the immune response. It is important to understand how these immune functions are regulated to exploit pDCs for therapeutic strategies. Indeed, numerous diseases could be ameliorated by enhancing either immunogenic or tolerogenic properties of pDCs, depending on the context. For example, it would be advantageous to modulate pDC production of IFN-I in autoimmune diseases such as lupus. On the other hand, it would be beneficial to promote pDC IFN-I production and T cell activation in tolerogenic tumor microenvironments.

While ontology studies have identified a clear developmental pathway of many immune cell types, the origin of pDCs remains unclear. This may reflect the convergence of different developmental paths (i.e., lymphoid and myeloid) into pDCs harboring similar phenotype and function.

Given the plasticity that pDCs demonstrate, a lymphoid versus myeloid origin might have considerably less weight than environmental signals in determining pDC function. Regardless of their origin, pDCs can become antigen-presenting cells with capacity to modulate immunogenicity and tolerogenicity. We consider this functional definition to be one of the strongest characteristics to classify a cell as a dendritic cell (Steinman and Idoyaga, 2010).

The DC compartment is tasked with initiating and modulating a wide array of immune responses. Identification of functionally specialized DC subsets has demonstrated one mechanism to achieve a division of labor. Now, the field must reconcile these ontologically defined subsets with the potential for functional plasticity, particularly under non-homeostatic conditions. These are two mechanisms that allow DCs to meet the challenges of their varied roles. Both aspects of DC biology must be addressed for the optimal design of vaccines (Palucka et al., 2010). In particular, further research is required to determine whether tDCs represent a cDC precursor, a distinct DC population or a pDC cell state. Independently of tDCs, pDCs demonstrate remarkable plasticity, indicating that development is not the sole source of their functional heterogeneity. This line of investigation has potential to unveil additional mechanisms that control the innate function of IFN-I production versus modulation of adaptive immune responses.

Future efforts to characterize pDC functions will benefit from placing more emphasis on studies performed with primary human pDCs, leveraging samples acquired from healthy donors and from varying disease states to infer function and mechanism. In particular, it will be imperative to study pDCs at the site of infection or pathology. The increasing feasibility of high-dimensional studies performed on low cell numbers holds promise for this type of work. Furthermore, advances in humanized mouse models for myeloid cells may allow mechanistic studies while better representing the biology of human pDCs. It is our hope that by better understanding how pDC function is modulated by the environment, the field can move toward the design of optimal pDC-based therapeutics to harness their versatility.

Acknowledgments

We thank the members of the Idoyaga Lab for their insight and support. We particularly thank Dr. Alcántara-Hernández for fruitful discussion and Dr. Gottfried for critical reading of the manuscript. R.L. is funded by the NSF GRFP. Work in the Idoyaga Lab is funded by NIH grants 1DP2AR069953 and 5R01CA219994, the Baxter and Freidenrich Foundations, and the Stanford Women's Health & Sex Differences in Medicine Center.

References

Adamus, T., Kortylewski, M., 2018. The revival of CpG oligonucleotide-based cancer immunotherapies. Contemp. Oncol. (Pozn) 22, 56–60.

Alcántara-Hernández, M., Leylek, R., Wagar, L.E., Engleman, E.G., Keler, T., Marinkovich, M.P., Davis, M.M., Nolan, G.P., Idoyaga, J., 2017. High-dimensional phenotypic mapping of human dendritic cells reveals interindividual variation and tissue specialization. Immunity 47. 1037–1050.e6.

Alculumbre, S., Raieli, S., Hoffmann, C., Chelbi, R., Danlos, F.-X., Soumelis, V., 2018a. Plasmacytoid pre-dendritic cells (pDC): from molecular pathways to function and disease association. Semin. Cell Dev. Biol. 86, 24–35.

Alculumbre, S.G., Saint-André, V., Di Domizio, J., Vargas, P., Sirven, P., Bost, P., Maurin, M., Maiuri, P., Wery, M., Roman, M.S., Savey, L., Touzot, M., Terrier, B., Saadoun, D., Conrad, C., Gilliet, M., Morillon, A., Soumelis, V., 2018b. Diversification of human plasmacytoid predendritic cells in response to a single stimulus. Nat. Immunol. 19, 63–75.

Amiel, J., Rio, M., de Pontual, L., Redon, R., Malan, V., Boddaert, N., Plouin, P., Carter, N.P., Lyonnet, S., Munnich, A., Colleaux, L., 2007. Mutations in TCF4, encoding a class I basic helix-loop-helix transcription factor, are responsible for Pitt-Hopkins syndrome, a severe epileptic encephalopathy associated with autonomic dysfunction. Am. J. Hum. Genet. 80, 988–993.

Asselin-Paturel, C., Boonstra, A., Dalod, M., Durand, I., Yessaad, N., Dezutter-Dambuyant, C., Vicari, A., O'Garra, A., Biron, C., Brière, F., Trinchieri, G., 2001. Mouse type I IFN-producing cells are immature APCs with plasmacytoid morphology. Nat. Immunol. 2, 1144–1150.

Assil, S., Coléon, S., Dong, C., Décembre, E., Sherry, L., Allatif, O., Webster, B., Dreux, M., 2019. Plasmacytoid dendritic cells and infected cells form an interferogenic synapse required for antiviral responses. Cell Host Microbe 25. 730–745.e6.

Bagadia, P., Huang, X., Liu, T.-T., Durai, V., Grajales-Reyes, G.E., Nitschké, M., Modrusan, Z., Granja, J.M., Satpathy, A.T., Briseño, C.G., Gargaro, M., Iwata, A., Kim, S., Chang, H.Y., Shaw, A.S., Murphy, T.L., Murphy, K.M., 2019. An Nfil3-Zeb2-Id2 pathway imposes Irf8 enhancer switching during cDC1 development. Nat. Immunol. 20, 1174–1185.

Bar-On, L., Birnberg, T., Lewis, K.L., Edelson, B.T., Bruder, D., Hildner, K., Buer, J., Murphy, K.M., Reizis, B., Jung, S., 2010. CX3CR1+ CD8alpha+ dendritic cells are a steady-state population related to plasmacytoid dendritic cells. Proc. Natl. Acad. Sci. U. S. A. 107, 14745–14750.

Barrat, F.J., Su, L., 2019. A pathogenic role of plasmacytoid dendritic cells in autoimmunity and chronic viral infection. J. Exp. Med. 216, 1974–1985.

Bauer, J., Dress, R.J., Schulze, A., Dresing, P., Ali, S., Deenen, R., Alferink, J., Scheu, S., 2016. Cutting edge: IFN-β expression in the spleen is restricted to a subpopulation of plasmacytoid dendritic cells exhibiting a specific immune modulatory transcriptome signature. J. Immunol. 196, 4447–4451.

Bekeredjian-Ding, I., Schäfer, M., Hartmann, E., Pries, R., Parcina, M., Schneider, P., Giese, T., Endres, S., Wollenberg, B., Hartmann, G., 2009. Tumour-derived prostaglandin E and transforming growth factor-beta synergize to inhibit plasmacytoid dendritic cell-derived interferon-alpha. Immunology 128, 439–450.

Bendriss-Vermare, N., Barthélémy, C., Durand, I., Bruand, C., Dezutter-Dambuyant, C., Moulian, N., Berrih-Aknin, S., Caux, C., Trinchieri, G., Brière, F., 2001. Human thymus contains IFN-alpha-producing CD11c(-), myeloid CD11c(+), and mature interdigitating dendritic cells. J. Clin. Invest. 107, 835–844.

Björck, P., 2001. Isolation and characterization of plasmacytoid dendritic cells from Flt3 ligand and granulocyte-macrophage colony-stimulating factor-treated mice. Blood 98, 3520–3526.

Björck, P., Leong, H.X., Engleman, E.G., 2011. Plasmacytoid dendritic cell dichotomy: identification of IFN-α producing cells as a phenotypically and functionally distinct subset. J. Immunol. 186, 1477–1485.

Blasius, A.L., Cella, M., Maldonado, J., Takai, T., Colonna, M., 2006a. Siglec-H is an IPC-specific receptor that modulates type I IFN secretion through DAP12. Blood 107, 2474–2476.

Blasius, A.L., Giurisato, E., Cella, M., Schreiber, R.D., Shaw, A.S., Colonna, M., 2006b. Bone marrow stromal cell antigen 2 is a specific marker of type I IFN-producing cells in the naive mouse, but a promiscuous cell surface antigen following IFN stimulation. J. Immunol. 177, 3260–3265.

Blom, B., Ho, S., Antonenko, S., Liu, Y.J., 2000. Generation of interferon alpha-producing predendritic cell (Pre-DC)2 from human CD34(+) hematopoietic stem cells. J. Exp. Med. 192, 1785–1796.

Bode, C., Fox, M., Tewary, P., Steinhagen, A., Ellerkmann, R.K., Klinman, D., Baumgarten, G., Hornung, V., Steinhagen, F., 2016. Human plasmacytoid dendritic cells elicit a type I interferon response by sensing DNA via the cGAS-STING signaling pathway. Eur. J. Immunol. 46, 1615–1621.

Brawand, P., Fitzpatrick, D.R., Greenfield, B.W., Brasel, K., Maliszewski, C.R., De Smedt, T., 2002. Murine plasmacytoid pre-dendritic cells generated from Flt3 ligand-supplemented bone marrow cultures are immature APCs. J. Immunol. 169, 6711–6719.

Breton, G., Lee, J., Zhou, Y.J., Schreiber, J.J., Keler, T., Puhr, S., Anandasabapathy, N., Schlesinger, S., Caskey, M., Liu, K., Nussenzweig, M.C., 2015. Circulating precursors of human CD1c+ and CD141+ dendritic cells. J. Exp. Med. 212, 401–413.

Brown, G., Ceredig, R., Tsapogas, P., 2018. The making of hematopoiesis: developmental ancestry and environmental nurture. Int. J. Mol. Sci. 19, 2122.

Bruni, D., Chazal, M., Sinigaglia, L., Chauveau, L., Schwartz, O., Desprès, P., Jouvenet, N., 2015. Viral entry route determines how human plasmacytoid dendritic cells produce type I interferons. Sci. Signal. 8, ra25.

Bryant, C., Fromm, P.D., Kupresanin, F., Clark, G., Lee, K., Clarke, C., Silveira, P.A., Suen, H., Brown, R., Newman, E., Cunningham, I., Ho, P.J., Gibson, J., Bradstock, K., Joshua, D., Hart, D.N., 2016. A CD2 high-expressing stress-resistant human plasmacytoid dendritic-cell subset. Immunol. Cell Biol. 94, 447–457.

Cella, M., Jarrossay, D., Facchetti, F., Alebardi, O., Nakajima, H., Lanzavecchia, A., Colonna, M., 1999. Plasmacytoid monocytes migrate to inflamed lymph nodes and produce large amounts of type I interferon. Nat. Med. 5, 919–923.

Cella, M., Facchetti, F., Lanzavecchia, A., Colonna, M., 2000. Plasmacytoid dendritic cells activated by influenza virus and CD40L drive a potent TH1 polarization. Nat. Immunol. 1, 305–310.

Cervantes-Barragan, L., Züst, R., Weber, F., Spiegel, M., Lang, K.S., Akira, S., Thiel, V., Ludewig, B., 2007. Control of coronavirus infection through plasmacytoid dendritic-cell-derived type I interferon. Blood 109, 1131–1137.

Cervantes-Barragan, L., Lewis, K.L., Firner, S., Thiel, V., Hugues, S., Reith, W., Ludewig, B., Reizis, B., 2012. Plasmacytoid dendritic cells control T-cell response to chronic viral infection. Proc. Natl. Acad. Sci. U. S. A. 109, 3012–3017.

Chappell, C.P., Giltiay, N.V., Draves, K.E., Chen, C., Hayden-Ledbetter, M.S., Shlomchik, M.J., Kaplan, D.H., Clark, E.A., 2014. Targeting antigens through blood dendritic cell antigen 2 on plasmacytoid dendritic cells promotes immunologic tolerance. J. Immunol. 192, 5789–5801.

Cisse, B., Caton, M.L., Lehner, M., Maeda, T., Scheu, S., Locksley, R., Holmberg, D., Zweier, C., den Hollander, N.S., Kant, S.G., Holter, W., Rauch, A., Zhuang, Y., Reizis, B., 2008. Transcription factor E2-2 is an essential and specific regulator of plasmacytoid dendritic cell development. Cell 135, 37–48.

Comeau, M.R., Van der Vuurst de Vries, A.-R., Maliszewski, C.R., Galibert, L., 2002. CD123bright plasmacytoid predendritic cells: progenitors undergoing cell fate conversion? J. Immunol. 169, 75–83.

Conrad, C., Di Domizio, J., Mylonas, A., Belkhodja, C., Demaria, O., Navarini, A.A., Lapointe, A.-K., French, L.E., Vernez, M., Gilliet, M., 2018. TNF blockade induces a dysregulated type I interferon response without autoimmunity in paradoxical psoriasis. Nat. Commun. 9, 25.

Contractor, N., Louten, J., Kim, L., Biron, C.A., Kelsall, B.L., 2007. Cutting edge: Peyer's patch plasmacytoid dendritic cells (pDCs) produce low levels of type I interferons: possible role for IL-10, TGFbeta, and prostaglandin E2 in conditioning a unique mucosal pDC phenotype. J. Immunol. 179, 2690–2694.

Corcoran, L., Ferrero, I., Vremec, D., Lucas, K., Waithman, J., O'Keeffe, M., Wu, L., Wilson, A., Shortman, K., 2003. The lymphoid past of mouse plasmacytoid cells and thymic dendritic cells. J. Immunol. 170, 4926–4932.

Crow, M.K., Olferiev, M., Kirou, K.A., 2019. Type I interferons in autoimmune disease. Annu. Rev. Pathol. 14, 369–393.

Cytlak, U., Resteu, A., Bogaert, D., Kuehn, H.S., Altmann, T., Gennery, A., Jackson, G., Kumanovics, A., Voelkerding, K.V., Prader, S., Dullaers, M., Reichenbach, J., Hill, H., Haerynck, F., Rosenzweig, S.D., Collin, M., Bigley, V., 2018. Ikaros family zinc finger 1 regulates dendritic cell development and function in humans. Nat. Commun. 9, 1239.

Dalod, M., Salazar-Mather, T.P., Malmgaard, L., Lewis, C., Asselin-Paturel, C., Brière, F., Trinchieri, G., Biron, C.A., 2002. Interferon alpha/beta and interleukin 12 responses to viral infections: pathways regulating dendritic cell cytokine expression in vivo. J. Exp. Med. 195, 517–528.

Dasgupta, S., Erturk-Hasdemir, D., Ochoa-Reparaz, J., Reinecker, H.-C., Kasper, D.L., 2014. Plasmacytoid dendritic cells mediate anti-inflammatory responses to a gut commensal molecule via both innate and adaptive mechanisms. Cell Host Microbe 15, 413–423.

Davison, L.M., Jørgensen, T.N., 2015. Sialic acid-binding immunoglobulin-type lectin H-positive plasmacytoid dendritic cells drive spontaneous lupus-like disease development in B6.Nba2 mice. Arthritis Rheumatol. 67, 1012–1022.

Di Pucchio, T., Chatterjee, B., Smed-Sörensen, A., Clayton, S., Palazzo, A., Montes, M., Xue, Y., Mellman, I., Banchereau, J., Connolly, J.E., 2008. Direct proteasome-independent cross-presentation of viral antigen by plasmacytoid dendritic cells on major histocompatibility complex class I. Nat. Immunol. 9, 551–557.

Douam, F., Ziegler, C.G.K., Hrebikova, G., Fant, B., Leach, R., Parsons, L., Wang, W., Gaska, J.M., Winer, B.Y., Heller, B., Shalek, A.K., Ploss, A., 2018. Selective expansion of myeloid and NK cells in humanized mice yields human-like vaccine responses. Nat. Commun. 9, 5031.

Dress, R.J., Dutertre, C.-A., Giladi, A., Schlitzer, A., Low, I., Shadan, N.B., Tay, A., Lum, J., Kairi, M.F.B.M., Hwang, Y.Y., Becht, E., Cheng, Y., Chevrier, M., Larbi, A., Newell, E.W., Amit, I., Chen, J., Ginhoux, F., 2019. Plasmacytoid dendritic cells develop from Ly6D + lymphoid progenitors distinct from the myeloid lineage. Nat. Immunol. 20, 852–864.

Drobits, B., Holcmann, M., Amberg, N., Swiecki, M., Grundtner, R., Hammer, M., Colonna, M., Sibilia, M., 2012. Imiquimod clears tumors in mice independent of adaptive immunity by converting pDCs into tumor-killing effector cells. J. Clin. Invest. 122, 575–585.

Duraes, F.V., Lippens, C., Steinbach, K., Dubrot, J., Brighouse, D., Bendriss-Vermare, N., Issazadeh-Navikas, S., Merkler, D., Hugues, S., 2015. pDC therapy induces recovery from EAE by recruiting endogenous pDC to sites of CNS inflammation. J. Autoimmun. 67, 8–18.

Durai, V., Bagadia, P., Briseño, C.G., Theisen, D.J., Iwata, A., Davidson, J.T., Gargaro, M., Fremont, D.H., Murphy, T.L., Murphy, K.M., 2018. Altered compensatory cytokine signaling underlies the discrepancy between Flt3-/- and Flt3l-/- mice. J. Exp. Med. 215, 1417–1435.

Durai, V., Bagadia, P., Granja, J.M., Satpathy, A.T., Kulkarni, D.H., Davidson, J.T., Wu, R., Patel, S.J., Iwata, A., Liu, T.-T., Huang, X., Briseño, C.G., Grajales-Reyes, G.E., Wöhner, M., Tagoh, H., Kee, B.L., Newberry, R.D., Busslinger, M., Chang, H.Y., Murphy, T.L., Murphy, K.M., 2019. Cryptic activation of an Irf8 enhancer governs cDC1 fate specification. Nat. Immunol. 20, 1161–1173.

Edwards, A.D., Diebold, S.S., Slack, E.M.C., Tomizawa, H., Hemmi, H., Kaisho, T., Akira, S., Reis e Sousa, C., 2003. Toll-like receptor expression in murine DC subsets: lack of TLR7 expression by CD8 alpha+ DC correlates with unresponsiveness to imidazoquinolines. Eur. J. Immunol. 33, 827–833.

Faget, J., Sisirak, V., Blay, J.-Y., Caux, C., Bendriss-Vermare, N., Ménétrier-Caux, C., 2013. ICOS is associated with poor prognosis in breast cancer as it promotes the amplification of immunosuppressive CD4(+) T cells by plasmacytoid dendritic cells. Oncoimmunology 2, e23185.

Fischer, C., Metsger, M., Bauch, S., Vidal, R., Böttcher, M., Grote, P., Kliem, M., Sauer, S., 2019. Signals trigger state-specific transcriptional programs to support diversity and homeostasis in immune cells. Sci. Signal. 12, eaao5820.

Fujimori, T., Grabiec, A.M., Kaur, M., Bell, T.J., Fujino, N., Cook, P.C., Svedberg, F.R., MacDonald, A.S., Maciewicz, R.A., Singh, D., Hussell, T., 2015. The Axl receptor tyrosine kinase is a discriminator of macrophage function in the inflamed lung. Mucosal Immunol. 8, 1021–1030.

Galli, S.J., Borregaard, N., Wynn, T.A., 2011. Phenotypic and functional plasticity of cells of innate immunity: macrophages, mast cells and neutrophils. Nat. Immunol. 12, 1035–1044.

Ghosh, H.S., Cisse, B., Bunin, A., Lewis, K.L., Reizis, B., 2010. Continuous expression of the transcription factor e2-2 maintains the cell fate of mature plasmacytoid dendritic cells. Immunity 33, 905–916.

Ghosh, H.S., Ceribelli, M., Matos, I., Lazarovici, A., Bussemaker, H.J., Lasorella, A., Hiebert, S.W., Liu, K., Staudt, L.M., Reizis, B., 2014. ETO family protein Mtg16 regulates the balance of dendritic cell subsets by repressing Id2. J. Exp. Med. 211, 1623–1635.

Gilliet, M., Boonstra, A., Paturel, C., Antonenko, S., Xu, X.-L., Trinchieri, G., O'Garra, A., Liu, Y.-J., 2002. The development of murine plasmacytoid dendritic cell precursors is differentially regulated by FLT3-ligand and granulocyte/macrophage colony-stimulating factor. J. Exp. Med. 195, 953–958.

Gilliet, M., Cao, W., Liu, Y.-J., 2008. Plasmacytoid dendritic cells: sensing nucleic acids in viral infection and autoimmune diseases. Nat. Rev. Immunol. 8, 594–606.

Glitzner, E., Korosec, A., Brunner, P.M., Drobits, B., Amberg, N., Schonthaler, H.B., Kopp, T., Wagner, E.F., Stingl, G., Holcmann, M., Sibilia, M., 2014. Specific roles for dendritic cell subsets during initiation and progression of psoriasis. EMBO Mol. Med. 6, 1312–1327.

Goubier, A., Dubois, B., Gheit, H., Joubert, G., Villard-Truc, F., Asselin-Paturel, C., Trinchieri, G., Kaiserlian, D., 2008. Plasmacytoid dendritic cells mediate oral tolerance. Immunity 29, 464–475.

Grouard, G., Rissoan, M.C., Filgueira, L., Durand, I., Banchereau, J., Liu, Y.J., 1997. The enigmatic plasmacytoid T cells develop into dendritic cells with interleukin (IL)-3 and CD40-ligand. J. Exp. Med. 185, 1101–1111.

Guiducci, C., Ott, G., Chan, J.H., Damon, E., Calacsan, C., Matray, T., Lee, K.-D., Coffman, R.L., Barrat, F.J., 2006. Properties regulating the nature of the plasmacytoid dendritic cell response to Toll-like receptor 9 activation. J. Exp. Med. 203, 1999–2008.

Gutschow, M.V., Mason, J.C., Lane, K.M., Maayan, I., Hughey, J.J., Bajar, B.T., Amatya, D.N., Valle, S.D., Covert, M.W., 2019. Combinatorial processing of bacterial and host-derived innate immune stimuli at the single-cell level. Mol. Biol. Cell 30, 282–292.

Haas, S., Trumpp, A., Milsom, M.D., 2018. Causes and consequences of hematopoietic stem cell heterogeneity. Cell Stem Cell 22, 627–638.

Hadeiba, H., Sato, T., Habtezion, A., Oderup, C., Pan, J., Butcher, E.C., 2008. CCR9 expression defines tolerogenic plasmacytoid dendritic cells able to suppress acute graft-versus-host disease. Nat. Immunol. 9, 1253–1260.

Hadeiba, H., Lahl, K., Edalati, A., Oderup, C., Habtezion, A., Pachynski, R., Nguyen, L., Ghodsi, A., Adler, S., Butcher, E.C., 2012. Plasmacytoid dendritic cells transport peripheral antigens to the thymus to promote central tolerance. Immunity 36, 438–450.

Helft, J., Anjos-Afonso, F., van der Veen, A.G., Chakravarty, P., Bonnet, D., Reis e Sousa, C., 2017. Dendritic cell lineage potential in human early hematopoietic progenitors. Cell Rep. 20, 529–537.

Hirahara, K., Poholek, A., Vahedi, G., Laurence, A., Kanno, Y., Milner, J.D., O'Shea, J.J., 2013. Mechanisms underlying helper T-cell plasticity: implications for immune-mediated disease. J. Allergy Clin. Immunol. 131, 1276–1287.

Hoeffel, G., Ripoche, A.-C., Matheoud, D., Nascimbeni, M., Escriou, N., Lebon, P., Heshmati, F., Guillet, J.-G., Gannagé, M., Caillat-Zucman, S., Casartelli, N., Schwartz, O., De la Salle, H., Hanau, D., Hosmalin, A., Marañón, C., 2007. Antigen crosspresentation by human plasmacytoid dendritic cells. Immunity 27, 481–492.

Hole, C.R., Wager, C.M.L., Castro-Lopez, N., Campuzano, A., Cai, H., Wozniak, K.L., Wang, Y., Wormley, F.L., 2019. Induction of memory-like dendritic cell responses in vivo. Nat. Commun. 10, 2955.

Hughes, G.C., Thomas, S., Li, C., Kaja, M.-K., Clark, E.A., 2008. Cutting edge: progesterone regulates IFN-alpha production by plasmacytoid dendritic cells. J. Immunol. 180, 2029–2033.

Idoyaga, J., Lubkin, A., Fiorese, C., Lahoud, M.H., Caminschi, I., Huang, Y., Rodriguez, A., Clausen, B.E., Park, C.G., Trumpfheller, C., Steinman, R.M., 2011. Comparable T helper 1 (Th1) and CD8 T-cell immunity by targeting HIV gag p24 to CD8 dendritic cells within antibodies to Langerin, DEC205, and Clec9A. Proc. Natl. Acad. Sci. U. S. A. 108, 2384–2389.

Idoyaga, J., Fiorese, C., Zbytnuik, L., Lubkin, A., Miller, J., Malissen, B., Mucida, D., Merad, M., Steinman, R.M., 2013. Specialized role of migratory dendritic cells in peripheral tolerance induction. J. Clin. Invest. 123, 844–854.

Ippolito, G.C., Dekker, J.D., Wang, Y.-H., Lee, B.-K., Shaffer, A.L., Lin, J., Wall, J.K., Lee, B.-S., Staudt, L.M., Liu, Y.-J., Iyer, V.R., Tucker, H.O., 2014. Dendritic cell fate is determined by BCL11A. Proc. Natl. Acad. Sci. U. S. A. 111, E998–1006.

Irla, M., Küpfer, N., Suter, T., Lissilaa, R., Benkhoucha, M., Skupsky, J., Lalive, P.H., Fontana, A., Reith, W., Hugues, S., 2010. MHC class II-restricted antigen presentation by plasmacytoid dendritic cells inhibits T cell-mediated autoimmunity. J. Exp. Med. 207, 1891–1905.

Ishikawa, H., Ma, Z., Barber, G.N., 2009. STING regulates intracellular DNA-mediated, type I interferon-dependent innate immunity. Nature 461, 788–792.

Ito, T., Kanzler, H., Duramad, O., Cao, W., Liu, Y.-J., 2006. Specialization, kinetics, and repertoire of type 1 interferon responses by human plasmacytoid predendritic cells. Blood 107, 2423–2431.

Ito, T., Yang, M., Wang, Y.-H., Lande, R., Gregorio, J., Perng, O.A., Qin, X.-F., Liu, Y.-J., Gilliet, M., 2007. Plasmacytoid dendritic cells prime IL-10-producing T regulatory cells by inducible costimulator ligand. J. Exp. Med. 204, 105–115.

Izaguirre, A., Barnes, B.J., Amrute, S., Yeow, W.-S., Megjugorac, N., Dai, J., Feng, D., Chung, E., Pitha, P.M., Fitzgerald-Bocarsly, P., 2003. Comparative analysis of IRF and IFN-alpha expression in human plasmacytoid and monocyte-derived dendritic cells. J. Leukoc. Biol. 74, 1125–1138.

Jaehn, P.S., Zaenker, K.S., Schmitz, J., Dzionek, A., 2008. Functional dichotomy of plasmacytoid dendritic cells: antigen-specific activation of T cells versus production of type I interferon. Eur. J. Immunol. 38, 1822–1832.

Jarrossay, D., Napolitani, G., Colonna, M., Sallusto, F., Lanzavecchia, A., 2001. Specialization and complementarity in microbial molecule recognition by human myeloid and plasmacytoid dendritic cells. Eur. J. Immunol. 31, 3388–3393.

Kadowaki, N., Antonenko, S., Lau, J.Y., Liu, Y.J., 2000. Natural interferon alpha/beta-producing cells link innate and adaptive immunity. J. Exp. Med. 192, 219–226.

Kalb, M.L., Glaser, A., Stary, G., Koszik, F., Stingl, G., 2012. TRAIL(+) human plasmacytoid dendritic cells kill tumor cells in vitro: mechanisms of imiquimod- and IFN-α-mediated antitumor reactivity. J. Immunol. 188, 1583–1591.

Kamogawa-Schifter, Y., Ohkawa, J., Namiki, S., Arai, N., Arai, K.-I., Liu, Y., 2005. Ly49Q defines 2 pDC subsets in mice. Blood 105, 2787–2792.

Kellogg, R.A., Tian, C., Etzrodt, M., Tay, S., 2017. Cellular decision making by non-integrative processing of TLR inputs. Cell Rep. 19, 125–135.

Kerkmann, M., Rothenfusser, S., Hornung, V., Towarowski, A., Wagner, M., Sarris, A., Giese, T., Endres, S., Hartmann, G., 2003. Activation with CpG-A and CpG-B oligonucleotides reveals two distinct regulatory pathways of type I IFN synthesis in human plasmacytoid dendritic cells. J. Immunol. 170, 4465–4474.

Kim, S., Kaiser, V., Beier, E., Bechheim, M., Guenthner-Biller, M., Ablasser, A., Berger, M., Endres, S., Hartmann, G., Hornung, V., 2014. Self-priming determines high type I IFN production by plasmacytoid dendritic cells. Eur. J. Immunol. 44, 807–818.

Kool, M., van Nimwegen, M., Willart, M.A.M., Muskens, F., Boon, L., Smit, J.J., Coyle, A., Clausen, B.E., Hoogsteden, H.C., Lambrecht, B.N., Hammad, H., 2009. An anti-inflammatory role for plasmacytoid dendritic cells in allergic airway inflammation. J. Immunol. 183, 1074–1082.

Krieg, A.M., 2008. Toll-like receptor 9 (TLR9) agonists in the treatment of cancer. Oncogene 27, 161–167.

Krieg, A.M., 2012. CpG still rocks! Update on an accidental drug. Nucleic Acid Ther. 22, 77–89.

Krug, A., French, A.R., Barchet, W., Fischer, J.A.A., Dzionek, A., Pingel, J.T., Orihuela, M.M., Akira, S., Yokoyama, W.M., Colonna, M., 2004. TLR9-dependent recognition of MCMV by IPC and DC generates coordinated cytokine responses that activate antiviral NK cell function. Immunity 21, 107–119.

Kumagai, Y., Kumar, H., Koyama, S., Kawai, T., Takeuchi, O., Akira, S., 2009. Cutting edge: TLR-dependent viral recognition along with type I IFN positive feedback signaling masks the requirement of viral replication for IFN-{alpha} production in plasmacytoid dendritic cells. J. Immunol. 182, 3960–3964.

Labidi-Galy, S.I., Treilleux, I., Goddard-Leon, S., Combes, J.-D., Blay, J.-Y., Ray-Coquard, I., Caux, C., Bendriss-Vermare, N., 2012. Plasmacytoid dendritic cells infiltrating ovarian cancer are associated with poor prognosis. Oncoimmunology 1, 380–382.

Lau, C.M., Nish, S.A., Yogev, N., Waisman, A., Reiner, S.L., Reizis, B., 2016. Leukemia-associated activating mutation of Flt3 expands dendritic cells and alters T cell responses. J. Exp. Med. 213, 415–431.

Lennert, K., Remmele, W., 1958. Karyometric research on lymph node cells in man. I. Germinoblasts, lymphoblasts & lymphocytes. Acta Haematol. 19, 99–113.

Li, X.-D., Wu, J., Gao, D., Wang, H., Sun, L., Chen, Z.J., 2013. Pivotal roles of cGAS-cGAMP signaling in antiviral defense and immune adjuvant effects. Science 341, 1390–1394.

Li, G., Cheng, M., Nunoya, J.-I., Cheng, L., Guo, H., Yu, H., Liu, Y.-J., Su, L., Zhang, L., 2014. Plasmacytoid dendritic cells suppress HIV-1 replication but contribute to HIV-1 induced immunopathogenesis in humanized mice. PLoS Pathog. 10, e1004291.

Liou, L.-Y., Blasius, A.L., Welch, M.J., Colonna, M., Oldstone, M.B.A., Zúñiga, E.I., 2008. In vivo conversion of BM plasmacytoid DC into CD11b+ conventional DC during virus infection. Eur. J. Immunol. 38, 3388–3394.

Liu, K., Waskow, C., Liu, X., Yao, K., Hoh, J., Nussenzweig, M., 2007. Origin of dendritic cells in peripheral lymphoid organs of mice. Nat. Immunol. 8, 578–583.

Liu, K., Victora, G.D., Schwickert, T.A., Guermonprez, P., Meredith, M.M., Yao, K., Chu, F.-F., Randolph, G.J., Rudensky, A.Y., Nussenzweig, M., 2009. In vivo analysis of dendritic cell development and homeostasis. Science 324, 392–397.

Loschko, J., Heink, S., Hackl, D., Dudziak, D., Reindl, W., Korn, T., Krug, A.B., 2011a. Antigen targeting to plasmacytoid dendritic cells via Siglec-H inhibits Th cell-dependent autoimmunity. J. Immunol. 187, 6346–6356.

Loschko, J., Schlitzer, A., Dudziak, D., Drexler, I., Sandholzer, N., Bourquin, C., Reindl, W., Krug, A.B., 2011b. Antigen delivery to plasmacytoid dendritic cells via BST2 induces protective T cell-mediated immunity. J. Immunol. 186, 6718–6725.

Loschko, J., Rieke, G.J., Schreiber, H.A., Meredith, M.M., Yao, K.-H., Guermonprez, P., Nussenzweig, M.C., 2016. Inducible targeting of cDCs and their subsets in vivo. J. Immunol. Methods 434, 32–38.

Macal, M., Jo, Y., Dallari, S., Chang, A.Y., Dai, J., Swaminathan, S., Wehrens, E.J., Fitzgerald-Bocarsly, P., Zúñiga, E.I., 2018. Self-renewal and Toll-like receptor signaling sustain exhausted plasmacytoid dendritic cells during chronic viral infection. Immunity 48. 730–744.e5.

Manh, T.-P.V., Alexandre, Y., Baranek, T., Crozat, K., Dalod, M., 2013. Plasmacytoid, conventional, and monocyte-derived dendritic cells undergo a profound and convergent genetic reprogramming during their maturation. Eur. J. Immunol. 43, 1706–1715.

Manz, M.G., Traver, D., Miyamoto, T., Weissman, I.L., Akashi, K., 2001. Dendritic cell potentials of early lymphoid and myeloid progenitors. Blood 97, 3333–3341.

Martín-Gayo, E., Sierra-Filardi, E., Corbí, A.L., Toribio, M.L., 2010. Plasmacytoid dendritic cells resident in human thymus drive natural Treg cell development. Blood 115, 5366–5375.

Matsui, T., Connolly, J.E., Michnevitz, M., Chaussabel, D., Yu, C. I., Glaser, C., Tindle, S., Pypaert, M., Freitas, H., Piqueras, B., Banchereau, J., Palucka, A.K., 2009. CD2 distinguishes two subsets of human plasmacytoid dendritic cells with distinct phenotype and functions. J. Immunol. 182, 6815–6823.

McGill, J., Van Rooijen, N., Legge, K.L., 2008. Protective influenza-specific CD8 T cell responses require interactions with dendritic cells in the lungs. J. Exp. Med. 205, 1635–1646.

McNab, F., Mayer-Barber, K., Sher, A., Wack, A., O'Garra, A., 2015. Type I interferons in infectious disease. Nat. Rev. Immunol. 15, 87–103.

Means, T.K., Latz, E., Hayashi, F., Murali, M.R., Golenbock, D.T., Luster, A.D., 2005. Human lupus autoantibody-DNA complexes activate DCs through cooperation of CD32 and TLR9. J. Clin. Invest. 115, 407–417.

Merrell, A.J., Stanger, B.Z., 2016. Adult cell plasticity in vivo: de-differentiation and transdifferentiation are back in style. Nat. Rev. Mol. Cell Biol. 17, 413–425.

Müller, U., Steinhoff, U., Reis, L.F., Hemmi, S., Pavlovic, J., Zinkernagel, R.M., Aguet, M., 1994. Functional role of type I and type II interferons in antiviral defense. Science 264, 1918–1921.

Munn, D.H., Sharma, M.D., Hou, D., Baban, B., Lee, J.R., Antonia, S.J., Messina, J.L., Chandler, P., Koni, P.A., Mellor, A.L., 2004. Expression of indoleamine 2,3-dioxygenase by plasmacytoid dendritic cells in tumor-draining lymph nodes. J. Clin. Invest. 114, 280–290.

Nagasawa, M., Schmidlin, H., Hazekamp, M.G., Schotte, R., Blom, B., 2008. Development of human plasmacytoid dendritic cells depends on the combined action of the basic helix-loop-helix factor E2-2 and the Ets factor Spi-B. Eur. J. Immunol. 38, 2389–2400.

Naik, S.H., Sathe, P., Park, H.-Y., Metcalf, D., Proietto, A.I., Dakic, A., Carotta, S., O'Keeffe, M., Bahlo, M., Papenfuss, A., Kwak, J.-Y., Wu, L., Shortman, K., 2007. Development of plasmacytoid and conventional dendritic cell subtypes from single precursor cells derived in vitro and in vivo. Nat. Immunol. 8, 1217–1226.

Nakano, H., Yanagita, M., Gunn, M.D., 2001. CD11c(+)B220(+)Gr-1(+) cells in mouse lymph nodes and spleen display characteristics of plasmacytoid dendritic cells. J. Exp. Med. 194, 1171–1178.

Nestle, F.O., Conrad, C., Tun-Kyi, A., Homey, B., Gombert, M., Boyman, O., Burg, G., Liu, Y.-J., Gilliet, M., 2005. Plasmacytoid predendritic cells initiate psoriasis through interferon-alpha production. J. Exp. Med. 202, 135–143.

Niederquell, M., Kurig, S., Fischer, J.A.A., Tomiuk, S., Swiecki, M., Colonna, M., Johnston, I.C.D., Dzionek, A., 2013. Sca-1 expression defines developmental stages of mouse pDCs that show functional heterogeneity in the endosomal but not lysosomal TLR9 response. Eur. J. Immunol. 43, 2993–3005.

Ochando, J.C., Homma, C., Yang, Y., Hidalgo, A., Garin, A., Tacke, F., Angeli, V., Li, Y., Boros, P., Ding, Y., Jessberger, R., Trinchieri, G., Lira, S.A., Randolph, G.J., Bromberg, J.S., 2006. Alloantigen-presenting plasmacytoid dendritic cells mediate tolerance to vascularized grafts. Nat. Immunol. 7, 652–662.

Omatsu, Y., Iyoda, T., Kimura, Y., Maki, A., Ishimori, M., Toyama-Sorimachi, N., Inaba, K., 2005. Development of murine plasmacytoid dendritic cells defined by increased expression of an inhibitory NK receptor, Ly49Q. J. Immunol. 174, 6657–6662.

Onai, N., Kurabayashi, K., Hosoi-Amaike, M., Toyama-Sorimachi, N., Matsushima, K., Inaba, K., Ohteki, T., 2013. A clonogenic progenitor with prominent plasmacytoid dendritic cell developmental potential. Immunity 38, 943–957.

Onai, N., Asano, J., Kurosaki, R., Kuroda, S., Ohteki, T., 2017. Flexible fate commitment of E2-2high common DC progenitors implies tuning in tissue microenvironments. Int. Immunol. 29, 443–456.

Palucka, A.K., Blanck, J.-P., Bennett, L., Pascual, V., Banchereau, J., 2005. Cross-regulation of TNF and IFN-alpha in autoimmune diseases. Proc. Natl. Acad. Sci. U. S. A. 102, 3372–3377.

Palucka, K., Banchereau, J., Mellman, I., 2010. Designing vaccines based on biology of human dendritic cell subsets. Immunity 33, 464–478.

Panda, S.K., Kolbeck, R., Sanjuan, M.A., 2017. Plasmacytoid dendritic cells in autoimmunity. Curr. Opin. Immunol. 44, 20–25.

Peippo, M., Ignatius, J., 2012. Pitt-Hopkins syndrome. Mol. Syndromol. 2, 171–180.

Pelayo, R., Hirose, J., Huang, J., Garrett, K.P., Delogu, A., Busslinger, M., Kincade, P.W., 2005. Derivation of 2 categories of plasmacytoid dendritic cells in murine bone marrow. Blood 105, 4407–4415.

Pulendran, B., Banchereau, J., Burkeholder, S., Kraus, E., Guinet, E., Chalouni, C., Caron, D., Maliszewski, C., Davoust, J., Fay, J., Palucka, K., 2000. Flt3-ligand and granulocyte colony-stimulating factor mobilize distinct human dendritic cell subsets in vivo. J. Immunol. 165, 566–572.

Raieli, S., Trichot, C., Korniotis, S., Pattarini, L., Soumelis, V., 2019. TLR1/2 orchestrate human plasmacytoid predendritic cell response to gram+ bacteria. PLoS Biol. 17, e3000209.

Reizis, B., 2019. Plasmacytoid dendritic cells: development, regulation, and function. Immunity 50, 37–50.

Reizis, B., Bunin, A., Ghosh, H.S., Lewis, K.L., Sisirak, V., 2011. Plasmacytoid dendritic cells: recent progress and open questions. Annu. Rev. Immunol. 29, 163–183.

Remer, K.A., Apetrei, C., Schwarz, T., Linden, C., Moll, H., 2007. Vaccination with plasmacytoid dendritic cells induces protection against infection with Leishmania major in mice. Eur. J. Immunol. 37, 2463–2473.

Res, P.C., Couwenberg, F., Vyth-Dreese, F.A., Spits, H., 1999. Expression of pTalpha mRNA in a committed dendritic cell precursor in the human thymus. Blood 94, 2647–2657.

Rissoan, M.C., Soumelis, V., Kadowaki, N., Grouard, G., Brière, F., de Waal Malefyt, R., Liu, Y.J., 1999. Reciprocal control of T helper cell and dendritic cell differentiation. Science 283, 1183–1186.

Rodrigues, P.F., Alberti-Servera, L., Eremin, A., Grajales-Reyes, G.E., Ivanek, R., Tussiwand, R., 2018. Distinct progenitor lineages contribute to the heterogeneity of plasmacytoid dendritic cells. Nat. Immunol. 19, 711–722.

Rowland, S.L., Riggs, J.M., Gilfillan, S., Bugatti, M., Vermi, W., Kolbeck, R., Unanue, E.R., Sanjuan, M.A., Colonna, M., 2014. Early, transient depletion of plasmacytoid dendritic cells ameliorates autoimmunity in a lupus model. J. Exp. Med. 211, 1977–1991.

Saitoh, S.-I., Abe, F., Kanno, A., Tanimura, N., Mori Saitoh, Y., Fukui, R., Shibata, T., Sato, K., Ichinohe, T., Hayashi, M., Kubota, K., Kozuka-Hata, H., Oyama, M., Kikko, Y., Katada, T., Kontani, K., Miyake, K., 2017. TLR7 mediated viral recognition results in focal type I interferon secretion by dendritic cells. Nat. Commun. 8, 1592.

Salio, M., Palmowski, M.J., Atzberger, A., Hermans, I.F., Cerundolo, V., 2004. CpG-matured murine plasmacytoid dendritic cells are capable of in vivo priming of functional CD8 T cell responses to endogenous but not exogenous antigens. J. Exp. Med. 199, 567–579.

Salvermoser, J., van Blijswijk, J., Papaioannou, N.E., Rambichler, S., Pasztoi, M., Pakalniškytė, D., Rogers, N.C., Keppler, S.J., Straub, T., Reis e Sousa, C., Schraml, B.U., 2018. Clec9a-mediated ablation of conventional dendritic cells suggests a lymphoid path to generating dendritic cells in vivo. Front. Immunol. 9, 699.

Sapoznikov, A., Fischer, J.A.A., Zaft, T., Krauthgamer, R., Dzionek, A., Jung, S., 2007. Organ-dependent in vivo priming of naive CD4+, but not CD8+, T cells by plasmacytoid dendritic cells. J. Exp. Med. 204, 1923–1933.

Sathe, P., Vremec, D., Wu, L., Corcoran, L., Shortman, K., 2013. Convergent differentiation: myeloid and lymphoid pathways to murine plasmacytoid dendritic cells. Blood 121, 11–19.

Sawai, C.M., Sisirak, V., Ghosh, H.S., Hou, E.Z., Ceribelli, M., Staudt, L.M., Reizis, B., 2013. Transcription factor Runx2 controls the development and migration of plasmacytoid dendritic cells. J. Exp. Med. 210, 2151–2159.

Schiavoni, G., Mattei, F., Sestili, P., Borghi, P., Venditti, M., Morse, H.C., Belardelli, F., Gabriele, L., 2002. ICSBP is essential for the development of mouse type I interferon-producing cells and for the generation and activation of CD8alpha(+) dendritic cells. J. Exp. Med. 196, 1415–1425.

Schlecht, G., Garcia, S., Escriou, N., Freitas, A.A., Leclerc, C., Dadaglio, G., 2004. Murine plasmacytoid dendritic cells induce effector/memory CD8+ T-cell responses in vivo after viral stimulation. Blood 104, 1808–1815.

Schlitzer, A., Loschko, J., Mair, K., Vogelmann, R., Henkel, L., Einwächter, H., Schiemann, M., Niess, J.-H., Reindl, W., Krug, A., 2011. Identification of CCR9- murine plasmacytoid DC precursors with plasticity to differentiate into conventional DCs. Blood 117, 6562–6570.

Schotte, R., Nagasawa, M., Weijer, K., Spits, H., Blom, B., 2004. The ETS transcription factor Spi-B is required for human plasmacytoid dendritic cell development. J. Exp. Med. 200, 1503–1509.

Scott, C.L., Soen, B., Martens, L., Skrypek, N., Saelens, W., Taminau, J., Blancke, G., Van Isterdael, G., Huylebroeck, D., Haigh, J., Saeys, Y., Guilliams, M., Lambrecht, B.N., Berx, G., 2016. The transcription factor Zeb2 regulates development of conventional and plasmacytoid DCs by repressing Id2. J. Exp. Med. 213, 897–911.

See, P., Dutertre, C.-A., Chen, J., Günther, P., McGovern, N., Irac, S.E., Gunawan, M., Beyer, M., Händler, K., Duan, K., Sumatoh, H.R.B., Ruffin, N., Jouve, M., Gea-Mallorquí, E., Hennekam, R.C.M., Lim, T., Yip, C.C., Wen, M., Malleret, B., Low, I., Shadan, N.B., Fen, C.F.S., Tay, A., Lum, J., Zolezzi, F., Larbi, A., Poidinger, M., Chan, J.K.Y., Chen, Q., Renia, L., Haniffa, M., Benaroch, P., Schlitzer, A., Schultze, J.L., Newell, E.W., Ginhoux, F., 2017. Mapping the human DC lineage through the integration of high-dimensional techniques. Science 356, eaag3009.

Segura, E., Wong, J., Villadangos, J.A., 2009. Cutting edge: B220+CCR9- dendritic cells are not plasmacytoid dendritic cells but are precursors of conventional dendritic cells. J. Immunol. 183, 1514–1517.

Segura, E., Durand, M., Amigorena, S., 2013. Similar antigen cross-presentation capacity and phagocytic functions in all freshly isolated human lymphoid organ-resident dendritic cells. J. Exp. Med. 210, 1035–1047.

Seillet, C., Laffont, S., Trémollières, F., Rouquié, N., Ribot, C., Arnal, J.-F., Douin-Echinard, V., Gourdy, P., Guéry, J.-C., 2012. The TLR-mediated response of plasmacytoid dendritic cells is positively regulated by estradiol in vivo through cell-intrinsic estrogen receptor α signaling. Blood 119, 454–464.

Shigematsu, H., Reizis, B., Iwasaki, H., Mizuno, S.-I., Hu, D., Traver, D., Leder, P., Sakaguchi, N., Akashi, K., 2004. Plasmacytoid dendritic cells activate lymphoid-specific genetic programs irrespective of their cellular origin. Immunity 21, 43–53.

Shinohara, M.L., Lu, L., Bu, J., Werneck, M.B.F., Kobayashi, K.S., Glimcher, L.H., Cantor, H., 2006. Osteopontin expression is essential for interferon-alpha production by plasmacytoid dendritic cells. Nat. Immunol. 7, 498–506.

Sichien, D., Scott, C.L., Martens, L., Vanderkerken, M., Van Gassen, S., Plantinga, M., Joeris, T., De Prijck, S., Vanhoutte, L., Vanheerswynghels, M., Van Isterdael, G., Toussaint, W., Madeira, F.B., Vergote, K., Agace, W.W., Clausen, B.E., Hammad, H., Dalod, M., Saeys, Y., Lambrecht, B.N., Guilliams, M., 2016. IRF8 transcription factor controls survival and function of terminally differentiated conventional and plasmacytoid dendritic cells, respectively. Immunity 45, 626–640.

Siegal, F.P., Kadowaki, N., Shodell, M., Fitzgerald-Bocarsly, P.A., Shah, K., Ho, S., Antonenko, S., Liu, Y.J., 1999. The nature of the principal type 1 interferon-producing cells in human blood. Science 284, 1835–1837.

Sisirak, V., Faget, J., Gobert, M., Goutagny, N., Vey, N., Treilleux, I., Renaudineau, S., Poyet, G., Labidi-Galy, S.I., Goddard-Leon, S., Durand, I., Le Mercier, I., Bajard, A., Bachelot, T., Puisieux, A., Puisieux, I., Blay, J.-Y., Ménétrier-Caux, C., Caux, C., Bendriss-Vermare, N., 2012. Impaired IFN-α production by plasmacytoid dendritic cells favors regulatory T-cell expansion that may contribute to breast cancer progression. Cancer Res. 72, 5188–5197.

Sisirak, V., Ganguly, D., Lewis, K.L., Couillault, C., Tanaka, L., Bolland, S., D'Agati, V., Elkon, K.B., Reizis, B., 2014. Genetic evidence for the role of plasmacytoid dendritic cells in systemic lupus erythematosus. J. Exp. Med. 211, 1969–1976.

Smit, J.J., Rudd, B.D., Lukacs, N.W., 2006. Plasmacytoid dendritic cells inhibit pulmonary immunopathology and promote clearance of respiratory syncytial virus. J. Exp. Med. 203, 1153–1159.

Snell, L.M., McGaha, T.L., Brooks, D.G., 2017. Type I interferon in chronic virus infection and cancer. Trends Immunol. 38, 542–557.

Steinman, R.M., Idoyaga, J., 2010. Features of the dendritic cell lineage. Immunol. Rev. 234, 5–17.

Sun, X., Hua, S., Chen, H.-R., Ouyang, Z., Einkauf, K., Tse, S., Ard, K., Ciaranello, A., Yawetz, S., Sax, P., Rosenberg, E.S., Lichterfeld, M., Yu, X.G., 2017. Transcriptional changes during naturally acquired zika virus infection render dendritic cells highly conducive to viral replication. Cell Rep. 21, 3471–3482.

Swiecki, M., Gilfillan, S., Vermi, W., Wang, Y., Colonna, M., 2010. Plasmacytoid dendritic cell ablation impacts early interferon responses and antiviral NK and CD8(+) T cell accrual. Immunity 33, 955–966.

Swiecki, M., Wang, Y., Gilfillan, S., Colonna, M., 2013. Plasmacytoid dendritic cells contribute to systemic but not local antiviral responses to HSV infections. PLoS Pathog. 9, e1003728.

Swiecki, M., Wang, Y., Riboldi, E., Kim, A.H.J., Dzutsev, A., Gilfillan, S., Vermi, W., Ruedl, C., Trinchieri, G., Colonna, M., 2014. Cell depletion in mice that express diphtheria toxin receptor under the control of SiglecH encompasses more than plasmacytoid dendritic cells. J. Immunol. 192, 4409–4416.

Takauji, R., Iho, S., Takatsuka, H., Yamamoto, S., Takahashi, T., Kitagawa, H., Iwasaki, H., Iida, R., Yokochi, T., Matsuki, T., 2002. CpG-DNA-induced IFN-alpha production involves p38 MAPK-dependent STAT1 phosphorylation in human plasmacytoid dendritic cell precursors. J. Leukoc. Biol. 72, 1011–1019.

Tel, J., Benitez-Ribas, D., Hoosemans, S., Cambi, A., Adema, G.J., Figdor, C.G., Tacken, P.J., de Vries, I.J.M., 2011. DEC-205 mediates antigen uptake and presentation by both resting and activated human plasmacytoid dendritic cells. Eur. J. Immunol. 41, 1014–1023.

Tel, J., van der Leun, A.M., Figdor, C.G., Torensma, R., de Vries, I.J.M., 2012. Harnessing human plasmacytoid dendritic cells as professional APCs. Cancer Immunol. Immunother. 61, 1279–1288.

Tel, J., Aarntzen, E.H.J.G., Baba, T., Schreibelt, G., Schulte, B.M., Benitez-Ribas, D., Boerman, O.C., Croockewit, S., Oyen, W.J.G., van Rossum, M., Winkels, G., Coulie, P.G., Punt, C.J.A., Figdor, C.G., de Vries, I.J.M., 2013. Natural human plasmacytoid dendritic cells induce antigen-specific T-cell responses in melanoma patients. Cancer Res. 73, 1063–1075.

Thomas, J.M., Pos, Z., Reinboth, J., Wang, R.Y., Wang, E., Frank, G.M., Lusso, P., Trinchieri, G., Alter, H.J., Marincola, F.M., Thomas, E., 2014. Differential responses of plasmacytoid dendritic cells to influenza virus and distinct viral pathogens. J. Virol. 88, 10758–10766.

Tomasello, E., Naciri, K., Chelbi, R., Bessou, G., Fries, A., Gressier, E., Abbas, A., Pollet, E., Pierre, P., Lawrence, T., Vu Manh, T.-P., Dalod, M., 2018. Molecular dissection of plasmacytoid dendritic cell activation in vivo during a viral infection. EMBO J. 37, e98836.

Treilleux, I., Blay, J.-Y., Bendriss-Vermare, N., Ray-Coquard, I., Bachelot, T., Guastalla, J.-P., Bremond, A., Goddard, S., Pin, J.-J., Barthelemy-Dubois, C., Lebecque, S., 2004. Dendritic cell infiltration and prognosis of early stage breast cancer. Clin. Cancer Res. 10, 7466–7474.

Tsujimura, H., Tamura, T., Ozato, K., 2003. Cutting edge: IFN consensus sequence binding protein/IFN regulatory factor 8 drives the development of type I IFN-producing plasmacytoid dendritic cells. J. Immunol. 170, 1131–1135.

Uematsu, S., Akira, S., 2007. Toll-like receptors and type I interferons. J. Biol. Chem. 282, 15319–15323.

Urosevic, M., Dummer, R., Conrad, C., Beyeler, M., Laine, E., Burg, G., Gilliet, M., 2005. Disease-independent skin recruitment and activation of plasmacytoid predendritic cells following imiquimod treatment. J. Natl. Cancer Inst. 97, 1143–1153.

Vermi, W., Soncini, M., Melocchi, L., Sozzani, S., Facchetti, F., 2011. Plasmacytoid dendritic cells and cancer. J. Leukoc. Biol. 90, 681–690.

Villadangos, J.A., Young, L., 2008. Antigen-presentation properties of plasmacytoid dendritic cells. Immunity 29, 352–361.

Villani, A.-C., Satija, R., Reynolds, G., Sarkizova, S., Shekhar, K., Fletcher, J., Griesbeck, M., Butler, A., Zheng, S., Lazo, S., Jardine, L., Dixon, D., Stephenson, E., Nilsson, E., Grundberg, I., McDonald, D., Filby, A., Li, W., De Jager, P.L., Rozenblatt-Rosen, O., Lane, A.A., Haniffa, M., Regev, A., Hacohen, N., 2017. Single-cell RNA-seq reveals new types of human blood dendritic cells, monocytes, and progenitors. Science 356, eaah4573.

Wang, J.P., Zhang, L., Madera, R.F., Woda, M., Libraty, D.H., 2012. Plasmacytoid dendritic cell interferon-α production to R-848 stimulation is decreased in male infants. BMC Immunol. 13, 35.

Wang, S., Campos, J., Gallotta, M., Gong, M., Crain, C., Naik, E., Coffman, R.L., Guiducci, C., 2016. Intratumoral injection of a CpG oligonucleotide reverts resistance to PD-1 blockade by expanding multifunctional CD8+ T cells. Proc. Natl. Acad. Sci. U. S. A. 113, E7240–E7249.

Webster, B., Werneke, S.W., Zafirova, B., This, S., Coléon, S., Décembre, E., Paidassi, H., Bouvier, I., Joubert, P.-E., Duffy, D., Walzer, T., Albert, M.L., Dreux, M., 2018. Plasmacytoid dendritic cells control dengue and Chikungunya virus infections via IRF7-regulated interferon responses. eLife 7, e34273.

Wilhelm, T.R., Taddeo, A., Winter, O., Schulz, A.R., Mälzer, J.-N., Domingo, C., Biesen, R., Alexander, T., Thiel, A., Radbruch, A., Hiepe, F., Gerl, V., 2016. Siglec-1-positive plasmacytoid dendritic cells (pDCs) in human peripheral blood: a semi-mature and myeloid-like subset imbalanced during protective and autoimmune responses. Clin. Immunol. 163, 42–51.

Wimmers, F., Subedi, N., van Buuringen, N., Heister, D., Vivié, J., Beeren-Reinieren, I., Woestenenk, R., Dolstra, H., Piruska, A., Jacobs, J.F.M., van Oudenaarden, A., Figdor, C.G., Huck, W.T.S., de Vries, I.J.M., Tel, J., 2018. Single-cell analysis reveals that stochasticity and paracrine signaling control interferon-alpha production by plasmacytoid dendritic cells. Nat. Commun. 9, 3317.

Wohn, C., Ober-Blöbaum, J.L., Haak, S., Pantelyushin, S., Cheong, C., Zahner, S.P., Onderwater, S., Kant, M., Weighardt, H., Holzmann, B., Reizis, B., Becher, B., Prens, E.P., Clausen, B.E., 2013. Langerin(neg) conventional dendritic cells produce IL-23 to drive psoriatic plaque formation in mice. Proc. Natl. Acad. Sci. U.S.A. 110, 10723–10728.

Wolf, A.I., Buehler, D., Hensley, S.E., Cavanagh, L.L., Wherry, E.J., Kastner, P., Chan, S., Weninger, W., 2009. Plasmacytoid dendritic cells are dispensable during primary influenza virus infection. J. Immunol. 182, 871–879.

Worbs, T., Hammerschmidt, S.I., Förster, R., 2017. Dendritic cell migration in health and disease. Nat. Rev. Immunol. 17, 30–48.

Wu, X., Satpathy, A.T., KC, W., Liu, P., Murphy, T.L., Murphy, K.M., 2013. Bcl11a controls Flt3 expression in early hematopoietic progenitors and is required for pDC development in vivo. PLoS One 8, e64800.

Wu, X., Briseño, C.G., Grajales-Reyes, G.E., Haldar, M., Iwata, A., Kretzer, N.M., KC, W., Tussiwand, R., Higashi, Y., Murphy, T.L., Murphy, K.M., 2016. Transcription factor Zeb2 regulates commitment to plasmacytoid dendritic cell and monocyte fate. Proc. Natl. Acad. Sci. U. S. A. 113, 14775–14780.

Yonkers, N.L., Rodriguez, B., Milkovich, K.A., Asaad, R., Lederman, M.M., Heeger, P.S., Anthony, D.D., 2007. TLR ligand-dependent activation of naive CD4 T cells by plasmacytoid dendritic cells is impaired in hepatitis C virus infection. J. Immunol. 178, 4436–4444.

Young, L.J., Wilson, N.S., Schnorrer, P., Proietto, A., ten Broeke, T., Matsuki, Y., Mount, A.M., Belz, G.T., O'Keeffe, M., Ohmura-Hoshino, M., Ishido, S.,

Stoorvogel, W., Heath, W.R., Shortman, K., Villadangos, J.A., 2008. Differential MHC class II synthesis and ubiquitination confers distinct antigen-presenting properties on conventional and plasmacytoid dendritic cells. Nat. Immunol. 9, 1244–1252.

Yu, X., Cai, B., Wang, M., Tan, P., Ding, X., Wu, J., Li, J., Li, Q., Liu, P., Xing, C., Wang, H.Y., Su, X.-Z., Wang, R.-F., 2016. Cross-regulation of two type I interferon signaling pathways in plasmacytoid dendritic cells controls anti-malaria immunity and host mortality. Immunity 45, 1093–1107.

Yun, T.J., Lee, J.S., Machmach, K., Shim, D., Choi, J., Wi, Y.J., Jang, H.S., Jung, I.-H., Kim, K., Yoon, W.K., Miah, M.A., Li, B., Chang, J., Bego, M.G., Pham, T.N.Q., Loschko, J., Fritz, J.H., Krug, A.B., Lee, S.-P., Keler, T., Guimond, J.V., Haddad, E., Cohen, E.A., Sirois, M.G., El-Hamamsy, I., Colonna, M., Oh, G.T., Choi, J.-H., Cheong, C., 2016. Indoleamine 2,3-dioxygenase-expressing aortic plasmacytoid dendritic cells protect against atherosclerosis by induction of regulatory T cells. Cell Metab. 23, 852–866.

Zhang, J., Raper, A., Sugita, N., Hingorani, R., Salio, M., Palmowski, M.J., Cerundolo, V., Crocker, P.R., 2006. Characterization of Siglec-H as a novel endocytic receptor expressed on murine plasmacytoid dendritic cell precursors. Blood 107, 3600–3608.

Zhang, H., Gregorio, J.D., Iwahori, T., Zhang, X., Choi, O., Tolentino, L.L., Prestwood, T., Carmi, Y., Engleman, E.G., 2017. A distinct subset of plasmacytoid dendritic cells induces activation and differentiation of B and T lymphocytes. Proc. Natl. Acad. Sci. U. S. A. 114, 1988–1993.

Zweier, C., Peippo, M.M., Hoyer, J., Sousa, S., Bottani, A., Clayton-Smith, J., Reardon, W., Saraiva, J., Cabral, A., Gohring, I., Devriendt, K., de Ravel, T., Bijlsma, E.K., Hennekam, R.C.M., Orrico, A., Cohen, M., Dreweke, A., Reis, A., Nurnberg, P., Rauch, A., 2007. Haploinsufficiency of TCF4 causes syndromal mental retardation with intermittent hyperventilation (Pitt-Hopkins syndrome). Am. J. Hum. Genet. 80, 994–1001.

CHAPTER FIVE

Dendritic cell extracellular vesicles

Joanna Kowal[a], **Mercedes Tkach**[b,*]
[a]Department of Oncology, Ludwig Institute for Cancer Research, University of Lausanne, Lausanne, Switzerland
[b]Institut Curie, PSL Research University, INSERM U932, Paris, France
*Corresponding author: e-mail address: mercedes.tkach@curie.fr

Contents

1. Introduction 214
2. Molecular identity of dendritic cell-derived extracellular vesicles 216
   2.1 Proteins 216
   2.2 Nucleic acids 219
   2.3 Lipids 220
3. Diversity of extracellular vesicles from dendritic cells 221
4. The journey of an extracellular vesicle: From its biogenesis to its final destination in the target cell 222
   4.1 Biogenesis of dendritic cell derived extracellular vesicles 222
   4.2 Regulation of extracellular vesicles release by external signals 225
   4.3 Interaction with recipient cells 226
5. Role of dendritic cell-derived extracellular vesicles as modulators of immune responses 227
   5.1 Antigen presentation by dendritic cell-derived vesicles 228
   5.2 Antigen-independent functions of dendritic cell-derived vesicles 232
   5.3 Dendritic cell extracellular vesicles and viruses 233
6. Clinical application of dendritic cell-derived extracellular vesicles 235
7. Concluding remarks and perspectives 236
Acknowledgments 237
References 238

Abstract

In addition to direct cell-to-cell contact, dendritic cells (DCs) can regulate the onset of adaptive immunity through the secretion of nano-sized membrane structures, called extracellular vesicles (EVs). This novel mode of communication between cells has added a new layer of complexity to the regulation of immune responses. DCs secrete into their environment different types of EVs containing immunomodulatory molecules that have distinct structural and biochemical properties depending on their intracellular site of origin. Exosomes are generated inside multivesicular bodies and are secreted when these compartments fuse with the plasma membrane, whereas microvesicles are formed and released by budding from the cells' plasma membrane. Once outside the cell of origin, these vesicles can reach target cells through membrane

International Review of Cell and Molecular Biology, Volume 349
ISSN 1937-6448
https://doi.org/10.1016/bs.ircmb.2019.08.005

receptor–ligand interactions, modifying their physiological state by the transfer of the EV content or by triggering cell signaling at the cells' surface. Particularly, EVs released by DCs contain major histocompatibility complex (MHC) class I and class II molecules able to activate cognate T cells and promote humoral responses. These activities motivated the use of DC-derived EVs in the treatment of cancer, infectious diseases and autoimmune disorders. The therapeutic potential of these vesicles led to the use of EVs from tumor antigen-loaded DCs in cancer clinical trials, although with limited clinical effects. In this review we will focus on the different EVs released by DCs, their composition and biogenesis, together with their proposed functions as immune regulators.

1. Introduction

Cells, from prokaryotic organisms to complex eukaryotic systems, are able to produce and secrete membrane-enclosed vesicles. Known as extracellular vesicles (EVs), they emerged as important players in cell-to-cell communication (Colombo et al., 2014; Yanez-Mo et al., 2015). EVs are limited by lipid bilayer with embedded transmembrane proteins that enclose soluble proteins and nucleic acids. Once outside the cell of origin, EVs can affect the physiological state of surrounding cells by the delivery of their cargo inside the acceptor cell but also by receptor-ligand interactions at the surface of the acceptor cell. EVs are present in several biological fluids from living organisms, particularly blood (Caby et al., 2005), semen (Aalberts et al., 2012; Gatti et al., 2005), urine (Kanno et al., 1995; Pisitkun et al., 2004) and tumor effusions (Clayton et al., 2008). There is a growing body of evidence attributing to EVs numerous physiological functions, such as regulation of immune responses, cell differentiation, tissue repair and neural communication (Yanez-Mo et al., 2015). EVs are also implicated in pathological processes such as viral infection or cancer development, progression and metastasis (Tkach and Théry, 2016).

High heterogeneity of EVs is believed to reflect the various intracellular processes that lead to EV biogenesis. Based on proposed EV origin, vesicles have been categorized in the literature into two main classes: (1) microvesicles or ectosomes which originate by direct outward budding from the cells' plasma membrane and (2) exosomes which are formed intracellularly within multivesicular endocytic compartments (multivesicular bodies, MVBs) and released into the extracellular space upon fusion of these MVBs with the plasma membrane (van Niel et al., 2018). Exosomes are small EVs with a size restricted to the diameter of vesicles found inside MVBs

(<150 nm in diameter) whereas microvesicles are not size-constrained and their diameter can vary (from 100 nm to >1 μm). As there is an overlap between the size distribution of EVs coming from endosomal origin and from the plasma membrane, it is very difficult to classify these two types of vesicles based on their size. The term "exosomes" has been extensively used in the literature to refer to the totality of small EVs that sediment at high speed ultracentrifugation (most commonly at 100,000 × *g*), even though the endosomal nature of the vesicles is poorly documented in most studies up to date. In this chapter, we use the generic term EVs, independently of the term used by the authors in the original article, unless there is a deep characterization of their endosomal origin.

Nonuniform nomenclature of EV subtypes across the literature and wide variety of EV isolation methods that are prone to user- and protocol-dependent variability all together impose the limitation on the interpretation of EV-based findings (Witwer and Théry, 2019). One cannot exclude that EVs with ascribed contradictory functions can be in fact preparations of EVs with different technical protocols that yield in the enrichment of distinct EV subtypes (discussed by Tkach et al., 2018). Thus, one must remain careful and attentive about the isolation methods used in different studies. The efforts are being made in the field of EVs to unify and compare methodology applied around the world to obtain EVs (EV-TRACK Consortium et al., 2017; Théry et al., 2018). Very recently new methodology that includes the reference of recombinant EVs was developed in order to aid the comparison of different EV preparation (Geeurickx et al., 2019). Hopefully, all these efforts, undertaken by the scientists across the world, will overcome some of technical hurdles related to EVs and result in easier and less controversial comparison of EV-based discoveries.

First studies reporting EVs were proposing that these vesicles were secreted from cells as a mechanism of waste disposal developed to eliminate unnecessary cellular components (Harding et al., 1983; Pan and Johnstone, 1983). Surprisingly, in the mid-1990s, the first evidence supporting the role of these nanosized structures as mediators of intercellular communication emerged, particularly as regulators of immune responses (Raposo et al., 1996). Raposo and colleagues showed that B cells released EVs bearing peptide-loaded major histocompatibility complex (MHC) class II molecules on their surface, able to promote T cells proliferation (Raposo et al., 1996). Shortly after, EVs from another professional antigen presenting cell, in this case from dendritic cells (DCs), loaded with tumor antigens were shown to promote cytotoxic T cell anti-tumor responses *in vivo* (Zitvogel et al., 1998).

In early 2000s, independent groups of researchers showed that RNA was enclosed within EVs and that could be functionally delivered into recipient cells (Ratajczak et al., 2006; Skog et al., 2008; Valadi et al., 2007). Taken together these seminal works uncovered a new spectrum of functions attributed to EVs and caught the attention of the worldwide scientific community. Since then, the high number of publications reporting functions of EVs in physiological and pathological circumstances have contributed to recognize EVs as intercellular mediators of cell-to-cell communication (Witwer and Théry, 2019). In this chapter, we make particular focus on the studies on EVs released by DCs, mainly on the vesicles released by mouse bone marrow-derived DCs (BMDCs), human primary monocyte-derived DCs (moDCs) and DC cell lines. We will discuss their biochemical properties, biogenesis, proposed functions as antigen presenting entities and antigen independent activities and finally their potential use in therapeutics applications.

2. Molecular identity of dendritic cell-derived extracellular vesicles

Extracellular vesicles are composed of a membrane lipid-bilayer and are packed with cytosol containing soluble proteins, small molecules and genetic information (Fig. 1). The vesicle membrane has the same geometry as plasma membrane of parent cells allowing external exposure of extracellular domains of transmembrane proteins to interact with their respective partners on target cells. However, the exact composition of EV lipid bilayer reflects the site of origin of EVs with endosomal membrane rafts enriched in endosome-derived EVs as exosomes (van Niel et al., 2018). Three main components of EV, meaning proteins, nucleic acids and lipids, will be discussed here.

2.1 Proteins

Since the attention given to EVs in the late 1990s, the protein content of EVs raised the curiosity of investigators. Applying several mass spectrometry and electron microscopy techniques allowed unveiling the theoretical composition of EVs (Chaput et al., 2006; Colombo et al., 2014; Thery et al., 1999), where proteins are proposed to be one of the key components. Proteins might be either enclosed within EV lumen, or included in the limiting membrane, or sometimes associated with the luminal-surface of EVs.

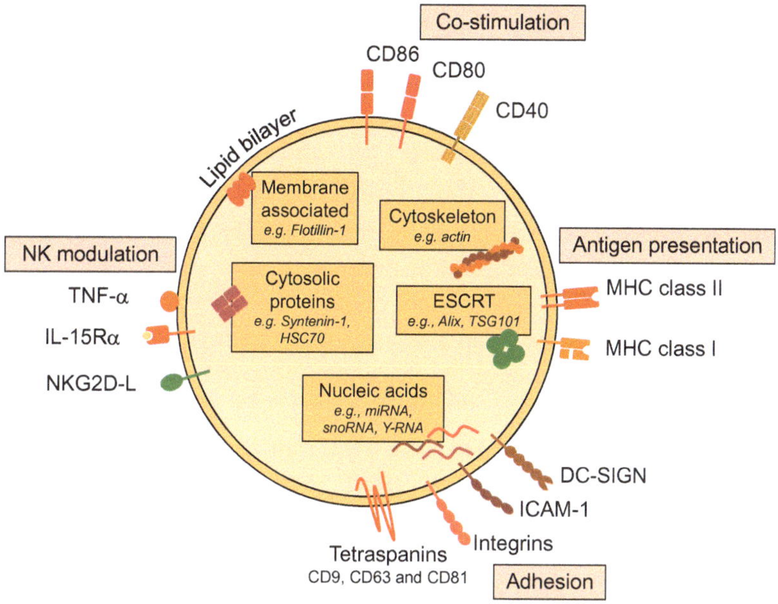

Fig. 1 Major components of dendritic cell-derived EVs. EVs are composed of a lipid bilayer, with an orientation of the limiting membrane similar to the orientation of the plasma membrane. Thus, extracellular domains of transmembrane proteins are exposed to the extracellular space, which allows them to interact with their respective partners. Soluble cytosolic proteins are enclosed within EVs, therefore they are protected from external proteinases. Finally, EVs carry genetic information such as various RNA molecules.

The first proteomic studies had analyzed the content of the bulk of small EVs isolated by high-speed ultracentrifugation. These had uncovered that those EVs, purified from different cell types or body fluids, are enriched in a specific set of proteins mainly present within the secreting cell in the plasma membrane, endosomal compartments and cytosol but not or rarely in other organelles (Bard et al., 2004; Gatti et al., 2005; Mears et al., 2004; Thery et al., 1999, 2001; van Niel et al., 2001; Wubbolts et al., 2003). However, more recently there had been an increased interest in the comparison of EVs of different sizes/density, released by tumor cells or DCs, which had suggested that larger EVs recovered at lower centrifugation speed (generally at 10,000 × *g*) contain other components, like mitochondria, endoplasmic reticulum and cytokeratins (Johnson et al., 2017; Keerthikumar et al., 2015; Kowal et al., 2016; Minciacchi et al., 2015). The restricted protein signature found within small EVs suggested that protein incorporation and sorting into these vesicles is a regulated process. Several families of proteins were shown to be highly abundant within vesicles in comparison with secreting cells, such as: heat shock proteins, ESCRT proteins, MHC complexes, tetraspanins, cytoskeleton proteins and others. It is likely that several abundant components of EVs are playing a role in their biogenesis and secretion. Several of these EV components are ubiquitous for EVs secreted by several cell types and are thus considered general EV markers (Minciacchi et al., 2015) suggesting that there is an EV secretion/biogenesis machinery shared between cells of different origins (Kowal et al., 2016; Mathivanan et al., 2010; Minciacchi et al., 2015).

Importantly, there are other EV components that are cell type-specific. This has been evidenced since the exact proteome of EVs generally reflects the cell of origin. Particularly, EVs released by DCs have been found to contain important amounts of MHC class I and class II molecules (Thery et al., 2009). Interestingly, immature DCs store MHC class II molecules within MVB luminal vesicles, but once these cells are activated, the luminal vesicles fuse with the MVBs, which then fuses with the cells' plasma membrane, thus increasing the MHC class II-peptide availability at the plasma membrane (ten Broeke et al., 2011). Indeed, this change in the subcellular protein distribution, together with changes in the expression profiles, are believed to play a role in the abundance of these protein markers in the EVs released upon maturation. Specifically, LPS-treated BMDCs secreted EVs contain increased levels of MHC class II and the co-stimulation marker CD86, together with increased ICAM-I (Admyre et al., 2006; Segura et al., 2005b). IFN-γ-treatment of murine BMDCs also promoted an increase in CD80 (Qazi et al., 2009). Similar results were observed with EVs released from IFN-γ-matured human moDCs, which contained increased co-stimulatory molecules (CD40, CD80 and CD86) and ICAM-I than EVs released from immature cells (Viaud et al., 2011). Moreover, in the presence of cognate T cells, DCs secrete higher amounts of MHC class II-bearing EVs (Buschow et al., 2009). This MHC class II^{+}-EV secretion up-regulation is even more striking, when a peptide specifically recognized by T cells is added to the DC:T cell co-culture. This MHC class II upregulation is dependent on the T-cell number present in the co-culture, showing the importance of the intensity of the signaling to the DCs (Buschow et al., 2009). Altogether, these results indicate that, in addition to the core of proteins that is maintained along different cell types, there is a specific protein signature of EV which is characteristic of the cell of origin and its physiological state.

2.2 Nucleic acids

Since the initial identification of nucleic acids enclosed within EVs in the mid-2000s, major progresses enabled to better describe the type of genetic messages shuttled within EVs, and put forward some fundamental understanding of their biology (Yanez-Mo et al., 2015). After proofs of presence of functional RNA (Ratajczak et al., 2006; Skog et al., 2008; Valadi et al., 2007) and due to the development of new generation sequencing, a closer look on RNA species within EVs became possible. Beside full length mRNA, EVs contain noncoding RNA that might regulate expression of genes in acceptor cells, such as microRNA (miRNA) (Mittelbrunn et al., 2011) and its immature form: pre-miRNA (Chen et al., 2010). Further confirmation of plethora of RNA species abundantly present in EVs was shown for: tRNA (Bellingham et al., 2012), vault-RNA, Y-RNA (Nolte-'T Hoen et al., 2012) and long noncoding RNA (Huang et al., 2013; Kogure and Patel, 2013). Alongside with RNA, mitochondrial DNA (Guescini et al., 2010; Sansone et al., 2017) and genomic DNA (Cai et al., 2013; Kahlert et al., 2014; Lazaro-Ibanez et al., 2014; Thakur et al., 2014) were also described as present in EV preparations.

DCs had been shown to release RNA enclosed within EVs which can play a role in the regulation of immune responses (Alexander et al., 2015; Montecalvo et al., 2012). Montecalvo and colleagues had compared the miRNA content of EVs purified from immature BMDCs to LPS-matured cells (Montecalvo et al., 2012). They suggested that these EVs were first internalized by target cells in endosomal compartments, followed by membrane fusion and luminal content delivery (Montecalvo et al., 2012). This enabled the transfer of miRNAs to the recipient cell, where they were shown to modulate gene expression by using miRNA luciferase reporter constructs (Montecalvo et al., 2012). Moreover, the EV derived miRNA repression of certain genes within the recipient cell had been linked to the regulation of inflammation, in particular by modifying the cellular response to endotoxins (Alexander et al., 2015). The sequencing of the RNA released within EVs upon cognate DC:T cell interactions led to the finding that EV shuttled several noncoding RNA species, such as tRNA and vRNA (Nolte-'T Hoen et al., 2012). More recently, the same group had shown that functionally distinct DC subsets release particular EV-associated RNA classes (Driedonks et al., 2018). The authors had analyzed by RNA deep sequencing all the RNA species released within EVs by immune-stimulatory (LPS-treated BMDCs) or immune-suppressive

DCs (Vit-D3-treated BMDCs) (Driedonks et al., 2018). In particular, EV-associated miRNA, snoRNA and Y-RNA were affected upon the different treatments of DCs, while other non-coding RNA types remained unchanged (Driedonks et al., 2018). Strikingly, the changes in the RNA species observed within EVs did not reflect the changes observed in the cellular RNA levels, suggesting that a selective set of RNAs got packaged within EVs (Driedonks et al., 2018; Shurtleff et al., 2016, 2017). However, the biological activity of these abundant small noncoding transcripts released in EVs remains to be determined.

Importantly, there are evidences in the literature suggesting that only a fraction of EVs encloses RNA (Chevillet et al., 2014). However, the final proof of RNA abundance within EV might be obtained from single EV studies in the future.

2.3 Lipids

Lipids play an important role in EV biology as they build its limiting membrane and protect the soluble cytosolic content enclosed within it. However, the lipid composition of EVs has been poorly explored in the literature, and even less studied in case of DC derived EVs. In general, it was shown that EVs from different cellular origins are highly enriched in cholesterol and sphingomyelin (Beloribi et al., 2012; Brouwers et al., 2013; Laulagnier et al., 2004; Llorente et al., 2013; Wubbolts et al., 2003). Particularly, sphingomyelin is highly represented in EVs from dendritic and mast cells (Laulagnier et al., 2004) and B-lymphocytes (Wubbolts et al., 2003). Ceramide was also shown as an important component of EVs (Trajkovic et al., 2008) participating in their generation. Lysobisphosphatidic acid, LBPA, a lipid important in the MVB biology, remains a controversial component of EVs detected in some reports (Denzer et al., 2000; Laulagnier et al., 2004) but not in others (Wubbolts et al., 2003). In a microenvironment of low pH, an accumulation of this lipid within the lipid layer of liposomes was shown to induce spontaneous vesicle budding *in vitro* and proposed to be one of the mechanisms driving ILV formation in MVB, therefore indirectly responsible for EV secretion (Matsuo et al., 2004). However, LBPA is clearly not enriched in EVs when its abundance is compared to secreting cells. It is proposed that LBPA remains on the limiting membrane of endosomes after the generation of ILVs is completed and vesicles have pinched off (Subra et al., 2007). Taken into consideration the entire lipid composition of EVs, it was proposed that they remain very rigid, what is reflected in their high insensitivity to detergents (Buschow et al., 2009; Osteikoetxea et al., 2015a). A particular lipid content of EVs suggested that their biogenesis occurs within highly specified membrane rafts (Subra et al., 2007; Tan et al., 2013). However, the different sensitivity of EV subtypes to detergents (Osteikoetxea et al., 2015b) and the different ratio between lipid and protein content between EV subtypes (Osteikoetxea et al., 2015a) together suggest that not all EVs derive from exactly the same lipid rafts. Even between EVs of the same size, but enriched in different protein markers (tetraspanins vs MHC II), slight differences in the lipid content were observed (Laulagnier et al., 2005).

Finally, DC-derived EVs contain phosphatidylserine (PS) (Laulagnier et al., 2004). PS can be found exposed at the surface of small and large EVs (Colombo et al., 2014) where it can bind other molecules, like the milk fat globule EGF/factor VIII (MFGE8), for the case of EVs released by immature BMDCs (Veron et al., 2005). Moreover, PS present on EV surface was shown to bind to Tim4 expressed in macrophages (Miyanishi et al., 2007; Nakai et al., 2016). This interaction had been suggested to be important for EV internalization by phagocytosis (Chairoungdua et al., 2010).

3. Diversity of extracellular vesicles from dendritic cells

The preparations of EVs tend to be a collection of morphologically and probably biogenetically different membrane-enclosed vesicles (Yuana et al., 2013). The heterogeneity of vesicles secreted by cells was observed very early on (Heijnen et al., 1999) and it was proposed that morphologically different EV types were originating from different cellular compartments. Several studies confirmed the coexistence of different EV subsets (Bobrie et al., 2012; Laulagnier et al., 2005; Minciacchi et al., 2015; Tauro et al., 2013; van Niel et al., 2001). The first evidences that DCs release heterogeneous EV populations were provided in the early studies on DC-derived EVs. Indeed, the electron microscopy images of the human moDCs by EM revealed clear presence of membrane-enclosed structures of various sizes at the cell surface (Zitvogel et al., 1998). The electron microscopy analysis revealed that the range of sizes of EVs isolated from moDCs conditioned medium is much wider than for similarly isolated EVs from HeLa CIITA (Colombo et al., 2013). By performing a combination of EV isolation methods, the classical differential ultracentrifugation protocol proposed by Thery et al. (2006) and a bottom-up floatation density gradient, we were able to identify four subpopulations of EVs emanating from DCs with different morphologies and different abundances (Kowal et al., 2016).

To further characterize the different membrane-enclosed structures within these four fractions, we performed extensive mass spectrometry analysis that led us to determine their protein content. Indeed, we had observed an overlapping molecule composition among these fractions but also we were able to identify several proteins specifically sorted to only one type of EVs (Kowal et al., 2016). For example, MHC molecules and other markers generally attributed to the endosomal derived exosomes, as flotillin-1 and HSC70, had been found to be present in all the different types of the isolated vesicles, irrespective of their size (either in small EVs as in the larger ones) (Kowal et al., 2016). Importantly, syntenin-1, the disintegrin and metalloproteinase domain-containing protein 10 (ADAM10), TSG101 and EHD4 were secreted with small EVs and absent in large ones. Conversely, actinin-4, a protein associated with the actin cytoskeleton was enriched in plasma-membrane derived large and medium size vesicles and almost absent in small ones (Kowal et al., 2016). Moreover, we performed an additional approach of separation of small EVs, where the EVs were distinguished thanks to the presence of different tetraspanins: CD63, CD9 and CD81. Comparing side-by-side the vesicles bearing the different tetraspanins on their surface to the vesicles devoided of these markers, we were able to propose a protein signature for the "bona-fide" exosomes from DCs: small vesicles that are coenriched in CD63, CD81 and CD9 and bear the proteins syntenin-1 and TSG101 (Kowal et al., 2016). We thus provided several evidences showing that heterogeneity of EVs is reflected by different sizes and, more importantly, protein content to some extent. However, their respective machineries of biogenesis and their respective relative abundance, for example in biological fluids, remains to be fully characterized.

4. The journey of an extracellular vesicle: From its biogenesis to its final destination in the target cell

4.1 Biogenesis of dendritic cell derived extracellular vesicles

Finding the molecular mechanisms that are exclusively involved in the formation and release of the different EVs is one of the major challenges in the EV field. We believe that the discovery of specific EV subtype biogenesis machineries may enable us firstly to interfere with their formation and subsequently to understand their physio/pathological functions *in vivo*. Such correlation would lead to a new or more precise definition of EV-subtype based on a function and not only on content/morphology.

In the past years great progress has been made to discover what pathways are involved in EV biogenesis in many cell types. As the generic pathways of EV biogenesis have been extensively reviewed elsewhere (Colombo et al., 2014; Kowal et al., 2014; Mathieu et al., 2019), here, we will only focus on discussing the current knowledge on the formation and release of DC derived EVs.

Exosomes had been described to be formed within endosomal compartments (Pan and Johnstone, 1983). Within MVBs, vesicles are generated by invagination of the limiting membrane of an endosome compartment leading to the formation of intraluminal vesicles (ILV), with exactly the same membrane orientation as the mother cell. These MVBs compartments with accumulated ILVs inside can either follow further maturation and fusion with the lysosomes for lysosomal degradation of their content, or alternatively, can be transported toward and fused with the plasma membrane to release their luminal vesicles to the extracellular space (reviewed by Colombo et al., 2014; Kowal et al., 2014). Cells can have different subpopulations of MVBs, with different morphology or lipid and protein composition (Klumperman and Raposo, 2014). When analyzing immature vs matured DCs undergoing cognate interactions with T cells, Buschow et al. have shown the presence of different MVBs in these cells (Buschow et al., 2009). Within immature DCs, ubiquitinated MHC class II molecules are sorted into MVBs for transfer to and degradation in lysosomes. However, upon maturation in the presence of antigen-specific T cells, another pathway is upregulated, which sorts MHC class II molecules together with CD9 to ILVs that are subsequently secreted as exosomes which are then found accumulated at the plasma membrane of the T cell (Buschow et al., 2009). The most-described molecular machinery involved in MVB and ILV formation is driven by the endosomal sorting complex required for transport (ESCRT). This complex has been widely reviewed in the literature (Christ et al., 2017; Hurley, 2015). It is mainly composed by four family of proteins (ESCRT-0, involved in cargo -I, -II and -III), involved in ubiquitinated cargo recognition and clustering, bud formation and vesicle scission (Hurley, 2015). Several ESCRT members were found present in EVs (Buschow et al., 2005; Mathivanan et al., 2010; Thery et al., 2001) and some of them have been associated to EV protein sorting and biogenesis in different cell types (Abrami et al., 2013; Baietti et al., 2012; Colombo et al., 2013). In particular, DC deficient for the ESCRT-0 protein HRS have impaired EVs release after being incubated with an antigen (Tamai et al., 2010). Another protein associated to the ESCRT-III complex, Alix, had been suggested to be involved in EV biogenesis and protein sorting into EVs in the breast cancer cells MCF-7 (Baietti et al., 2012). Moreover, Alix silencing in moDCs increased the overall level of MHC class II expression in cells, DC maturation and decreased the secretion of $CD63^+$/MHC class II^+ EVs (Colombo et al., 2013). ESCRT proteins had been involved in several cellular processes, as plasma membrane repair or autophagosome formation, and therefore, this complex might be involved in the formation of other type of vesicles, and not only bona fide exosomes. We had observed that the inhibition of some particular proteins by lentiviral-based transduction of moDCs led to an increased mortality of these cells. Thus, analyzing exclusively the vesicles coming from live cells present in the culture supernatant of gene silenced moDCs is a big challenge (J. Kowal and C. Thery, unpublished data). Thus, further work on DCs generated from mice genetically engineered to conditionally delete these proteins could overcome these hurdles.

Other molecules, not associated to the ESCRT complex had been also implicated in DC-derived EV biogenesis. For example, BMDCs from mice deficient in the tumor suppressor-activated pathway 6 (TSAP6), a molecule present in the trans-Golgi network, endosomal compartment and plasma membrane, released fewer EVs than their wild-type counterparts (Lespagnol et al., 2008).

In addition to the endosome-derived vesicles, cells release vesicles of different sizes originating from the plasma membrane. Membrane budding that led to vesicle budding and separation was proposed to occur at specialized membrane sites, which can function as platforms of associated lipids and proteins that enable this process. Outward vesicles release was firstly described as a product of plasma membrane remodeling and budding (Trams et al., 1981). It was proposed later on that secretion is induced by signals such as calcium signaling (Heijnen et al., 1999), starvation (Antonyak et al., 2011), hypoxia (Wang et al., 2014) and extracellular ATP (Pizzirani et al., 2007). Furthermore, it was proposed that they might bud within plasma membrane endosome-like domains that are described to be enriched in lipid rafts (Booth et al., 2006). Recently, Vps4 has been associated with the formation of DC-derived EVs from plasma membrane (Choi et al., 2018). Vps4a knocked-down BMDCs showed, by electron microscopy, a limited capacity to shed EVs from their membrane, which resulted in a decrease number of EVs bigger than 500 nm (Choi et al., 2018).

It is important to highlight that the molecular machineries described above to be involved in the release of EVs from cells are also important regulators of other intracellular processes. For example, ESCRT machinery was described as crucial for the repair of plasma membrane (Jimenez et al., 2014) and nuclear envelope (Raab et al., 2016). ESCRT-associated protein ALIX that was shown important for EV release (Colombo et al., 2013) is also crucial for the release of viral particles (Bissig et al., 2013). Interestingly, viral secretion shares several molecular machineries with EV secretion as for example Rab27a was described as regulator of both exosome and virus release (Gerber et al., 2015; Ostrowski et al., 2010). Thus, the interventions leading to the inhibition of EV secretion may also interrupt other important intracellular processes and impose the off-target effects that may be difficult to disconnect from the effects of EVs inhibition as such.

4.2 Regulation of extracellular vesicles release by external signals

In the majority of EV studies, vesicles were recovered at the steady-state, i.e., from cells not submitted to a determined stimulus to trigger EV secretion. Thus, constitutive secretion of EV is known to occur and has been a subject of detailed investigation. Importantly, one has to keep in mind that there is a balance between what cells secrete and what they reincorporate (Colombo et al., 2014). Therefore, the amount of EVs present in the conditioned medium highly depends on this equilibrium (Colombo et al., 2014). Moreover, the nature of these EVs may depend on extracellular signals that the secreting cells receive (Yanez-Mo et al., 2015). Indeed, several studies indicate that secretion might be modulated and/or triggered by external signals. Particularly, changes in the vesicles' properties and/or numbers were observed when secreting DCs were challenged with external stimuli. For example, the controversies about DC vesicles after TLR stimulation with LPS endures as some studies reported a decrease of EV secretion by mature DCs, evaluated by overall secretion of a set of EV-associated protein markers (Segura et al., 2005b; Thery et al., 2002; Zitvogel et al., 1998), or a total protein content in the EV preparation (Montecalvo et al., 2012), while others noticed an increase in EV secretion (Nolte-'T Hoen et al., 2013). In the latter study, changes within EV subpopulations as well as a clear expansion of the bulk population were shown by single vesicle flow cytometry analysis of EV separated by their density. Mature DCs secreted higher numbers of EVs achieving 1.14 g/mL densities when compared with immature ones. At the same time, the vesicle number in both the extremely dense and the lightest fractions remained almost unchanged (Nolte-'T Hoen et al., 2013). The discrepancies in evaluation of EV secretion by LPS-pulsed DCs might be caused by different protocols of EV recovery that were carried out in these studies, as recent studies clearly demonstrate that the chosen methodology potentially biases the outcome of EV (Takov et al., 2019; Tauro et al., 2012; Van Deun et al., 2014).

Interestingly, the secretion of large vesicles (Muralidharan-Chari et al., 2009) was often triggered by external stimuli. In particular, human moDCs express in their plasma membrane receptors for extracellular nucleotides, as the P2X7 receptor, which upon stimulation by extracellular ATP caused the rapid and massive shedding of EVs of very heterogeneous sizes (from small EVs of 100 nm up to 1000 nm in diameter) (Pizzirani et al., 2007). This vesicle release occurs from both immature and mature DCs, however the composition of the EVs was different upon maturation. Vesicles from LPS-matured moDCs treated with extracellular ATP contained high levels of the immunoregulatory cytokine IL-1β (Pizzirani et al., 2007).

Consequently, these data altogether suggest that both small and large vesicle secretion could be altered upon stimulation of the secreting cell. To study it in a more comprehensive way, we had compared side-by-side the different types of EVs that are released upon human moDC IFN-γ treatment (Tkach et al., 2017). IFN-γ-matured DCs released similar number of vesicles than immature DCs (Tkach et al., 2017). However, while analyzing the size distribution of these vesicles by nanoparticle tracking analysis, we noticed that the fraction containing the large vesicles coming from the plasma membrane contained now a higher proportion of small EVs (Tkach et al., 2017). Indeed, this was consistent with morphological changes observed in the matured DC by scanning electron microscopy, which displayed a smooth surface, devoid of the plasma membrane ruffles observed in immature cells (Tkach et al., 2017). Thus, different stimulation of DCs can alter qualitatively the EVs which they release.

4.3 Interaction with recipient cells

Once vesicles have been released outside of the secreting cells, they influence their environment by different manners. First, they can interact either with neighboring cells or with cells at long distances from its original place of secretion. Distant interactions of EVs are enabled by their entrance into the blood circulation, or other body fluids, like lymph (Milasan et al., 2016), semen (Aalberts et al., 2013), milk (Admyre et al., 2007). Indeed, EVs were shown to be able to travel to distant organs where they interact with recipient cells (Pucci et al., 2016; Ridder et al., 2014). Once the EV reaches a target cell, there are different mechanisms by which an EV can interact with it: (1) EVs can be immobilized on the recipient cell's surface through the interaction of specific receptors present both at the EV and plasma membrane (Denzer et al., 2000; Morelli et al., 2004; Tian et al., 2013); (2) EVs can trigger signaling by engaging cell surface receptors (Raposo et al., 1996; Zitvogel et al., 1998); (3) EVs can fuse directly with the plasma membrane of target cells and release their content into the cytosol (Mathieu et al., 2019; Montecalvo et al., 2012; Obregon et al., 2006); (4) EVs can be endocytosed or micropinocytosed (Escrevente et al., 2011; Feng et al., 2010; Fitzner et al., 2011; Mallegol et al., 2007; Montecalvo et al., 2008; Morelli et al., 2004). Regarding the release of EV content, it has been proposed that the EV cargo can be released from the vesicles once they are in contact with a plasma membrane and exposed to acidic pH, suggesting that EV content delivery occurs within acidic endo/lysosomes (Bonsergent and Lavieu, 2019).

In particular, it had been proposed different mechanisms of interaction between EVs derived from DCs with their target cells. Segura and colleagues had proposed that DC-derived EVs bearing ICAM-1 at their surface can be captured by other DCs through its binding with LFA-1 (Segura et al., 2005b). Moreover, it has been shown that other molecules also play a role in this interaction, as the αv and β3 integrins or tetraspanins CD9 and CD81 (Morelli et al., 2004). DC-derived EVs can be rapidly internalized into early endosomes of recipient DCs (Morelli et al., 2004). In addition, these EVs can fuse to target cells and deliver their content into the cytoplasm of the acceptor cells, transferring both proteins and RNA (Montecalvo et al., 2012).

5. Role of dendritic cell-derived extracellular vesicles as modulators of immune responses

One of the major functions attributed to DC-derived EVs has been their capacity to activate T cells. Since the finding that antigen presenting cells (APC) derived EVs bear MHC molecules, both MHC class I and II, the interest in their activity as small antigen presenting entities has been widely studied. Importantly, several other functions had been attributed to these vesicles, as for example their capacity to activate NK cells. Here, we will summarize the proposed roles of DC-derived EVs on the induction of adaptive responses through their capacity to activate antigen-specific T cells together with the antigen-independent functions of these EVs.

5.1 Antigen presentation by dendritic cell-derived vesicles

After the report in 1996 (Raposo et al., 1996) that B cell-derived EVs were able to stimulate T-cell responses via their MHC II-antigen complexes, Zitvogel and colleagues investigated vesicles derived from another professional APC: DCs. In particular, it was demonstrated that EVs derived from BMDCs were shown to play an important role in regulating the adaptive immunity by triggering both $CD8^+$ and $CD4^+$ T lymphocyte responses *in vivo* (Zitvogel et al., 1998). Indeed, the vesicles from peptide-loaded BMDCs were able to trigger antigen-specific immune responses *in vivo* against established tumors. In both aggressive mastocytoma P815 and mammary carcinoma TS/A, the administration of tumor antigen-pulsed BMDC-derived EVs induced syngeneic and antigen-specific T cell responses that led to tumor growth retardation with stronger effect than the administration of BMDCs themselves (Zitvogel et al., 1998). Subsequently, it was shown that APC-derived EVs might transfer MHC-antigen complexes to T cells (Arnold and Mannie, 1999; Bedford et al., 1999; Patel et al., 1999) and *vice versa* (Nolte-'T Hoen et al., 2004). There were described different models for EV-mediated antigen presentation to T cells (Fig. 2). On the one hand, DCs can capture DC-derived EVs, keeping them on their surface where they could directly present the peptide-loaded MHC complexes to $CD4^+$ and $CD8^+$ T cells. This mechanism would allow to amplify T cell responses through cross-dressing of bystander DCs that had not been in direct contact with the antigen (Andre et al., 2004; Montecalvo et al., 2008; Segura et al., 2005b; Thery et al., 2002; Vincent-Schneider et al., 2002). Particularly, EVs bearing functional tumor-derived peptide-loaded MHC class I complexes were shown to be transferred to DCs to promote the expansion of melanoma specific cytotoxic $CD8^+$ T lymphocytes in the lymph node of transgenic HLA-A2 mice (Andre et al., 2004). *In vivo* activation of $CD8^+$ T cell responses by the injection of DC-derived EVs was only observed when the host DCs where matured with CpG (Chaput et al., 2004). This suggests that co-stimulatory molecules present on the recipient DC could be important for the efficient presentation of the EV-associated antigen. Moreover, EV-derived MHC-antigen complexes can be presented more efficiently to T cells when obtained from matured DCs than those coming from immature cells (Montecalvo et al., 2008; Segura et al., 2005b). This could be explained by the increased levels of ICAM-1 on EVs from mature cells (Segura et al., 2005b) a molecule involved in DC and T interactions. The importance of this cross-dressing mechanism for the induction of $CD8^+$ T cell

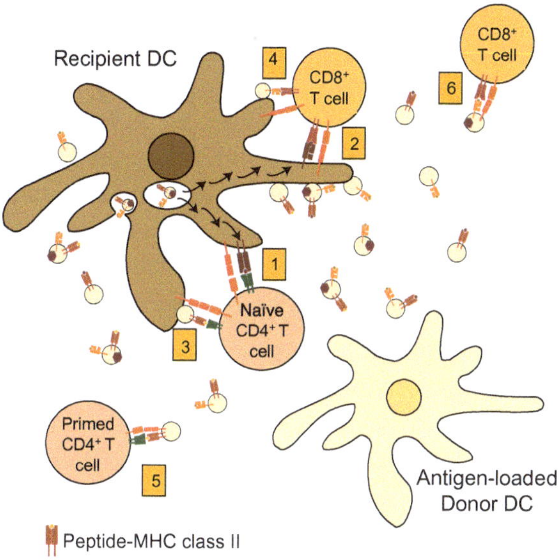

Fig. 2 Activation of antigen-specific T cell immune responses by dendritic cell-derived EVs. Donor DCs pulsed with a specific antigen release EVs that bear peptides from the antigen on MHC class I and/or MHC class II which can be transferred to recipient DCs to indirectly activate T cells. Donor peptide-MHC complexes can be presented to naïve $CD4^+$ T cells (1) or to $CD8^+$ T lymphocytes (2), by a mechanism known as crossdressing. Alternatively, EVs are internalized and antigen peptides from the EVs are transferred to MHC molecules of the recipient cell, which then activate $CD4^+$ (3) and $CD8^+$ (4) lymphocytes. Finally, EVs bearing peptide-MHC complexes and costimulatory molecules can directly activate primed $CD4^+$ (5) or $CD8^+$ (6) T cells.

responses was also evidenced for the presentation of allopeptides. After transplantation, few DCs from the graft migrated to lymphoid organs where they could stimulate alloreactive T cells by transferring clusters of EVs bearing functional MHC molecules to resident DCs from the host (Liu et al., 2016). This allowed the activation of memory T cells that could infiltrate the graft, leading to transplant rejection (Liu et al., 2016).

Another model of EV antigen delivery toward T cells had proposed that recipient DCs internalize EVs containing full antigens or peptides, process and present them on endogenous MHC molecules (Montecalvo et al., 2008; Morelli et al., 2004). The injection of EVs from LPS-matured DCs pulsed with the model antigen OVA were more efficient at eliciting specific $CD4^+$ T cell responses than EVs bearing MHC complexes directly loaded with antigen peptides (Qazi et al., 2009). Particularly, immunization with EVs from OVA-loaded DCs promoted a $CD4^+$ T helper (Th) 1 response dependent on B cell activation (Qazi et al., 2009). The importance of this Th1 response was later demonstrated by the same group, who showed that these vesicles, and not those from peptide loaded DCs, promoted a $CD8^+$ T cell response *in vivo* dependent on $CD4^+$ T and B cells (Naslund et al., 2013). Furthermore, evidence that the whole antigen, and not the preformed MHC-peptides complexes, was being transferred through EVs toward antigen-presenting cells was subsequently demonstrated by the use of EVs purified from DCs lacking MHC class I and of allogeneic EVs (Hiltbrunner et al., 2016). Indeed, the injection of allogeneic EVs containing full antigens was shown to induce a more potent and long-lasting Ag-specific immunity that syngeneic EVs, suggesting their potential use in vaccination strategies (Larssen et al., 2019).

Finally, DC-derived EVs had been shown to directly induce primary T cell or T cell clone activation *in vitro* and *in vivo* (Admyre et al., 2006; Luketic et al., 2007; Nolte-'T Hoen et al., 2009; Segura et al., 2007; Tkach et al., 2017; Utsugi-Kobukai et al., 2003). This latter mechanism has been mainly proposed for the activation of primed T cells or memory cells, and seems to be poorly relevant for the stimulation of naïve T cells (Admyre et al., 2006; Nolte-'T Hoen et al., 2009; Thery et al., 2002; Tkach et al., 2017). Several molecules had been suggested to be involved in this interaction (Fig. 3) (Admyre et al., 2006; Nolte-'T Hoen et al., 2009; Segura et al., 2007; Tkach et al., 2017). For the case of direct $CD8^+$ T cell

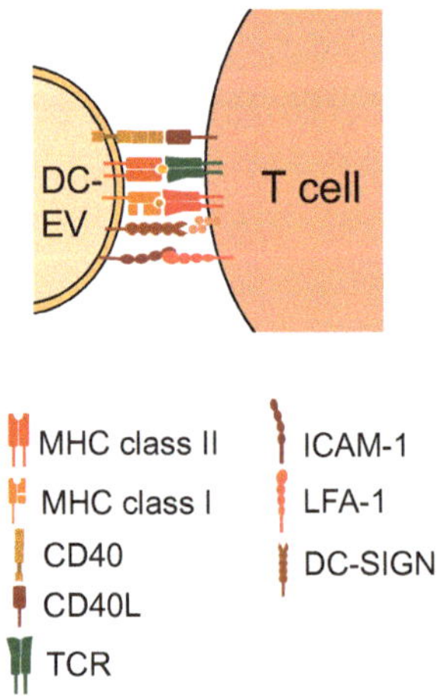

Fig. 3 Molecules involved in the direct activation of T cells induced by dendritic cell derived EVs. Several molecules had been shown to be important for the DC-EV activation of primed T cells. In particular, adhesion molecules, as ICAM-1, were shown to be indispensable for the interaction between the EVs and the target cells.

activation, it was shown that EVs purified from mature human moDCs were more potent at stimulating, in an MHC class I-dependent manner, IFN-γ secretion than the EVs from immature cells (Admyre et al., 2006). Moreover, EVs released by LPS-matured human moDCs upon contact with activated bystander T-cells promote stronger peptide-specific $CD8^+$ T-cell responses than those released upon culture with non-activated bystander T-cells (Lindenbergh et al., 2019). Altogether, this suggests that molecules present on mature DC-derived EVs participate in co-stimulating T cells (Admyre et al., 2007) and that this EV composition can be further modified upon noncognate interaction between DCs and T cells to support subsequent antigen-specific T cell activation (Lindenbergh et al., 2019).

In addition to all these studies analyzing the effect of EVs obtained by high-speed ultracentrifugation from the supernatant of DCs on T cells, some reports together with ours, had considered the fact that DCs release very heterogeneous EV populations, and had extended their analysis to other type of released vesicles (Liu et al., 2016; Tkach et al., 2017; Wahlund et al., 2017). In particular, we separated the EV subtypes from human moDCs as previously described (Kowal et al., 2016) and evaluated whether the capacity of the small EVs to activate T cells was a property unique for a given EV subtype or was a more common feature shared among different EVs. Notably, all the different EV subtypes from immature moDCs were able to directly induce an MHC class II-dependent activation and proliferation of human blood-derived allogeneic total $CD4^+$ T cells, but not naïve T cells (Tkach et al., 2017). Strikingly, activation of the T cells induced by the different EVs resulted in distinctive functional outcomes: T cells activated by medium and small EVs were oriented towards a Th1 phenotype whereas large EVs skewed T cells towards a Th2 phenotype (Tkach et al., 2017). Moreover, we had identified several of the molecules responsible for this differential activation. CD40 (*TNFRSF5*) and DC-SIGN (*CD209*) are enriched in small EVs compared to the other types, and are partially responsible for the Th1-polarizing capacity of these EVs (Tkach et al., 2017). On the other hand, CD80, equally distributed among the different types of EVs, was involved in the Th2-activating capacity of the large EVs (Tkach et al., 2017). Additionally, when the different small EVs were separated by floatation into density gradients, we observed that the exosome-enriched fraction was proportionally less efficient at inducing Th1 cytokines than the non-exosomal fraction. Thus, our results suggested that *bona fide* exosomes are not responsible for the Th1-inducing activity (Tkach et al., 2017).

In addition to the mentioned mechanisms of transfer of peptide-bound MHC class molecules or antigens to DCs, DC-derived EVs antigen presenting complexes have been shown to be transferred to $CD4^+$ T cells during antigen-specific immune synapse formation (Buschow et al., 2009; Nolte-'T Hoen et al., 2009; Wetzel et al., 2005). This transfer of MHC was dependent on LFA-1 on the T cells and on ICAM-1 on the T cells (Nolte-'T Hoen et al., 2009). It had been proposed that these antigen-presenting $CD4^+$ T cells maintained the activated status of T cells independently of the presence of DCs (Zhou et al., 2005), or, conversely, primed antigen specific $CD8^+$ T cells (Lindenbergh et al., 2019). Moreover, these T cells impaired the recruitment of memory T cells following a re-exposure to the antigen, while not affecting naïve T cell recruitment, thus allowing an enlargement of the T cell repertoire and limiting clonal expansion (Helft et al., 2008).

Finally, it was suggested that perivascular DCs capture circulatory antigens and transfer them to mast cells through vesicles released by the plasma membrane (Choi et al., 2018). This cooperation between mast cells and DCs was thought to play an important role in the trigger of anaphylaxis, since once the IgE-bound mast cells contacted an allergen on the surface of a DC-derived EV, they degranulated releasing their inflammatory mediators (Choi et al., 2018).

Another property described for EVs released by antigen-loaded DCs is their capacity to promote antigen-specific humoral responses when injected to mice (Aline et al., 2004; Colino and Snapper, 2006). Exosomes from diphtheria toxoid (DT)-pulsed mature and immature BMDCs, which bore undetectable levels of DT, induced a specific IgM and IgG response *in vivo* (Colino and Snapper, 2006). Moreover, injection of EVs released by DCs pulsed with *Toxoplasma gondii* proteins led to a protective immune response against the parasite infection through the induction of humoral and cellular immunity (Aline et al., 2004). Similar to this strategy, vaccination with *Leishmania major* antigen-loaded EVs from DCs elicited a protective immune response against *L. major* through activation of splenic DCs (Schnitzer et al., 2010). Altogether, these results highlight the potential use of EVs as a cell-free vaccine against infectious diseases.

5.2 Antigen-independent functions of dendritic cell-derived vesicles

EVs derived from DCs, both from immature or from mature cells, contain several ligands for NK cell-activating receptors and other molecules that are important to regulate NK cell function (Reiners et al., 2014). On the one hand, NK cell activation can be triggered by the binding of BAT3, a natural ligand for the NK cell receptor NKp30, present on EVs from heat shocked induced DCs (Simhadri et al., 2008) or through the combination of NKG2D ligands and the IL-15 receptor α-chain on the EVs surface (Viaud et al., 2009). On the other hand, DC derived EVs contain on their surface several transmembrane proteins from the TNF superfamily, as TNF, FasL and Trail, which can directly bind NK cell receptors. Particularly, it has been shown that DC-derived EVs can activate NK cells and stimulate them to secrete IFN-γ via TNF on the EV surface (Munich et al., 2012). Moreover, this capacity was highly increased when the EVs were purified from matured DCs (Munich et al., 2012). On the top of these activities as NK cell modulators, EVs from LPS-matured DCs have been shown to be uptaken by epithelial cells, stimulating their activation and cytokine and chemokine secretion, thereby amplifying the magnitude of innate immune responses.

More recently, a novel function of EVs derived from mature DCs as stimulators of monocyte differentiation towards immature DC has been proposed (Schierer et al., 2018). In particular, several cytokines and chemokines had been found in DC-derived EVs isolated by ultracentrifugation and floatation into density gradient (Schierer et al., 2018). It had been suggested that this array of factors present on the EVs derived from mature DC, as for example, GM-CSF, promote the differentiation of monocytes into inflammatory DCs, and particularly, to an immature DC with high expression of the slan protein (Schierer et al., 2018).

5.3 Dendritic cell extracellular vesicles and viruses

In addition to the plethora of functions described for the EVs released by DCs as antigen presenting entities or as NK cell stimulators, DC-derived vesicles had been shown to play an important role upon viral infection. In the early 2000, it was proposed that retroviruses could hijack the host intracellular vesicle biogenesis and trafficking pathway to release EVs containing both retroviral and host molecules able to infect target cells (Gould et al., 2003). In accordance to this hypothesis, it was demonstrated that HIV-1 exploits EVs dissemination pathway from DCs allowing the trans-infection of $CD4^+$ T cells (Izquierdo-Useros et al., 2009; Wiley and Gummuluru, 2006). Immature DCs were shown to constitutively release endocytosed HIV-1 virions to the extracellular milieu in association with

DC-derived EVs (Wiley and Gummuluru, 2006). These released virions were 10-fold more infectious per particle than cell-free virus particles (Wiley and Gummuluru, 2006). In this study, HIV particles were endocytosed by immature DCs to CD9- and CD63-rich compartments, where exosomes are produced by inward budding and then exocytosed together with other endogenous vesicles (Wiley and Gummuluru, 2006). In addition, Izquierdo-Useros et al. suggested that HIV-1 exploits the pathway used by mature DCs to trans-disseminate functional peptide-MHC complexes present on uptaken DC-derived EVs (Izquierdo-Useros et al., 2009). The matured DCs would capture EVs or HIV-1 or viral like particles (VLPs) through a common mechanism and store them in CD81-enriched compartments. Previously internalized particles are then released allowing the trans-infection of neighboring cells as $CD4^{+}$ T cells (Izquierdo-Useros et al., 2009). More recently, another group confirmed the increased infectivity of HIV-1 upon being internalized by DCs and released together with DC-derived EVs compared to cell-free HIV-1 and analyzed the composition of these vesicles (Kulkarni and Prasad, 2017). Compared to steady-state DC-derived EVs, those released by HIV-infected DCs contained higher amounts of fibronectin and galectin-3, which were shown to mediate the higher infectivity of these particles (Kulkarni and Prasad, 2017). While these studies focused on the EV-mediated infectivity and speeding of virus, as well as the molecular composition of particles released upon viral infection, one would be curious to assess whether EVs are able to mount immune responses against the virus or be a reservoir of viral antigens. Following this line of thoughts, EVs produced by endothelial cells infected with cytomegalovirus were shown to transfer viral antigens towards DC, which was followed by $CD4^{+}$ T cell activation (Walker et al., 2009). On the other hand, EVs secreted by DCs infected with the lymphocytic choriomeningitis virus (lCMv) failed to mount a cytotoxic T cell response (Coppieters et al., 2009). In addition, EVs produced by Hepatitis C virus-infected hepatocytes were shown to contain viral RNA that can be transferred to plasmacytoid DCs, triggering a strong interferon production contributing to the innate antiviral response (Dreux et al., 2012). Altogether, these studies point that immunogenicity of EVs released by cells undergoing infection seems to be highly dependent on the virus itself but also on both the secreting and target cells, which requires further investigation. The transfer of immune stimulating signals contained within EVs released by viral infected cells emerged as an exciting topic and had been reviewed elsewhere (Kouwaki et al., 2017).

The majority of the studies on DC-derived EVs in the context of viral infection had been performed upon HIV uptake by DCs. However, more recently, Martins et al. reported that EVs derived from moDC cells infected with Dengue virus were enriched in mRNAs and miRNAs associated to immune responses, compared to EVs from control cells. Interestingly, the RNA composition was dependent on the viral strain used to infect DCs, highlighting the different response of DCs upon infection with a mild or a severe strain (Martins et al., 2018). Additionally, together with the EVs, dengue infectious particles that had low or no ability to bind antibodies against the envelope protein of the virus, had been detected in EV preparations (Martins et al., 2018). Thus, as shown for HIV, the notion that the EV pathway is being hijacked by viruses to favor their spreading might be also true for other viruses.

6. Clinical application of dendritic cell-derived extracellular vesicles

With increasing numbers of studies reporting functions of EVs in pathological and physiological contexts, they emerged as potential therapeutic agents. Additionally, their high abundance in biological fluids and tumor effusions suggested their possible application as biomarkers to detect the onset of several diseases. The possibility to diagnose disease, especially cancer, through the presence of EVs in blood circulation was recently evidenced as sufficiently sensitive and reliable to represent a possible path to follow (Melo et al., 2015).

Because APC-derived EVs, especially from DCs, acted as MHC molecules carriers and activated several components of the immune system (Thery et al., 2009) they emerged as potent stimulators of immune responses and their potential was investigated in clinical trials involving melanoma and non-small cell lung cancer patients. Autologous DC-derived EVs were isolated after cells pulsing with a mixture of tumor antigen-derived MHC-binding peptides and administered to the patients without major toxic side effects (Escudier et al., 2005; Morse et al., 2005). These studies showed the feasibility of applying vesicle-based immune-stimulatory vaccines. The final results highlighted, however, the need for the optimization of this technique as they led to some clinical responses, but modest antigen-specific T cell activation (Escudier et al., 2005; Morse et al., 2005), with some cases of specific NK cell response (Viaud et al., 2009). Subsequently, efforts were made to design new and more effective DC-derived "exosome" based vaccines (called Dex) (Viaud et al., 2011), based on *in vitro* observations that EVs from mature DCs induced more efficient T cell activation than those from immature DCs (Andre et al., 2004; Segura et al., 2005a). This second generation of Dex has been used in a new phase II clinical trial in advanced non-small cell lung cancer patients (Besse et al., 2016), which again did not lead to antigen-specific T cell responses, but enhanced, in some patients, NK cells activation with minimal toxicity. These studies have proven the feasibility of an EV-based vaccine and its potential in boosting an anti-tumor immune response, although they were disappointing in terms of clinical and antigen-specific immune responses.

Finally, DC-derived vesicles therapeutic potential had also been explored for the treatment of autoimmune diseases. Vesicles purified from genetically modified DCs to express immunosuppressive molecules, as TGF-β1, had been used as therapy against inflammatory bowel disease (Cai et al., 2012) and in experimental autoimmune encephalomyelitis (Yu et al., 2013). When injected into mice, these EVs promoted immunosuppressive responses decreasing signs of disease (Cai et al., 2012; Yu et al., 2013). Moreover, vesicles purified from DCs expressing cytokines, as IL-4 or IL-10, or immunosuppressive molecules, as FasL, had anti-inflammatory capacities and ameliorated arthritis progression in mouse models (Kim et al., 2005, 2006, 2007). Despite the encouraging preclinical data on the immunosuppressive activity of these EVs in the context of autoimmunity, there has not been any advance up-to-date in the use of DC-derived EVs in the treatment of these diseases in humans.

The development of efficient EV-mimetics for therapeutics delivery certainly require the engineering of more potent immunomodulatory EVs together with the discovery of the core machinery required for EV targeting, uptake and content delivery within acceptor cells.

7. Concluding remarks and perspectives

Here, we had exposed part of the rich evidence of the unique features and functions of EVs released by DCs, making emphasis on the fact that they contain a specific cargo that enables immune system modulation. Taking together this information, EVs released by DCs had been considered a mechanism by which the immune system can amplify immune responses, by transmitting immune modulating signals far from their site of production and without the need of the parental DCs to sustain stimulation of other immune cells. However, the majority of the research presented in this chapter had been performed with EVs isolated from DCs that had been differentiated in *in vitro* systems. Moreover, these EVs had been mainly used to stimulate immune cells *in vitro* or had been injected into mice to evaluate their activity *in vivo*. Despite few reports describing the release of DC-derived EVs *in vivo* (Liu et al., 2016), there is a lack of information regarding the functionality of these vesicles once they are released by DCs *in vivo*, especially during the steady state or inflammatory circumstances as infectious and autoimmune diseases or cancer. To achieve better understanding of DC-derived EVs functions, further understanding of the pathways involved specifically in the biogenesis of these small particles is needed. Once we will be able to interfere with their biogenesis and/or secretion, we will be in position to evaluate their activity *in situ*, disclosing their interacting partners and their capacity to modulate immune responses at long distances.

Moreover, another limitation in the studying of EVs is there low abundance. To isolate and investigate EVs by current methodology there is a requirement for large number of secreting cells in order to analyze them properly. It constrains the study of EVs released by cells of low frequency, as for example, already differentiated DCs that can be isolated directly from human blood circulation or from mice. Furthermore, in the past years, distinct DC subsets were identified with their respective specific functions regarding the mounting of the immune response (Segura, 2016). As shown in previous reports, EVs reflect the secreting cells, therefore one can expect that EVs released by distinct DCs could have specialized functions. Thus, there is a need to develop methodology that would enable the study of EVs in small-scale, allowing the challenging analysis of vesicles released by DC subsets. Once this crucial tool developed, it would allow the comprehensive understanding of their potential role as regulators of the immune responses. Moreover, this knowledge is crucial to further exploit the therapeutic promise of DC-derived EVs in future clinical investigations.

Acknowledgments

Some parts of the PhD thesis of J.K. were used to write this chapter. Work performed in the authors' groups was funded by Institut National de la Santé et de la Recherche Médicale (INSERM), Institut Curie (including PhD fellowship to J.K.), the IDEX and Labex programs (ANR-10-IDEX-0001-02 PSL* and ANR-11-LABX-0043), Fondation ARC (SL220120605293 grant and post-doc grant to M.T.) and Institut National du Cancer (INCA-11548). We apologize to numerous colleagues for being unable to extensively quote their work. We thank Dr. Clotilde Théry and Dr. Gregory Lavieu for helpful discussions and comments on the manuscript.

References

Aalberts, M., Van Dissel-Emiliani, F.M., Van Adrichem, N.P., Van Wijnen, M., Wauben, M.H., Stout, T.A., Stoorvogel, W., 2012. Identification of distinct populations of prostasomes that differentially express prostate stem cell antigen, annexin A1 and GLIPR2 in humans. Biol. Reprod. 86, 82.

Aalberts, M., Stout, T.A., Stoorvogel, W., 2013. Prostasomes: extracellular vesicles from the prostate. Reproduction 147, R1–R14.

Abrami, L., Brandi, L., Moayeri, M., Brown, M.J., Krantz, B.A., Leppla, S.H., Van Der Goot, F.G., 2013. Hijacking multivesicular bodies enables long-term and exosome-mediated long-distance action of anthrax toxin. Cell Rep. 5, 986–996.

Admyre, C., Johansson, S.M., Paulie, S., Gabrielsson, S., 2006. Direct exosome stimulation of peripheral human T cells detected by ELISPOT. Eur. J. Immunol. 36, 1772–1781.

Admyre, C., Johansson, S.M., Qazi, K.R., Filen, J.J., Lahesmaa, R., Norman, M., Neve, E.P., Scheynius, A., Gabrielsson, S., 2007. Exosomes with immune modulatory features are present in human breast milk. J. Immunol. 179, 1969–1978.

Alexander, M., Hu, R., Runtsch, M.C., Kagele, D.A., Mosbruger, T.L., Tolmachova, T., Seabra, M.C., Round, J.L., Ward, D.M., O'connell, R.M., 2015. Exosome-delivered microRNAs modulate the inflammatory response to endotoxin. Nat. Commun. 6, 7321.

Aline, F., Bout, D., Amigorena, S., Roingeard, P., Dimier-Poisson, I., 2004. *Toxoplasma gondii* antigen-pulsed-dendritic cell-derived exosomes induce a protective immune response against *T. gondii* infection. Infect. Immun. 72, 4127–4137.

Andre, F., Chaput, N., Schartz, N.E., Flament, C., Aubert, N., Bernard, J., Lemonnier, F., Raposo, G., Escudier, B., Hsu, D.H., Tursz, T., Amigorena, S., Angevin, E., Zitvogel, L., 2004. Exosomes as potent cell-free peptide-based vaccine. I. Dendritic cell-derived exosomes transfer functional MHC class I/peptide complexes to dendritic cells. J. Immunol. 172, 2126–2136.

Antonyak, M.A., Li, B., Boroughs, L.K., Johnson, J.L., Druso, J.E., Bryant, K.L., Holowka, D.A., Cerione, R.A., 2011. Cancer cell-derived microvesicles induce transformation by transferring tissue transglutaminase and fibronectin to recipient cells. Proc. Natl. Acad. Sci. U. S. A. 108, 4852–4857.

Arnold, P.Y., Mannie, M.D., 1999. Vesicles bearing MHC class II molecules mediate transfer of antigen from antigen-presenting cells to CD4+ T cells. Eur. J. Immunol. 29, 1363–1373.

Baietti, M.F., Zhang, Z., Mortier, E., Melchior, A., Degeest, G., Geeraerts, A., Ivarsson, Y., Depoortere, F., Coomans, C., Vermeiren, E., Zimmermann, P., David, G., 2012. Syndecan-syntenin-ALIX regulates the biogenesis of exosomes. Nat. Cell Biol. 14, 677–685.

Bard, M.P., Hegmans, J.P., Hemmes, A., Luider, T.M., Willemsen, R., Severijnen, L.A., Van Meerbeeck, J.P., Burgers, S.A., Hoogsteden, H.C., Lambrecht, B.N., 2004. Proteomic analysis of exosomes isolated from human malignant pleural effusions. Am. J. Respir. Cell Mol. Biol. 31, 114–121.

Bedford, P., Garner, K., Knight, S.C., 1999. MHC class II molecules transferred between allogeneic dendritic cells stimulate primary mixed leukocyte reactions. Int. Immunol. 11, 1739–1744.

Bellingham, S.A., Coleman, B.M., Hill, A.F., 2012. Small RNA deep sequencing reveals a distinct miRNA signature released in exosomes from prion-infected neuronal cells. Nucleic Acids Res. 40, 10937–10949.

Beloribi, S., Ristorcelli, E., Breuzard, G., Silvy, F., Bertrand-Michel, J., Beraud, E., Verine, A., Lombardo, D., 2012. Exosomal lipids impact notch signaling and induce death of human pancreatic tumoral SOJ-6 cells. PLoS One 7, e47480.

Besse, B., Charrier, M., Lapierre, V., Dansin, E., Lantz, O., Planchard, D., Le Chevalier, T., Livartoski, A., Barlesi, F., Laplanche, A., Ploix, S., Vimond, N., Peguillet, I., Théry, C., Lacroix, L., Zoernig, I., Dhodapkar, K., Dhodapkar, M., Viaud, S., Soria, J.-C., Reiners, K.S., Pogge Von Strandmann, E., Vély, F., Rusakiewicz, S., Eggermont, A., Pitt, J.M., Zitvogel, L., Chaput, N., 2016. Dendritic cell-derived exosomes as maintenance immunotherapy after first line chemotherapy in NSCLC. Oncoimmunology 5, e1071008.

Bissig, C., Lenoir, M., Velluz, M.-C., Kufareva, I., Abagyan, R., Overduin, M., Gruenberg, J., 2013. Viral infection controlled by a calcium-dependent lipid-binding module in ALIX. Dev. Cell 25, 364–373.

Bobrie, A., Colombo, M., Krumeich, S., Raposo, G., Thery, C., 2012. Diverse subpopulations of vesicles secreted by different intracellular mechanisms are present in exosome preparations obtained by differential ultracentrifugation. J. Extracell. Vesicles 1, 18397.

Bonsergent, E., Lavieu, G., 2019. Content release of extracellular vesicles in a cell-free extract. FEBS Lett. 593, 1983–1992.

Booth, A.M., Fang, Y., Fallon, J.K., Yang, J.M., Hildreth, J.E., Gould, S.J., 2006. Exosomes and HIV Gag bud from endosome-like domains of the T cell plasma membrane. J. Cell Biol. 172, 923–935.

Brouwers, J.F., Aalberts, M., Jansen, J.W., Van Niel, G., Wauben, M.H., Stout, T.A., Helms, J.B., Stoorvogel, W., 2013. Distinct lipid compositions of two types of human prostasomes. Proteomics 13, 1660–1666.

Buschow, S.I., Liefhebber, J.M., Wubbolts, R., Stoorvogel, W., 2005. Exosomes contain ubiquitinated proteins. Blood Cells Mol. Dis. 35, 398–403.

Buschow, S.I., Nolte-'T Hoen, E.N., Van Niel, G., Pols, M.S., Ten Broeke, T., Lauwen, M., Ossendorp, F., Melief, C.J., Raposo, G., Wubbolts, R., Wauben, M.H., Stoorvogel, W., 2009. MHC II in dendritic cells is targeted to lysosomes or T cell-induced exosomes via distinct multivesicular body pathways. Traffic 10, 1528–1542.

Caby, M.P., Lankar, D., Vincendeau-Scherrer, C., Raposo, G., Bonnerot, C., 2005. Exosomal-like vesicles are present in human blood plasma. Int. Immunol. 17, 879–887.

Cai, Z., Zhang, W., Yang, F., Yu, L., Yu, Z., Pan, J., Wang, L., Cao, X., Wang, J., 2012. Immunosuppressive exosomes from TGF-beta1 gene-modified dendritic cells attenuate Th17-mediated inflammatory autoimmune disease by inducing regulatory T cells. Cell Res. 22, 607–610.

Cai, J., Han, Y., Ren, H., Chen, C., He, D., Zhou, L., Eisner, G.M., Asico, L.D., Jose, P.A., Zeng, C., 2013. Extracellular vesicle-mediated transfer of donor genomic DNA to recipient cells is a novel mechanism for genetic influence between cells. J. Mol. Cell Biol. 5, 227–238.

Chairoungdua, A., Smith, D.L., Pochard, P., Hull, M., Caplan, M.J., 2010. Exosome release of {beta}-catenin: a novel mechanism that antagonizes Wnt signaling. J. Cell Biol. 190, 1079–1091.

Chaput, N., Schartz, N.E., Andre, F., Taieb, J., Novault, S., Bonnaventure, P., Aubert, N., Bernard, J., Lemonnier, F., Merad, M., Adema, G., Adams, M., Ferrantini, M., Carpentier, A.F., Escudier, B., Tursz, T., Angevin, E., Zitvogel, L., 2004. Exosomes as potent cell-free peptide-based vaccine. II. Exosomes in CpG adjuvants efficiently prime naive Tc1 lymphocytes leading to tumor rejection. J. Immunol. 172, 2137–2146.

Chaput, N., Flament, C., Viaud, S., Taieb, J., Roux, S., Spatz, A., Andre, F., Lepecq, J.B., Boussac, M., Garin, J., Amigorena, S., Thery, C., Zitvogel, L., 2006. Dendritic cell derived-exosomes: biology and clinical implementations. J. Leukoc. Biol. 80, 471–478.

Chen, T.S., Lai, R.C., Lee, M.M., Choo, A.B., Lee, C.N., Lim, S.K., 2010. Mesenchymal stem cell secretes microparticles enriched in pre-microRNAs. Nucleic Acids Res. 38, 215–224.

Chevillet, J.R., Kang, Q., Ruf, I.K., Briggs, H.A., Vojtech, L.N., Hughes, S.M., Cheng, H.H., Arroyo, J.D., Meredith, E.K., Gallichotte, E.N., Pogosova-Agadjanyan, E.L., Morrissey, C., Stirewalt, D.L., Hladik, F., Yu, E.Y., Higano, C.S., Tewari, M., 2014. Quantitative and stoichiometric analysis of the microRNA content of exosomes. Proc. Natl. Acad. Sci. U. S. A. 111, 14888–14893.

Choi, H.W., Suwanpradid, J., Kim, I.H., Staats, H.F., Haniffa, M., Macleod, A.S., Abraham, S.N., 2018. Perivascular dendritic cells elicit anaphylaxis by relaying allergens to mast cells via microvesicles. Science 362, eaao0666.

Christ, L., Raiborg, C., Wenzel, E.M., Campsteijn, C., Stenmark, H., 2017. Cellular functions and molecular mechanisms of the ESCRT membrane-scission machinery. Trends Biochem. Sci. 42, 42–56.

Clayton, A., Mitchell, J.P., Court, J., Linnane, S., Mason, M.D., Tabi, Z., 2008. Human tumor-derived exosomes down-modulate NKG2D expression. J. Immunol. 180, 7249–7258.

Colino, J., Snapper, C.M., 2006. Exosomes from bone marrow dendritic cells pulsed with diphtheria toxoid preferentially induce type 1 antigen-specific IgG responses in naive recipients in the absence of free antigen. J. Immunol. 177, 3757–3762.

Colombo, M., Moita, C., Van Niel, G., Kowal, J., Vigneron, J., Benaroch, P., Manel, N., Moita, L.F., Thery, C., Raposo, G., 2013. Analysis of ESCRT functions in exosome biogenesis, composition and secretion highlights the heterogeneity of extracellular vesicles. J. Cell Sci. 126, 5553–5565.

Colombo, M., Raposo, G., Thery, C., 2014. Biogenesis, secretion, and intercellular interactions of exosomes and other extracellular vesicles. Annu. Rev. Cell Dev. Biol. 30, 255–289.

Coppieters, K., Barral, A.M., Juedes, A., Wolfe, T., Rodrigo, E., Thery, C., Amigorena, S., Von Herrath, M.G., 2009. No significant CTL cross-priming by dendritic cell-derived exosomes during murine lymphocytic choriomeningitis virus infection. J. Immunol. 182, 2213–2220.

Denzer, K., Van Eijk, M., Kleijmeer, M.J., Jakobson, E., De Groot, C., Geuze, H.J., 2000. Follicular dendritic cells carry MHC class II-expressing microvesicles at their surface. J. Immunol. 165, 1259–1265.

Dreux, M., Garaigorta, U., Boyd, B., Decembre, E., Chung, J., Whitten-Bauer, C., Wieland, S., Chisari, F.V., 2012. Short-range exosomal transfer of viral RNA from infected cells to plasmacytoid dendritic cells triggers innate immunity. Cell Host Microbe 12, 558–570.

Driedonks, T.A.P., Van Der Grein, S.G., Ariyurek, Y., Buermans, H.P.J., Jekel, H., Chow, F.W.N., Wauben, M.H.M., Buck, A.H., 'T Hoen, P.A.C., Nolte-'T Hoen, E.N.M., 2018. Immune stimuli shape the small non-coding transcriptome of extracellular vesicles released by dendritic cells. Cell. Mol. Life Sci. 75, 3857–3875.

Escrevente, C., Keller, S., Altevogt, P., Costa, J., 2011. Interaction and uptake of exosomes by ovarian cancer cells. BMC Cancer 11, 108.

Escudier, B., Dorval, T., Chaput, N., Andre, F., Caby, M.P., Novault, S., Flament, C., Leboulaire, C., Borg, C., Amigorena, S., Boccaccio, C., Bonnerot, C., Dhellin, O., Movassagh, M., Piperno, S., Robert, C., Serra, V., Valente, N., Le Pecq, J.B., Spatz, A., Lantz, O., Tursz, T., Angevin, E., Zitvogel, L., 2005. Vaccination of metastatic melanoma patients with autologous dendritic cell (DC) derived-exosomes: results of the first phase I clinical trial. J. Transl. Med. 3, 10.

EV-TRACK Consortium, Van Deun, J., Mestdagh, P., Agostinis, P., Akay, Ö., Anand, S., Anckaert, J., Martinez, Z.A., Baetens, T., Beghein, E., Bertier, L., Berx, G., Boere, J., Boukouris, S., Bremer, M., Buschmann, D., Byrd, J.B., Casert, C., Cheng, L., Cmoch, A., Daveloose, D., De Smedt, E., Demirsoy, S., Depoorter, V., Dhondt, B., Driedonks, T.A.P., Dudek, A., Elsharawy, A., Floris, I., Foers, A.D., Gärtner, K.,

Garg, A.D., Geeurickx, E., Gettemans, J., Ghazavi, F., Giebel, B., Kormelink, T.G., Hancock, G., Helsmoortel, H., Hill, A.F., Hyenne, V., Kalra, H., Kim, D., Kowal, J., Kraemer, S., Leidinger, P., Leonelli, C., Liang, Y., Lippens, L., Liu, S., Lo Cicero, A., Martin, S., Mathivanan, S., Mathiyalagan, P., Matusek, T., Milani, G., Monguió-Tortajada, M., Mus, L.M., Muth, D.C., Németh, A., Nolte-'T Hoen, E.N.M., O'driscoll, L., Palmulli, R., Pfaffl, M.W., Primdal-Bengtson, B., Romano, E., Rousseau, Q., Sahoo, S., Sampaio, N., Samuel, M., Scicluna, B., Soen, B., Steels, A., Swinnen, J.V., Takatalo, M., Thaminy, S., Théry, C., Tulkens, J., Van Audenhove, I., Van Der Grein, S., Van Goethem, A., Van Herwijnen, M.J., Van Niel, G., Van Roy, N., Van Vliet, A.R., Vandamme, N., Vanhauwaert, S., Vergauwen, G., Verweij, F., Wallaert, A., Wauben, M., Witwer, K.W., Zonneveld, M.I., De Wever, O., Vandesompele, J., Hendrix, A., 2017. EV-TRACK: transparent reporting and centralizing knowledge in extracellular vesicle research. Nat. Methods 14, 228.

Feng, D., Zhao, W.L., Ye, Y.Y., Bai, X.C., Liu, R.Q., Chang, L.F., Zhou, Q., Sui, S.F., 2010. Cellular internalization of exosomes occurs through phagocytosis. Traffic 11, 675–687.

Fitzner, D., Schnaars, M., Van Rossum, D., Krishnamoorthy, G., Dibaj, P., Bakhti, M., Regen, T., Hanisch, U.K., Simons, M., 2011. Selective transfer of exosomes from oligodendrocytes to microglia by macropinocytosis. J. Cell Sci. 124, 447–458.

Gatti, J.L., Metayer, S., Belghazi, M., Dacheux, F., Dacheux, J.L., 2005. Identification, proteomic profiling, and origin of ram epididymal fluid exosome-like vesicles. Biol. Reprod. 72, 1452–1465.

Geeurickx, E., Tulkens, J., Dhondt, B., Van Deun, J., Lippens, L., Vergauwen, G., Heyrman, E., De Sutter, D., Gevaert, K., Impens, F., Miinalainen, I., Van Bockstal, P.-J., De Beer, T., Wauben, M.H.M., Nolte-'T-Hoen, E.N.M., Bloch, K., Swinnen, J.V., Van Der Pol, E., Nieuwland, R., Braems, G., Callewaert, N., Mestdagh, P., Vandesompele, J., Denys, H., Eyckerman, S., De Wever, O., Hendrix, A., 2019. The generation and use of recombinant extracellular vesicles as biological reference material. Nat. Commun. 10, 3288.

Gerber, P.P., Cabrini, M., Jancic, C., Paoletti, L., Banchio, C., Von Bilderling, C., Sigaut, L., Pietrasanta, L.I., Duette, G., Freed, E.O., De Saint Basile, G., Moita, C.F., Moita, L.F., Amigorena, S., Benaroch, P., Geffner, J., Ostrowski, M., 2015. Rab27a controls HIV-1 assembly by regulating plasma membrane levels of phosphatidylinositol 4,5-bisphosphate. J. Cell Biol. 209 (3), 435–452.

Gould, S.J., Booth, A.M., Hildreth, J.E., 2003. The Trojan exosome hypothesis. Proc. Natl. Acad. Sci. U. S. A. 100, 10592–10597.

Guescini, M., Genedani, S., Stocchi, V., Agnati, L.F., 2010. Astrocytes and glioblastoma cells release exosomes carrying mtDNA. J. Neural Transm. 117, 1–4.

Harding, C., Heuser, J., Stahl, P., 1983. Receptor-mediated endocytosis of transferrin and recycling of the transferrin receptor in rat reticulocytes. J. Cell Biol. 97, 329–339.

Heijnen, H.F., Schiel, A.E., Fijnheer, R., Geuze, H.J., Sixma, J.J., 1999. Activated platelets release two types of membrane vesicles: microvesicles by surface shedding and exosomes derived from exocytosis of multivesicular bodies and alpha-granules. Blood 94, 3791–3799.

Helft, J., Jacquet, A., Joncker, N.T., Grandjean, I., Dorothee, G., Kissenpfennig, A., Malissen, B., Matzinger, P., Lantz, O., 2008. Antigen-specific T-T interactions regulate CD4 T-cell expansion. Blood 112, 1249–1258.

Hiltbrunner, S., Larssen, P., Eldh, M., Martinez-Bravo, M.-J., Wagner, A.K., Karlsson, M.C.I., Gabrielsson, S., 2016. Exosomal cancer immunotherapy is independent of MHC molecules on exosomes. Oncotarget 7, 38707–38717.

Huang, X., Yuan, T., Tschannen, M., Sun, Z., Jacob, H., Du, M., Liang, M., Dittmar, R.L., Liu, Y., Kohli, M., Thibodeau, S.N., Boardman, L., Wang, L., 2013. Characterization of human plasma-derived exosomal RNAs by deep sequencing. BMC Genomics 14, 319.

Hurley, J.H., 2015. ESCRTs are everywhere. EMBO J. 34, 2398.
Izquierdo-Useros, N., Naranjo-Gomez, M., Archer, J., Hatch, S.C., Erkizia, I., Blanco, J., Borras, F.E., Puertas, M.C., Connor, J.H., Fernandez-Figueras, M.T., Moore, L., Clotet, B., Gummuluru, S., Martinez-Picado, J., 2009. Capture and transfer of HIV-1 particles by mature dendritic cells converges with the exosome-dissemination pathway. Blood 113, 2732–2741.
Jimenez, A.J., Maiuri, P., Lafaurie-Janvore, J., Divoux, S., Piel, M., Perez, F., 2014. ESCRT machinery is required for plasma membrane repair. Science 343, 1247136.
Johnson, S.M., Dempsey, C., Parker, C., Mironov, A., Bradley, H., Saha, V., 2017. Acute lymphoblastic leukaemia cells produce large extracellular vesicles containing organelles and an active cytoskeleton. J. Extracell. Vesicles 6, 1294339.
Kahlert, C., Melo, S.A., Protopopov, A., Tang, J., Seth, S., Koch, M., Zhang, J., Weitz, J., Chin, L., Futreal, A., Kalluri, R., 2014. Identification of double-stranded genomic DNA spanning all chromosomes with mutated KRAS and p53 DNA in the serum exosomes of patients with pancreatic cancer. J. Biol. Chem. 289, 3869–3875.
Kanno, K., Sasaki, S., Hirata, Y., Ishikawa, S.-E., Fushimi, K., Nakanishi, S., Bichet, D.G., Marumo, F., 1995. Urinary excretion of aquaporin-2 in patients with diabetes insipidus. N. Engl. J. Med. 332, 1540–1545.
Keerthikumar, S., Gangoda, L., Liem, M., Fonseka, P., Atukorala, I., Ozcitti, C., Mechler, A., Adda, C.G., Ang, C.S., Mathivanan, S., 2015. Proteogenomic analysis reveals exosomes are more oncogenic than ectosomes. Oncotarget 6, 15375–15396.
Kim, S.H., Lechman, E.R., Bianco, N., Menon, R., Keravala, A., Nash, J., Mi, Z., Watkins, S.C., Gambotto, A., Robbins, P.D., 2005. Exosomes derived from IL-10-treated dendritic cells can suppress inflammation and collagen-induced arthritis. J. Immunol. 174, 6440–6448.
Kim, S.H., Bianco, N., Menon, R., Lechman, E.R., Shufesky, W.J., Morelli, A.E., Robbins, P.D., 2006. Exosomes derived from genetically modified DC expressing FasL are anti-inflammatory and immunosuppressive. Mol. Ther. 13, 289–300.
Kim, S.H., Bianco, N.R., Shufesky, W.J., Morelli, A.E., Robbins, P.D., 2007. Effective treatment of inflammatory disease models with exosomes derived from dendritic cells genetically modified to express IL-4. J. Immunol. 179, 2242–2249.
Klumperman, J., Raposo, G., 2014. The complex ultrastructure of the endolysosomal system. Cold Spring Harb. Perspect. Biol. 6, a016857.
Kogure, T., Patel, T., 2013. Isolation of extracellular nanovesicle microRNA from liver cancer cells in culture. Methods Mol. Biol. 1024, 11–18.
Kouwaki, T., Okamoto, M., Tsukamoto, H., Fukushima, Y., Oshiumi, H., 2017. Extracellular vesicles deliver host and virus RNA and regulate innate immune response. Int. J. Mol. Sci. 18, 666.
Kowal, J., Tkach, M., Thery, C., 2014. Biogenesis and secretion of exosomes. Curr. Opin. Cell Biol. 29C, 116–125.
Kowal, J., Arras, G., Colombo, M., Jouve, M., Morath, J.P., Primdal-Bengtson, B., Dingli, F., Loew, D., Tkach, M., Théry, C., 2016. Proteomic comparison defines novel markers to characterize heterogeneous populations of extracellular vesicle subtypes. Proc. Natl. Acad. Sci. U. S. A. 113, E968–E977.
Kulkarni, R., Prasad, A., 2017. Exosomes derived from HIV-1 infected DCs mediate viral trans-infection via fibronectin and galectin-3. Sci. Rep. 7, 14787.
Larssen, P., Veerman, R.E., Akpinar, G.G., Hiltbrunner, S., Karlsson, M.C.I., Gabrielsson, S., 2019. Allogenicity boosts extracellular vesicle–induced antigen-specific immunity and mediates tumor protection and long-term memory in vivo. J. Immunol. 203, 825.
Laulagnier, K., Motta, C., Hamdi, S., Roy, S., Fauvelle, F., Pageaux, J.F., Kobayashi, T., Salles, J.P., Perret, B., Bonnerot, C., Record, M., 2004. Mast cell- and dendritic cell-

derived exosomes display a specific lipid composition and an unusual membrane organization. Biochem. J. 380, 161–171.

Laulagnier, K., Vincent-Schneider, H., Hamdi, S., Subra, C., Lankar, D., Record, M., 2005. Characterization of exosome subpopulations from RBL-2H3 cells using fluorescent lipids. Blood Cells Mol. Dis. 35, 116–121.

Lazaro-Ibanez, E., Sanz-Garcia, A., Visakorpi, T., Escobedo-Lucea, C., Siljander, P., Ayuso-Sacido, A., Yliperttula, M., 2014. Different gDNA content in the subpopulations of prostate cancer extracellular vesicles: apoptotic bodies, microvesicles, and exosomes. Prostate 74, 1379–1390.

Lespagnol, A., Duflaut, D., Beekman, C., Blanc, L., Fiucci, G., Marine, J.C., Vidal, M., Amson, R., Telerman, A., 2008. Exosome secretion, including the DNA damage-induced p53-dependent secretory pathway, is severely compromised in TSAP6/Steap3-null mice. Cell Death Differ. 15, 1723.

Lindenbergh, M.F.S., Koerhuis, D.G.J., Borg, E.G.F., Van'T Veld, E.M., Driedonks, T.A.P., Wubbolts, R., Stoorvogel, W., Boes, M., 2019. Bystander T-Cells support clonal T-cell activation by controlling the release of dendritic cell-derived immune-stimulatory extracellular vesicles. Front. Immunol. 10, 448.

Liu, Q., Rojas-Canales, D.M., Divito, S.J., Shufesky, W.J., Stolz, D.B., Erdos, G., Sullivan, M.L.G., Gibson, G.A., Watkins, S.C., Larregina, A.T., Morelli, A.E., 2016. Donor dendritic cell-derived exosomes promote allograft-targeting immune response. J. Clin. Invest. 126, 2805–2820.

Llorente, A., Skotland, T., Sylvanne, T., Kauhanen, D., Rog, T., Orlowski, A., Vattulainen, I., Ekroos, K., Sandvig, K., 2013. Molecular lipidomics of exosomes released by PC-3 prostate cancer cells. Biochim. Biophys. Acta 1831, 1302–1309.

Luketic, L., Delanghe, J., Sobol, P.T., Yang, P., Frotten, E., Mossman, K.L., Gauldie, J., Bramson, J., Wan, Y., 2007. Antigen presentation by exosomes released from peptide-pulsed dendritic cells is not suppressed by the presence of active CTL. J. Immunol. 179, 5024–5032.

Mallegol, J., Van Niel, G., Lebreton, C., Lepelletier, Y., Candalh, C., Dugave, C., Heath, J.K., Raposo, G., Cerf-Bensussan, N., Heyman, M., 2007. T84-intestinal epithelial exosomes bear MHC class II/peptide complexes potentiating antigen presentation by dendritic cells. Gastroenterology 132, 1866–1876.

Martins, S.D.T., Kuczera, D., Lötvall, J., Bordignon, J., Alves, L.R., 2018. Characterization of dendritic cell-derived extracellular vesicles during dengue virus infection. Front. Microbiol. 9, 1792.

Mathieu, M., Martin-Jaular, L., Lavieu, G., Théry, C., 2019. Specificities of secretion and uptake of exosomes and other extracellular vesicles for cell-to-cell communication. Nat. Cell Biol. 21, 9–17.

Mathivanan, S., Lim, J.W., Tauro, B.J., Ji, H., Moritz, R.L., Simpson, R.J., 2010. Proteomic analysis of A33-immunoaffinity-purified exosomes released from the human colon tumor cell line LIM1215 reveals a tissue-specific protein signature. Mol. Cell. Proteomics 9, 197–208.

Matsuo, H., Chevallier, J., Mayran, N., Le Blanc, I., Ferguson, C., Faure, J., Blanc, N.S., Matile, S., Dubochet, J., Sadoul, R., Parton, R.G., Vilbois, F., Gruenberg, J., 2004. Role of LBPA and Alix in multivesicular liposome formation and endosome organization. Science 303, 531–534.

Mears, R., Craven, R.A., Hanrahan, S., Totty, N., Upton, C., Young, S.L., Patel, P., Selby, P.J., Banks, R.E., 2004. Proteomic analysis of melanoma-derived exosomes by two-dimensional polyacrylamide gel electrophoresis and mass spectrometry. Proteomics 4, 4019–4031.

Melo, S.A., Luecke, L.B., Kahlert, C., Fernandez, A.F., Gammon, S.T., Kaye, J., Lebleu, V.S., Mittendorf, E.A., Weitz, J., Rahbari, N., Reissfelder, C., Pilarsky, C.,

Fraga, M.F., Piwnica-Worms, D., Kalluri, R., 2015. Glypican-1 identifies cancer exosomes and detects early pancreatic cancer. Nature 523, 177–182.

Milasan, A., Tessandier, N., Tan, S., Brisson, A., Boilard, E., Martel, C., 2016. Extracellular vesicles are present in mouse lymph and their level differs in atherosclerosis. J. Extracell. Vesicles 5, 31427.

Minciacchi, V.R., You, S., Spinelli, C., Morley, S., Zandian, M., Aspuria, P.J., Cavallini, L., Ciardiello, C., Reis Sobreiro, M., Morello, M., Kharmate, G., Jang, S.C., Kim, D.K., Hosseini-Beheshti, E., Tomlinson Guns, E., Gleave, M., Gho, Y.S., Mathivanan, S., Yang, W., Freeman, M.R., Di Vizio, D., 2015. Large oncosomes contain distinct protein cargo and represent a separate functional class of tumor-derived extracellular vesicles. Oncotarget 6, 11327–11341.

Mittelbrunn, M., Gutierrez-Vazquez, C., Villarroya-Beltri, C., Gonzalez, S., Sanchez-Cabo, F., Gonzalez, M.A., Bernad, A., Sanchez-Madrid, F., 2011. Unidirectional transfer of microRNA-loaded exosomes from T cells to antigen-presenting cells. Nat. Commun. 2, 282.

Miyanishi, M., Tada, K., Koike, M., Uchiyama, Y., Kitamura, T., Nagata, S., 2007. Identification of Tim4 as a phosphatidylserine receptor. Nature 450, 435–439.

Montecalvo, A., Shufesky, W.J., Stolz, D.B., Sullivan, M.G., Wang, Z., Divito, S.J., Papworth, G.D., Watkins, S.C., Robbins, P.D., Larregina, A.T., Morelli, A.E., 2008. Exosomes as a short-range mechanism to spread alloantigen between dendritic cells during T cell allorecognition. J. Immunol. 180, 3081–3090.

Montecalvo, A., Larregina, A.T., Shufesky, W.J., Stolz, D.B., Sullivan, M.L., Karlsson, J.M., Baty, C.J., Gibson, G.A., Erdos, G., Wang, Z., Milosevic, J., Tkacheva, O.A., Divito, S.J., Jordan, R., Lyons-Weiler, J., Watkins, S.C., Morelli, A.E., 2012. Mechanism of transfer of functional microRNAs between mouse dendritic cells via exosomes. Blood 119, 756–766.

Morelli, A.E., Larregina, A.T., Shufesky, W.J., Sullivan, M.L., Stolz, D.B., Papworth, G.D., Zahorchak, A.F., Logar, A.J., Wang, Z., Watkins, S.C., Falo Jr., L.D., Thomson, A.W., 2004. Endocytosis, intracellular sorting, and processing of exosomes by dendritic cells. Blood 104, 3257–3266.

Morse, M.A., Garst, J., Osada, T., Khan, S., Hobeika, A., Clay, T.M., Valente, N., Shreeniwas, R., Sutton, M.A., Delcayre, A., Hsu, D.H., Le Pecq, J.B., Lyerly, H.K., 2005. A phase I study of dexosome immunotherapy in patients with advanced non-small cell lung cancer. J. Transl. Med. 3, 9.

Munich, S., Sobo-Vujanovic, A., Buchser, W.J., Beer-Stolz, D., Vujanovic, N.L., 2012. Dendritic cell exosomes directly kill tumor cells and activate natural killer cells via TNF superfamily ligands. Oncoimmunology 1, 1074–1083.

Muralidharan-Chari, V., Clancy, J., Plou, C., Romao, M., Chavrier, P., Raposo, G., D'souza-Schorey, C., 2009. ARF6-regulated shedding of tumor cell-derived plasma membrane microvesicles. Curr. Biol. 19, 1875–1885.

Nakai, W., Yoshida, T., Diez, D., Miyatake, Y., Nishibu, T., Imawaka, N., Naruse, K., Sadamura, Y., Hanayama, R., 2016. A novel affinity-based method for the isolation of highly purified extracellular vesicles. Sci. Rep. 6, 33935.

Naslund, T.I., Gehrmann, U., Qazi, K.R., Karlsson, M.C., Gabrielsson, S., 2013. Dendritic cell-derived exosomes need to activate both T and B cells to induce antitumor immunity. J. Immunol. 190, 2712–2719.

Nolte-'T Hoen, E.N., Wagenaar-Hilbers, J.P., Peters, P.J., Gadella, B.M., Van Eden, W., Wauben, M.H., 2004. Uptake of membrane molecules from T cells endows antigen-presenting cells with novel functional properties. Eur. J. Immunol. 34, 3115–3125.

Nolte-'T Hoen, E.N., Buschow, S.I., Anderton, S.M., Stoorvogel, W., Wauben, M.H., 2009. Activated T-cells recruit exosomes secreted by dendritic cells via LFA-1. Blood 113, 1977–1981.

Nolte-'T Hoen, E.N., Buermans, H.P., Waasdorp, M., Stoorvogel, W., Wauben, M.H., T Hoen, P.A., 2012. Deep sequencing of RNA from immune cell-derived vesicles uncovers the selective incorporation of small non-coding RNA biotypes with potential regulatory functions. Nucleic Acids Res. 40, 9272–9285.
Nolte-'T Hoen, E.N., Van Der Vlist, E.J., De Boer-Brouwer, M., Arkesteijn, G.J., Stoorvogel, W., Wauben, M.H., 2013. Dynamics of dendritic cell-derived vesicles: high-resolution flow cytometric analysis of extracellular vesicle quantity and quality. J. Leukoc. Biol. 93, 395–402.
Obregon, C., Rothen-Rutishauser, B., Gitahi, S.K., Gehr, P., Nicod, L.P., 2006. Exovesicles from human activated dendritic cells fuse with resting dendritic cells, allowing them to present alloantigens. Am. J. Pathol. 169, 2127–2136.
Osteikoetxea, X., Balogh, A., Szabo-Taylor, K., Nemeth, A., Szabo, T.G., Paloczi, K., Sodar, B., Kittel, A., Gyorgy, B., Pallinger, E., Matko, J., Buzas, E.I., 2015a. Improved characterization of EV preparations based on protein to lipid ratio and lipid properties. PLoS One 10, e0121184.
Osteikoetxea, X., Sódar, B., Németh, A., Szabó-Taylor, K., Pálóczi, K., Vukman, K.V., Tamási, V., Balogh, A., Kittel, Á., Pállinger, É., Buzás, E.I., 2015b. Differential detergent sensitivity of extracellular vesicle subpopulations. Org. Biomol. Chem. 13, 9775–9782.
Ostrowski, M., Carmo, N.B., Krumeich, S., Fanget, I., Raposo, G., Savina, A., Moita, C.F., Schauer, K., Hume, A.N., Freitas, R.P., Goud, B., Benaroch, P., Hacohen, N., Fukuda, M., Desnos, C., Seabra, M.C., Darchen, F., Amigorena, S., Moita, L.F., Thery, C., 2010. Rab27a and Rab27b control different steps of the exosome secretion pathway. Nat. Cell Biol. 12, 19–30.
Pan, B.T., Johnstone, R.M., 1983. Fate of the transferrin receptor during maturation of sheep reticulocytes in vitro: selective externalization of the receptor. Cell 33, 967–978.
Patel, D.M., Arnold, P.Y., White, G.A., Nardella, J.P., Mannie, M.D., 1999. Class II MHC/peptide complexes are released from APC and are acquired by T cell responders during specific antigen recognition. J. Immunol. 163, 5201–5210.
Pisitkun, T., Shen, R.F., Knepper, M.A., 2004. Identification and proteomic profiling of exosomes in human urine. Proc. Natl. Acad. Sci. U. S. A. 101, 13368–13373.
Pizzirani, C., Ferrari, D., Chiozzi, P., Adinolfi, E., Sandona, D., Savaglio, E., Di Virgilio, F., 2007. Stimulation of P2 receptors causes release of IL-1beta-loaded microvesicles from human dendritic cells. Blood 109, 3856–3864.
Pucci, F., Garris, C., Lai, C.P., Newton, A., Pfirschke, C., Engblom, C., Alvarez, D., Sprachman, M., Evavold, C., Magnuson, A., Von Andrian, U.H., Glatz, K., Breakefield, X.O., Mempel, T.R., Weissleder, R., Pittet, M.J., 2016. SCS macrophages suppress melanoma by restricting tumor-derived vesicle-B cell interactions. Science 352, 242–246.
Qazi, K.R., Gehrmann, U., Domange Jordo, E., Karlsson, M.C., Gabrielsson, S., 2009. Antigen-loaded exosomes alone induce Th1-type memory through a B-cell-dependent mechanism. Blood 113, 2673–2683.
Raab, M., Gentili, M., De Belly, H., Thiam, H.-R., Vargas, P., Jimenez, A.J., Lautenschlaeger, F., Voituriez, R., Lennon-Duménil, A.-M., Manel, N., Piel, M., 2016. ESCRT III repairs nuclear envelope ruptures during cell migration to limit DNA damage and cell death. Science 352, 359.
Raposo, G., Nijman, H.W., Stoorvogel, W., Liejendekker, R., Harding, C.V., Melief, C.J., Geuze, H.J., 1996. B lymphocytes secrete antigen-presenting vesicles. J. Exp. Med. 183, 1161–1172.
Ratajczak, J., Miekus, K., Kucia, M., Zhang, J., Reca, R., Dvorak, P., Ratajczak, M.Z., 2006. Embryonic stem cell-derived microvesicles reprogram hematopoietic progenitors: evidence for horizontal transfer of mRNA and protein delivery. Leukemia 20, 847–856.

Reiners, K.S., Dassler, J., Coch, C., Pogge Von Strandmann, E., 2014. Role of exosomes released by dendritic cells and/or by tumor targets: regulation of NK cell plasticity. Front. Immunol. 5, 91.

Ridder, K., Keller, S., Dams, M., Rupp, A.K., Schlaudraff, J., Turco, D.D., Starmann, J., Macas, J., Karpova, D., Devraj, K., Depboylu, C., Landfried, B., Arnold, B., Plate, K.H., Hoglinger, G., Sultmann, H., Altevogt, P., Momma, S., 2014. Extracellular vesicle-mediated transfer of genetic information between the hematopoietic system and the brain in response to inflammation. PLoS Biol. 12, e1001874.

Sansone, P., Savini, C., Kurelac, I., Chang, Q., Amato, L.B., Strillacci, A., Stepanova, A., Iommarini, L., Mastroleo, C., Daly, L., Galkin, A., Thakur, B.K., Soplop, N., Uryu, K., Hoshino, A., Norton, L., Bonafé, M., Cricca, M., Gasparre, G., Lyden, D., Bromberg, J., 2017. Packaging and transfer of mitochondrial DNA via exosomes regulate escape from dormancy in hormonal therapy-resistant breast cancer. Proc. Natl. Acad. Sci. U. S. A. 114, E9066.

Schierer, S., Ostalecki, C., Zinser, E., Lamprecht, R., Plosnita, B., Stich, L., Dörrie, J., Lutz, M.B., Schuler, G., Baur, A.S., 2018. Extracellular vesicles from mature dendritic cells (DC) differentiate monocytes into immature DC. Life Sci. Alliance 1, e201800093.

Schnitzer, J.K., Berzel, S., Fajardo-Moser, M., Remer, K.A., Moll, H., 2010. Fragments of antigen-loaded dendritic cells (DC) and DC-derived exosomes induce protective immunity against *Leishmania* major. Vaccine 28, 5785–5793.

Segura, E., 2016. Review of mouse and human dendritic cell subsets. Methods Mol. Biol. 1423, 3–15.

Segura, E., Amigorena, S., Thery, C., 2005a. Mature dendritic cells secrete exosomes with strong ability to induce antigen-specific effector immune responses. Blood Cells Mol. Dis. 35, 89–93.

Segura, E., Nicco, C., Lombard, B., Veron, P., Raposo, G., Batteux, F., Amigorena, S., Thery, C., 2005b. ICAM-1 on exosomes from mature dendritic cells is critical for efficient naive T-cell priming. Blood 106, 216–223.

Segura, E., Guerin, C., Hogg, N., Amigorena, S., Thery, C., 2007. CD8+ dendritic cells use LFA-1 to capture MHC-peptide complexes from exosomes in vivo. J. Immunol. 179, 1489–1496.

Shurtleff, M.J., Temoche-Diaz, M.M., Karfilis, K.V., Ri, S., Schekman, R., 2016. Y-box protein 1 is required to sort microRNAs into exosomes in cells and in a cell-free reaction. elife 5, e19276.

Shurtleff, M.J., Yao, J., Qin, Y., Nottingham, R.M., Temoche-Diaz, M.M., Schekman, R., Lambowitz, A.M., 2017. Broad role for YBX1 in defining the small noncoding RNA composition of exosomes. Proc. Natl. Acad. Sci. U. S. A. 114, E8987.

Simhadri, V.R., Reiners, K.S., Hansen, H.P., Topolar, D., Simhadri, V.L., Nohroudi, K., Kufer, T.A., Engert, A., Pogge Von Strandmann, E., 2008. Dendritic cells release HLA-B-associated transcript-3 positive exosomes to regulate natural killer function. PLoS One 3, e3377.

Skog, J., Wurdinger, T., Van Rijn, S., Meijer, D.H., Gainche, L., Curry Jr., W.T., Carter, B.S., Krichevsky, A.M., Breakefield, X.O., 2008. Glioblastoma microvesicles transport RNA and proteins that promote tumour growth and provide diagnostic biomarkers. Nat. Cell Biol. 10, 1470–1476.

Subra, C., Laulagnier, K., Perret, B., Record, M., 2007. Exosome lipidomics unravels lipid sorting at the level of multivesicular bodies. Biochimie 89, 205–212.

Takov, K., Yellon, D.M., Davidson, S.M., 2019. Comparison of small extracellular vesicles isolated from plasma by ultracentrifugation or size-exclusion chromatography: yield, purity and functional potential. J. Extracell. Vesicles 8, 1560809.

Tamai, K., Tanaka, N., Nakano, T., Kakazu, E., Kondo, Y., Inoue, J., Shiina, M., Fukushima, K., Hoshino, T., Sano, K., Ueno, Y., Shimosegawa, T., Sugamura, K., 2010. Exosome secretion of dendritic cells is regulated by Hrs, an ESCRT-0 protein. Biochem. Biophys. Res. Commun. 399, 384–390.

Tan, S.S., Yin, Y., Lee, T., Lai, R.C., Yeo, R.W., Zhang, B., Choo, A., Lim, S.K., 2013. Therapeutic MSC exosomes are derived from lipid raft microdomains in the plasma membrane. J. Extracell. Vesicles 2, 22614.

Tauro, B.J., Greening, D.W., Mathias, R.A., Ji, H., Mathivanan, S., Scott, A.M., Simpson, R.J., 2012. Comparison of ultracentrifugation, density gradient separation, and immunoaffinity capture methods for isolating human colon cancer cell line LIM1863-derived exosomes. Methods 56, 293–304.

Tauro, B.J., Greening, D.W., Mathias, R.A., Mathivanan, S., Ji, H., Simpson, R.J., 2013. Two distinct populations of exosomes are released from LIM1863 colon carcinoma cell-derived organoids. Mol. Cell. Proteomics 12, 587–598.

Ten Broeke, T., Van Niel, G., Wauben, M.H.M., Wubbolts, R., Stoorvogel, W., 2011. Endosomally stored MHC class II does not contribute to antigen presentation by dendritic cells at inflammatory conditions. Traffic 12, 1025–1036.

Thakur, B.K., Zhang, H., Becker, A., Matei, I., Huang, Y., Costa-Silva, B., Zheng, Y., Hoshino, A., Brazier, H., Xiang, J., Williams, C., Rodriguez-Barrueco, R., Silva, J.M., Zhang, W., Hearn, S., Elemento, O., Paknejad, N., Manova-Todorova, K., Welte, K., Bromberg, J., Peinado, H., Lyden, D., 2014. Double-stranded DNA in exosomes: a novel biomarker in cancer detection. Cell Res. 24, 766–769.

Thery, C., Regnault, A., Garin, J., Wolfers, J., Zitvogel, L., Ricciardi-Castagnoli, P., Raposo, G., Amigorena, S., 1999. Molecular characterization of dendritic cell-derived exosomes. Selective accumulation of the heat shock protein hsc73. J. Cell Biol. 147, 599–610.

Thery, C., Boussac, M., Veron, P., Ricciardi-Castagnoli, P., Raposo, G., Garin, J., Amigorena, S., 2001. Proteomic analysis of dendritic cell-derived exosomes: a secreted subcellular compartment distinct from apoptotic vesicles. J. Immunol. 166, 7309–7318.

Thery, C., Duban, L., Segura, E., Veron, P., Lantz, O., Amigorena, S., 2002. Indirect activation of naive CD4+ T cells by dendritic cell-derived exosomes. Nat. Immunol. 3, 1156–1162.

Théry, C., Amigorena, S., Raposo, G., Clayton, A., 2006. Isolation and characterization of exosomes from cell culture supernatants and biological fluids. Curr. Protoc. Cell Biol. 30, 3.22.1–3.22.29.

Thery, C., Ostrowski, M., Segura, E., 2009. Membrane vesicles as conveyors of immune responses. Nat. Rev. Immunol. 9, 581–593.

Théry, C., Witwer, K.W., Aikawa, E., Alcaraz, M.J., Anderson, J.D., Andriantsitohaina, R., Antoniou, A., Arab, T., Archer, F., Atkin-Smith, G.K., Ayre, D.C., Bach, J.-M., Bachurski, D., Baharvand, H., Balaj, L., Baldacchino, S., Bauer, N.N., Baxter, A.A., Bebawy, M., Beckham, C., Bedina Zavec, A., Benmoussa, A., Berardi, A.C., Bergese, P., Bielska, E., Blenkiron, C., Bobis-Wozowicz, S., Boilard, E., Boireau, W., Bongiovanni, A., Borràs, F.E., Bosch, S., Boulanger, C.M., Breakefield, X., Breglio, A.M., Brennan, M.Á., Brigstock, D.R., Brisson, A., Broekman, M.L.D., Bromberg, J.F., Bryl-Górecka, P., Buch, S., Buck, A.H., Burger, D., Busatto, S., Buschmann, D., Bussolati, B., Buzás, E.I., Byrd, J.B., Camussi, G., Carter, D.R.F., Caruso, S., Chamley, L.W., Chang, Y.-T., Chen, C., Chen, S., Cheng, L., Chin, A.R., Clayton, A., Clerici, S.P., Cocks, A., Cocucci, E., Coffey, R.J., Cordeiro-Da -Silva, A., Couch, Y., Coumans, F.A.W., Coyle, B., Crescitelli, R., Criado, M.F., D'souza-Schorey, C., Das, S., Datta Chaudhuri, A., De Candia, P., De Santana, E.F., De Wever, O., Del Portillo, H.A., Demaret, T.,

Deville, S., Devitt, A., Dhondt, B., Di Vizio, D., Dieterich, L.C., Dolo, V., Dominguez Rubio, A.P., Dominici, M., Dourado, M.R., Driedonks, T.A.P., Duarte, F.V., Duncan, H.M., Eichenberger, R.M., Ekström, K., El Andaloussi, S., Elie-Caille, C., Erdbrügger, U., Falcón-Pérez, J.M., Fatima, F., Fish, J.E., Flores-Bellver, M., Försönits, A., Frelet-Barrand, A., et al., 2018. Minimal information for studies of extracellular vesicles 2018 (MISEV2018): a position statement of the International Society for Extracellular Vesicles and update of the MISEV2014 guidelines. J. Extracell. Vesicles 7, 1535750.

Tian, T., Zhu, Y.L., Hu, F.H., Wang, Y.Y., Huang, N.P., Xiao, Z.D., 2013. Dynamics of exosome internalization and trafficking. J. Cell. Physiol. 228, 1487–1495.

Tkach, M., Théry, C., 2016. Communication by extracellular vesicles: where we are and where we need to go. Cell 164, 1226–1232.

Tkach, M., Kowal, J., Zucchetti, A.E., Enserink, L., Jouve, M., Lankar, D., Saitakis, M., Martin-Jaular, L., Théry, C., 2017. Qualitative differences in T-cell activation by dendritic cell-derived extracellular vesicle subtypes. EMBO J. 36, 3012–3028.

Tkach, M., Kowal, J., Théry, C., 2018. Why the need and how to approach the functional diversity of extracellular vesicles. Philos. Trans. R. Soc., B 373, 20160479.

Trajkovic, K., Hsu, C., Chiantia, S., Rajendran, L., Wenzel, D., Wieland, F., Schwille, P., Brugger, B., Simons, M., 2008. Ceramide triggers budding of exosome vesicles into multivesicular endosomes. Science 319, 1244–1247.

Trams, E.G., Lauter, C.J., Salem Jr., N., Heine, U., 1981. Exfoliation of membrane ecto-enzymes in the form of micro-vesicles. Biochim. Biophys. Acta 645, 63–70.

Utsugi-Kobukai, S., Fujimaki, H., Hotta, C., Nakazawa, M., Minami, M., 2003. MHC class I-mediated exogenous antigen presentation by exosomes secreted from immature and mature bone marrow derived dendritic cells. Immunol. Lett. 89, 125–131.

Valadi, H., Ekstrom, K., Bossios, A., Sjostrand, M., Lee, J.J., Lotvall, J.O., 2007. Exosome-mediated transfer of mRNAs and microRNAs is a novel mechanism of genetic exchange between cells. Nat. Cell Biol. 9, 654–659.

Van Deun, J., Mestdagh, P., Sormunen, R., Cocquyt, V., Vermaelen, K., Vandesompele, J., Bracke, M., De Wever, O., Hendrix, A., 2014. The impact of disparate isolation methods for extracellular vesicles on downstream RNA profiling. J. Extracell. Vesicles 3, 24858.

Van Niel, G., Raposo, G., Candalh, C., Boussac, M., Hershberg, R., Cerf-Bensussan, N., Heyman, M., 2001. Intestinal epithelial cells secrete exosome-like vesicles. Gastroenterology 121, 337–349.

Van Niel, G., D'angelo, G., Raposo, G., 2018. Shedding light on the cell biology of extracellular vesicles. Nat. Rev. Mol. Cell Biol. 19, 213.

Veron, P., Segura, E., Sugano, G., Amigorena, S., Thery, C., 2005. Accumulation of MFG-E8/lactadherin on exosomes from immature dendritic cells. Blood Cells Mol. Dis. 35, 81–88.

Viaud, S., Terme, M., Flament, C., Taieb, J., Andre, F., Novault, S., Escudier, B., Robert, C., Caillat-Zucman, S., Tursz, T., Zitvogel, L., Chaput, N., 2009. Dendritic cell-derived exosomes promote natural killer cell activation and proliferation: a role for NKG2D ligands and IL-15Ralpha. PLoS One 4, e4942.

Viaud, S., Ploix, S., Lapierre, V., Thery, C., Commere, P.H., Tramalloni, D., Gorrichon, K., Virault-Rocroy, P., Tursz, T., Lantz, O., Zitvogel, L., Chaput, N., 2011. Updated technology to produce highly immunogenic dendritic cell-derived exosomes of clinical grade: a critical role of interferon-gamma. J. Immunother. 34, 65–75.

Vincent-Schneider, H., Stumptner-Cuvelette, P., Lankar, D., Pain, S., Raposo, G., Benaroch, P., Bonnerot, C., 2002. Exosomes bearing HLA-DR1 molecules need dendritic cells to efficiently stimulate specific T cells. Int. Immunol. 14, 713–722.

Wahlund, C.J.E., Güclüler, G., Hiltbrunner, S., Veerman, R.E., Näslund, T.I., Gabrielsson, S., 2017. Exosomes from antigen-pulsed dendritic cells induce stronger antigen-specific immune responses than microvesicles in vivo. Sci. Rep. 7, 17095.

Walker, J.D., Maier, C.L., Pober, J.S., 2009. Cytomegalovirus-infected human endothelial cells can stimulate allogeneic CD4+ memory T cells by releasing antigenic exosomes. J. Immunol. 182, 1548–1559.

Wang, T., Gilkes, D.M., Takano, N., Xiang, L., Luo, W., Bishop, C.J., Chaturvedi, P., Green, J.J., Semenza, G.L., 2014. Hypoxia-inducible factors and RAB22A mediate formation of microvesicles that stimulate breast cancer invasion and metastasis. Proc. Natl. Acad. Sci. U. S. A. 111, E3234–E3242.

Wetzel, S.A., Mckeithan, T.W., Parker, D.C., 2005. Peptide-specific intercellular transfer of MHC class II to CD4+ T cells directly from the immunological synapse upon cellular dissociation. J. Immunol. 174, 80–89.

Wiley, R.D., Gummuluru, S., 2006. Immature dendritic cell-derived exosomes can mediate HIV-1 trans infection. Proc. Natl. Acad. Sci. U. S. A. 103, 738–743.

Witwer, K.W., Théry, C., 2019. Extracellular vesicles or exosomes? On primacy, precision, and popularity influencing a choice of nomenclature. J. Extracell. Vesicles 8, 1648167.

Wubbolts, R., Leckie, R.S., Veenhuizen, P.T., Schwarzmann, G., Mobius, W., Hoernschemeyer, J., Slot, J.W., Geuze, H.J., Stoorvogel, W., 2003. Proteomic and biochemical analyses of human B cell-derived exosomes. Potential implications for their function and multivesicular body formation. J. Biol. Chem. 278, 10963–10972.

Yanez-Mo, M., Siljander, P.R., Andreu, Z., Zavec, A.B., Borras, F.E., Buzas, E.I., Buzas, K., Casal, E., Cappello, F., Carvalho, J., Colas, E., Cordeiro-Da Silva, A., Fais, S., Falcon-Perez, J.M., Ghobrial, I.M., Giebel, B., Gimona, M., Graner, M., Gursel, I., Gursel, M., Heegaard, N.H., Hendrix, A., Kierulf, P., Kokubun, K., Kosanovic, M., Kralj-Iglic, V., Kramer-Albers, E.M., Laitinen, S., Lasser, C., Lener, T., Ligeti, E., Line, A., Lipps, G., Llorente, A., Lotvall, J., Mancek-Keber, M., Marcilla, A., Mittelbrunn, M., Nazarenko, I., Nolte-'T Hoen, E.N., Nyman, T.A., O'driscoll, L., Olivan, M., Oliveira, C., Pallinger, E., Del Portillo, H.A., Reventos, J., Rigau, M., Rohde, E., Sammar, M., Sanchez-Madrid, F., Santarem, N., Schallmoser, K., Ostenfeld, M.S., Stoorvogel, W., Stukelj, R., Van Der Grein, S.G., Vasconcelos, M.H., Wauben, M.H., De Wever, O., 2015. Biological properties of extracellular vesicles and their physiological functions. J. Extracell. Vesicles 4, 27066.

Yu, L., Yang, F., Jiang, L., Chen, Y., Wang, K., Xu, F., Wei, Y., Cao, X., Wang, J., Cai, Z., 2013. Exosomes with membrane-associated TGF-beta1 from gene-modified dendritic cells inhibit murine EAE independently of MHC restriction. Eur. J. Immunol. 43, 2461–2472.

Yuana, Y., Sturk, A., Nieuwland, R., 2013. Extracellular vesicles in physiological and pathological conditions. Blood Rev. 27, 31–39.

Zhou, J., Tagaya, Y., Tolouei-Semnani, R., Schlom, J., Sabzevari, H., 2005. Physiological relevance of antigen presentasome (APS), an acquired MHC/costimulatory complex, in the sustained activation of $CD4^+$ T cells in the absence of APCs. Blood 105, 3238.

Zitvogel, L., Regnault, A., Lozier, A., Wolfers, J., Flament, C., Tenza, D., Ricciardi-Castagnoli, P., Raposo, G., Amigorena, S., 1998. Eradication of established murine tumors using a novel cell-free vaccine: dendritic cell-derived exosomes. Nat. Med. 4, 594–600.

CHAPTER SIX

Ex vivo dendritic cell generation—A critical comparison of current approaches

Patrick Han[a], Douglas Hanlon[b], Olga Sobolev[b], Rabib Chaudhury[a], Richard L. Edelson[b,*]

[a]Department of Chemical and Environmental Engineering, School of Engineering and Applied Science, Yale University, New Haven, CT, United States
[b]Department of Dermatology, School of Medicine, Yale University, New Haven, CT, United States
*Corresponding author: e-mail address: redelson@yale.edu

Contents

1. Cytokine signals 253
2. Physiological signals for DC differentiation 260
 2.1 DC generation via inflammatory and tissue signaling 262
 2.2 DC generation via extracorporeal photopheresis (ECP) 265
3. Gene engineering tools 267
 3.1 Antigen expression vectors (Signal 1) 268
 3.2 Adjuvant for post-differentiation maturation (Signal 2) 274
 3.3 Enhanced cytokine release/conditioning (Signal 3) 277
 3.4 Transcriptional (re)programming of DC 280
4. Clinical trials of DC-based vaccines 281
 4.1 Early clinical trials with cytokine-derived mDCs 282
 4.2 Naturally occurring DCs, neoantigens, and combination therapies 283
 4.3 Prospective clinical applications of ECP-driven DC therapies 287
5. Concluding remarks 289
Conflict of interest 290
References 290

Abstract

Dendritic cells (DCs) are professional antigen-presenting cells, required for the initiation of naïve and memory T cell responses and regulation of adaptive immunity. The discovery of DCs in 1973, which culminated in the Nobel Prize in Physiology or Medicine in 2011 for Ralph Steinman and colleagues, initially focused on the identification of adherent mononuclear cell fractions with uniquely stellate dendritic morphology, followed by key discoveries of their critical immunologic role in initiating and maintaining antigen-specific immunity and tolerance. The medical promise of marshaling these key capabilities of DCs for therapeutic modulation of antigen-specific immune responses has guided decades of research in hopes to achieve genuine physiologic partnership with

International Review of Cell and Molecular Biology, Volume 349
ISSN 1937-6448
https://doi.org/10.1016/bs.ircmb.2019.10.003

the immune system. The potential uses of DCs in immunotherapeutic applications include cancer, infectious diseases, and autoimmune disorders; thus, methods for rapid and reliable large-scale production of DCs have been of great academic and clinical interest. However, difficulties in obtaining DCs from lymphoid and peripheral tissues, low numbers and poor survival in culture, have led to advancements in *ex vivo* production of DCs, both for probing molecular details of DC function as well as for experimenting with their clinical utility. Here, we review the development of a diverse array of DC production methodologies, ranging from cytokine-based strategies to genetic engineering tools devised for enhancing DC-specific immunologic functions. Further, we explore the current state of DC therapies in clinic, as well as emerging insights into physiologic production of DCs inspired by existing therapies.

Dendritic cells (DCs) are professional antigen-presenting cells (APCs), required for the initiation of naïve and memory T cell responses and regulation of adaptive immunity (Steinman, 2012). They were initially discovered within the adherent mononuclear cell fraction from mouse spleens (Mosier, 1967), and distinguished from other mononuclear cells by their unique stellate dendritic morphology (Steinman and Cohn, 1973, 1974). DCs were subsequently shown to be present as resident cells in both lymphoid organs and in peripheral tissues, such as the Langerhans cells in the epidermis, and to function as uniquely potent stimulators of T cell effector responses (Schuler et al., 1985; Schuler and Steinman, 1985; Steiner et al., 1985).

The ability to isolate and culture dendritic cells from lymphoid and peripheral tissues was a key first step that enabled the discovery of their physiologic function. However, molecular understanding of how DCs develop their unique properties was hampered by the low numbers of cells obtainable from tissues, and by their poor survival in culture, mainly due to spontaneous maturation (Lutz et al., 1999; Markowicz and Engleman, 1990; Schuler and Steinman, 1985; Steinman and Cohn, 1973, 1974). While immature DCs avidly take up antigen, mature DCs lose this capacity and instead gain the ability to potently stimulate T cells. DC maturation in culture, a spontaneous and uncontrolled process, presented a barrier to studying the mechanisms of DC antigen capture, processing, and presentation (Schuler and Steinman, 1985; Steinman and Cohn, 1973, 1974). Investigating human DCs posed additional challenges of accessibility and cell number, since the only readily available source are rare DC populations in the blood, comprising $<1\%$ of circulating leukocytes (Fearnley et al., 1999).

Despite their rarity in physiology, DCs have become the subject of interest in immunologic and biomedical research for their role in controlling adaptive immunity and maintaining tolerance. As APCs, the primary role of DCs is to uptake, process, and present antigen. Their unique ability to efficiently present antigen and activate helper T cells, as well as cytotoxic T cells through cross-presentation, earned DCs the title as "professional" APCs. Characterizations of DC function in T cell activation have identified three key signals mediated by DCs: (1) antigen presentation, (2) costimulatory signals, and (3) cytokine secretion. Signal 1 refers to antigen presentation on DC surface in the form of a peptide-MHC (class I or II) complex. Signal 2 refers to co-stimulation markers presented by DCs in the form of surface ligands (CD80, CD86, PD-1, CTLA-4, etc.), the specific combination of which was determined by DC maturation either toward immunity or tolerance. Signal 3 refers to the release of instructive cytokines by DCs to polarize T cells toward different effector phenotypes (Th1, Th2, Th17, Treg, etc.). Together, the exceptional ability for DCs to prime naïve T cells and, thus, trigger robust T cell-mediated immune responses has made DCs a target of immunotherapeutic approaches to control the mobilization of cellular immunity and tolerance. In the following sections, we will discuss the key approaches currently developed to generate and modify DCs *ex vivo* for the purposes of harnessing their immunomodulatory capabilities, followed by an overview of their therapeutic use in clinical trials.

1. Cytokine signals

Work on granulocyte-macrophage colony-stimulating factor (GM-CSF) inspired the early development of DC culture methods. GM-CSF was initially identified in tissue culture supernatants conditioned with lung tissue of mice following lipopolysaccharide (LPS) injection, by its ability to stimulate the differentiation of mouse bone marrow cells into macrophage and granulocyte colonies (Burgess et al., 1977; Burgess and Metcalf, 1980). This growth factor was then demonstrated to enable longer survival of tissue-derived mouse and human DCs in culture (Markowicz and Engleman, 1990; Witmer-Pack et al., 1987). The seminal finding that GM-CSF acts on the myeloid progenitor cells in a dose-dependent manner, and that an intermediate dose of GM-CSF favors *ex vivo* DC production from the bone marrow (Inaba et al., 1992, 1993b), subsequently enabled the classic 10–12 days culture methods for robust differentiation of large numbers of immature DCs from murine bone marrow progenitors using this cytokine (Inaba et al., 1993a; Lutz et al., 1999).

Similarly to mouse bone marrow cell culture method, GM-CSF culture of human hematopoietic stem cells (HSCs), either derived from the bone marrow or mobilized into the blood and isolated as lineage-negative CD34+ cells, enabled differentiation of human stem cell precursors into DCs (Bernhard et al., 1995; Reid et al., 1992). The human bone marrow culture methods additionally required tumor necrosis factor alpha (TNFa), which upregulated GM-CSF receptor expression and inhibited expression of receptors for other lineage-specific factors (Caux et al., 1990). Since human bone marrow HSCs are difficult to obtain routinely, and circulating HSCs are few, an alternative source of human DC progenitors was found in umbilical cord blood (UCB), which is especially rich in hematopoietic precursor cells (Broxmeyer et al., 1989). In addition, in the 1990s UCB was already being successfully used in stem cell transplantation (Ballen et al., 2013; Broxmeyer et al., 1991), making it an attractive source of DC precursors for potential clinical use. Cord blood therefore became the preferred source of HSC-differentiated human DCs, with best-established culture protocols (Balan et al., 2009; Caux et al., 1996).

Stem cell factor (SCF/c-kit ligand) and FMS-like tyrosine kinase 3 ligand (Flt3L) are notable common additions to human and mouse HSC DC cultures. These early-acting cytokines share similar tyrosine kinase receptors and are similarly expressed on early bone marrow progenitor cells (Lyman and Jacobsen, 1998). Both increase DC production from either human bone marrow or from mobilized HSC without affecting DC differentiation (Siena et al., 1995; Szabolcs et al., 1995), and the stimulating effect is additive when the factors are combined (Siena et al., 1995). Similarly to human HSC cultures, in mice SCF was found to cooperate with GM-CSF when inducing DC differentiation from the bone marrow (Zhang et al., 1997), whereas Flt3L was able to completely substitute for GM-CSF (Brasel et al., 2000; Naik et al., 2005), although the GM-CSF-derived DCs and Flt3L-derived DCs differed somewhat in their phenotypes (Xu et al., 2007). The discovery that *in vivo* Flt3L administration significantly expands endogenous DC populations of both mice and humans (Maraskovsky et al., 1996, 2000) prompted development of hybrid *in vivo* and *ex vivo* human DC culture methods, where DCs isolated from Flt3L-treated individuals undergo subsequent *ex vivo* manipulation (Fong et al., 2001).

Despite advances in culture methods, human HSCs remained a rare and specialized source of DCs, and well-controlled production of sufficient quantities of human DCs for research and clinical applications continued to be elusive. These issues were addressed with the development of a landmark *ex vivo* DC culture method (Sallusto and Lanzavecchia, 1994; Thurner et al., 1999), that allowed for differentiation of large numbers of immature DCs from readily available human blood monocytes by 5–7 days of culture with GM-CSF and interleukin 4 (IL-4), and for subsequent controlled DC maturation upon addition of inflammatory stimuli for a further 2–3 days of culture. In addition to GM-CSF's established role in promoting DC differentiation and viability, IL-4 was known to inhibit monocyte and macrophage differentiation from bone marrow, thus promoting DC differentiation (Jansen et al., 1989), and was also speculated to maintain DCs in an immature state by antagonizing the maturational effect of inflammatory agents (Sallusto and Lanzavecchia, 1994). The GM-CSF/IL-4 culture method became the experimental platform for much of our understanding of human dendritic cell biology, and solidified the role of DCs as mediators of both immunity and tolerance (Amigorena, 2018; Banchereau and Steinman, 1998; Steinman, 2012; Steinman et al., 2003). It also provided the basis for numerous dendritic cell clinical trials, primarily in cancer immunotherapy (Palucka et al., 2011; Palucka and Banchereau, 2012, 2013).

Substitutions to the GM-CSF/IL-4 monocyte culture method have been attempted since its publication, but are not as widely used as the original. In monocyte DC cultures, GM-CSF can be replaced by interleukin 3 (IL-3) (Ebner et al., 2002; Lutz, 2004), a closely related molecule that is expressed from the same locus as GM-CSF (Nicola, 1989). The resulting DCs are phenotypically similar to GM-CSF/IL-4 DCs, but differ in function, promoting more IL-4 and IL-5 secretion and less IL-12 and interferon gamma (IFNg) from stimulated T cells (Ebner et al., 2002). Similar Th2 bias was observed in DCs derived from human monocytes using IL-3 and IFNb culture (Buelens et al., 2002). IL-4, in its turn, can be replaced by a closely related Th2 cytokine IL-13 for monocyte DC generation (Morse et al., 1999; Piemonti et al., 1995). While the phenotype of GM-CSF/IL-13 DCs was found to be similar to that of GM-CSF/IL-4 cells, conflicting reports about the relative efficiency of the substitution protocol (Ahn and Agrawal, 2005; Sato et al., 1999) limited its use. Monocyte DCs differentiated with GM-CSF and interferon alpha (IFNa) secrete large amounts of inflammatory cytokines to initiate Th1 T cell responses (Mohty et al., 2003) and have been tested in cancer immunotherapy clinical trials (Farkas and Kemény, 2011). In contrast, monocyte DCs differentiated with GM-CSF and IL-15 were better inducers of Th17 T cell response (Harris, 2011).

A very successful modification of the GM-CSF/IL-4 protocol arose from the observation that 1 day of human monocyte culture with GM-CSF and IL-4, followed by 1 day of stimulation with proinflammatory factors (TNFa, IL-1b, IL-6, or PGE2), is sufficient to produce "FastDC," which are highly comparable to DCs derived by the original 10-day GM-CSF/IL-4 culture method, but much less labor-intensive and more efficient to manufacture (Dauer et al., 2003, 2005). "FastDC" are now extensively used for research purposes, and are being introduced into immunotherapy trials (Bürdek et al., 2010; Kvistborg et al., 2009; Schaller and Sampson, 2017). Fig. 1 outlines the evolution of DC culture methods. Shorter versions of other monocyte DC culture methods have also been successfully explored, for instance 2-day human monocyte culture with GM-CSF and IL-15 followed by stimulation (Anguille et al., 2009).

In conjunction with culture methods that can provide a rich source of immature DCs, there is a great variety of factors being used to promote DC maturation, and to polarize their phenotypes toward immunization or tolerance. DC maturation is triggered by environmental cues, including pathogen-associated molecular patterns (PAMPs, such as lipopolysaccharide (LPS)) and damage-associated molecular patterns (DAMPs, such as IL-1b), cytokines and secreted factors reporting on tissue homeostasis (such as tumor necrosis factor (TNF) and prostaglandin E2 (PGE2)), and immune

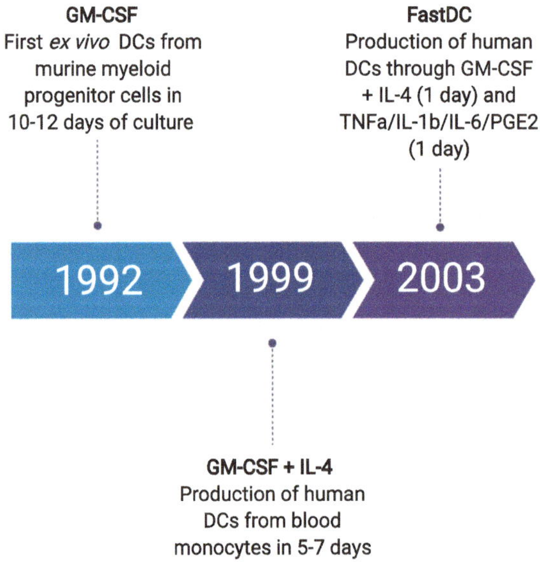

Fig. 1 Summary of DC culture innovations.

co-stimulation (such as T cell ligation of the CD40 receptor on DCs) (Cerboni et al., 2013; e Sousa, 2006; Schaefer, 2014).

The early human monocyte-derived DC culture methods mainly focused on producing immune-stimulatory DCs suitable for use in anti-cancer therapy trials. They employed a standard DC maturation mixture consisting of TNFa, IL-1b, IL-6, and PGE2 (e Sousa, 2006; Jonuleit et al., 1997; Nair et al., 2012); the same cocktail was used for FastDC maturation (Dauer et al., 2003). However, cells derived by this method ultimately proved ineffective for cancer immunotherapy (de Vries et al., 2003; Kalinski and Okada, 2010; Schadendorf et al., 2006; Vilella et al., 2004), prompting its re-examination, and the recognition that when considering DC manufacture for anti-cancer applications, with the desired outcome of stimulating cytotoxic T cell response against tumor cells, this classic mixture has some deficits. For instance, PGE2 was found to block the production of IL-12 p70 by DCs, thus skewing subsequent T differentiation toward Th2 instead of Th1 effector function (Kaliński et al., 1998, 2001). DC maturation with TNFa in a mouse model was also implicated in impaired inflammatory cytokine production, and even found to be protective against autoimmunity (Menges et al., 2002). IL-6 was found to suppress DC maturation in mouse models (Park et al., 2004) and its overproduction was linked to a functional DC defect in multiple myeloma (Ratta et al., 2002).

Alternative stimulatory cocktails that would selectively promote DC induction of cytotoxic T cell response were therefore developed, incorporating known Th1-skewing agents. These typically include a combination of type I and type II interferons, and pathogen-derived molecules and toll-like receptor (TLR) agonists. Successful polarization of monocyte-derived DCs toward a Th1-promoting phenotype was demonstrated with an IL-1b/TNFa/IFNg cocktail (Vieira et al., 2000) which could be further potentiated with polyinosinic:polycytidylic acid (polyI:C, a dsRNA analog and TLR3 ligand) (Mailliard et al., 2004); with an IL-1b/TNFa/IFNg/PGE2/PoyI:C/R848 (Resiquimod, a TLR7/8 ligand) mixture (Zobywalski et al., 2007); with a combination of IFNg and monophosphoryl lipid A (MPLA; a chemically modified version of lipopolysaccharide, LPS, a TLR4 ligand) (Ten Brinke et al., 2007); with IFNa/PGE2/OK432 (a *Streptococcus pyrogenes* preparation) (Sakakibara et al., 2006); and other similar preparations (Peng et al., 2005). While same direct *ex vivo* comparisons between these alternative maturation methods have been carried out (Vopenkova et al., 2012), clinical effectiveness comparisons are still lacking.

Although much of the work on DC maturation focused on deriving immunizing DCs, progress was also made on developing monocyte-derived DCs that could stably suppress inflammatory T cell responses and promote regulatory T cell activation, for use in autoimmunity and transplantation settings. The main features of tolerogenic DCs are low expression of co stimulatory molecules, high production of anti-inflammatory cytokines, particularly IL-10, and the reciprocal low production of IL-12 and other proinflammatory factors (Banchereau and Steinman, 1998; Garcia-Gonzalez et al., 2016; Steinman et al., 2003). It is worth mentioning here that there is still some debate in the field whether tolerogenic DC are a genuine lineage, or a less mature DC activation state (e Sousa, 2006; Takenaka and Quintana, 2017). In keeping with the latter hypothesis, some of the tolerogenic DC manufacturing techniques achieve their goal by blocking known DC activation pathways, such as nuclear factor-kappa B (NF-kB) (Xie et al., 2005).

Three main classes of compounds used for promoting tolerogenic DCs are anti-inflammatory cytokines, natural immunosuppressive compounds, and pharmaceutical agents known to inhibit key DC activation pathways. Anti-inflammatory cytokines shown to promote tolerogenic DC production include IL-10 itself (Steinbrink et al., 1997), IL-6 (Hegde et al., 2004; Park et al., 2004), IL-21 (Brandt et al., 2003), IL-27 (Mascanfroni et al., 2013), TNFa (Menges et al., 2002), and TGF-b (Fogel-Petrovic et al., 2007; Song et al., 2014).

The leading natural compounds used for tolerogenic DC production are 1α,25-dihydroxyvitamin D3, the active form of Vitamin D3 (Harry et al., 2010; Penna and Adorini, 2000), and Vitamin A, which certain classes of DCs can metabolize into retinoic acid (RA) and thus promote regulatory T cell differentiation, a pathway particularly relevant in the gut lamina propria (Sun et al., 2007). Other natural compounds tested include resveratrol (Švajger et al., 2010), curcumin (Rogers et al., 2010), sulforaphane (Geisel et al., 2014), andrographolide (Iruretagoyena et al., 2006), and vasoactive intestinal peptide (Chorny et al., 2005). Although the mechanisms of action of these supplements are not always clear, some have been shown to inhibit NF-kB (Iruretagoyena et al., 2006), an important driver of DC activation (Xie et al., 2005).

Pharmaceutical agents used to create tolerogenic DCs either target known DC activation pathways, such as Janus kinase (JAK) inhibitor tofacitinib and calcineurin inhibitors tacrolimus (Kubo et al., 2014) and cyclosporine (Lee et al., 1999), mTOR pathway inhibitor rapamycin

BOX 1 Summary of cytokine cocktails used in DC maturation

Immunogenic

FastDC Maturation
- TNFa/IL-1b/IL-6/PGE2
- Ineffective in clinical trials
- Skew T cells towards Th2 instead of Th1
- TNFa implicated in impaired inflammatory cytokine production

Th1-Promoting DC
- IL-1b/TNFa/IFNg
- Potentiated with polyI:C

Toleragenic

Anti-Inflammatory Cytokines
- IL-10/IL-6/IL-21/IL-27/TNFa/TGFb

Natural Compounds
- Vitamins D3 and A, resveratrol, curcumin, sulforaphane, andrographolide, vasoactive intestinal peptide
- Some have been shown to inhibit NF-kb

Pharmaceutical Agents
- JAK inhibitor tofacitinib, calcineurin inhibitors tacro limus and cyclosporine, mTOR inhibitor rapamycin, PKC inhibitors, dexamethasone, aspirin, rosiglitazone, mycophenolic acid

(Fedoric and Krishnan, 2008; Hackstein et al., 2003), and protein kinase C (PKC) inhibitors (Matsumoto et al., 2013), or employ known anti-inflammatory drugs such as corticosteroid dexamethasone (Xia et al., 2005), aspirin (Buckland et al., 2006; Buckland and Lombardi, 2009), rosiglitazone (Iruretagoyena et al., 2006), and mycophenolic acid (Kazma et al., 2013). Box 1 provides a summary of the various cocktails used for DC maturation for both proinflammatory DCs (immunogenic) and tolerogenic DCs.

Clinical studies using tolerogenic DCs are relatively recent, and tend to use a combination of the above compounds, sometimes followed by a short activating signal. For instance, dexamethasone and Vitamin A culture, followed by 1-day stimulation with the classic stimulatory cocktail of TNFa, IL-1b, IL-6, and PGE2 have been used to create peritoneally injected tolerogenic DCs used in a Phase I Crohn's disease trial (Jauregui-Amezaga et al., 2015), while DCs manufactured using dexamethasone and Vitamin D3 and briefly stimulated with MPLA were injected into the knee joint of rheumatoid arthritis patients (Bell et al., 2017). The clinical effectiveness of these methods, and comparative advantages and disadvantages among them, remain to be determined.

Although monocyte-derived human DCs are now easy to obtain and polarize to the desired modalities in culture, their relationship to DC populations arising *in vivo* is still poorly understood (Collin and Bigley, 2018; Geissmann et al., 2010). While cytokine-derived DC and classical DC (cDC) share some transcriptomic similarities in genes such as those associated with antigen cross-presentation capability (Briseño et al., 2016; Haniffa et al., 2015; Miller et al., 2012; Villani et al., 2017), they have been shown to be functionally distinct, especially in their capacity to induce effector responses *in vitro* (Osugi et al., 2002) and *in vivo* (Kalinski et al., 2011). Notably, as immunotherapies using cytokine-derived DCs have met with disappointing clinical outcomes (Anguille et al., 2014; Rosenberg et al., 2004; Schüler and Blankenstein, 2003), the biologic integrity of *ex vivo* cytokine-maturation has been questioned even by the originators of this methodology (Steinman and Mellman, 2004). Current hypotheses attribute cDC therapeutic dysfunctions to a combination of factors, namely restricted life-cycle and migration to the draining LNs (Tacken et al., 2007), loss of responsiveness and "exhaustion" against *in vivo* cytokine cues, and limited capacity for effector activation (Camporeale et al., 2003; Langenkamp et al., 2000; Watchmaker et al., 2010), implicating the method of production in cytokine-derived DCs' shortcomings concerning physiologic performance *in vivo*. Given the drawbacks in deploying cytokine cocktails and concentrations seldom seen in nature, there has been a push in the field toward understanding and mimicking DC differentiation *in vivo*. This drive toward understanding physiological signals is the subject of the next section.

2. Physiological signals for DC differentiation

The discovery of DC in the 1973 by Ralph Steinman of Rockefeller University, followed by their scientific elucidation by him and his worldwide acolytes, earned him the Nobel Prize in Physiology or Medicine in 2011. The title of that award is appropriate, since the *medical promise* it has long portended is that therapeutic marshaling of the key capability of DC, initiation of specific T cell responses to antigenic peptides can be distinguished from all man-made immune interventions, because they would operate through a genuine *physiologic* partnership with the immune system itself.

Far and away, the normal immune system is the single best therapy in existence, or likely even possible. Discretely, and without our momentary awareness, but with uncanny efficiency, our armamentarium of billions of clones antigen-specific T cells protect our tissues from microbial and cellular invaders, distinguish self from non-self and eliminate cancerous cells in their tracks. Out of necessity, our efforts to treat those disorders resulting from immunologic errors typically require administration of blunt instruments, in the form of non-specific immune enhancers or suppressors, or biologic constructs. While these agents can be extremely beneficial, they are by their very nature stopgap, are efforts to circumvent the immune system rather than collaborate with it, lack the immune system's exquisite specificity and commonly produce substantial adverse side reactions.

The tantalizing potential of DC, after arming with the relevant antigen(s), to provide a fully natural assist to the normal immune system to fill its own periodic functional gaps ran into a limiting roadblock. The method by which readily accessible blood monocytes are physiologically signaled to become DC was unknown. To circumvent this problem, laboratory approaches to monocyte-to-DC differentiation involving non-physiologic *ex vivo* cytokine cocktails, including IL-4 and GM-CSF, were devised. This approach provided sufficient DC for detailed laboratory study, around which the scientific field developed. However, human administration of these artificially induced DC regularly produced disappointing clinical results.

In contrast to long-term culture in non-physiological cytokine cocktails, an alternative approach to effective DC production might be to first elucidate the physiologic mechanisms of *in vivo* APC generation from monocytes, and then to co-opt these mechanisms for experimental and therapeutic purposes. Since rapid response to infection or injury is a hallmark of both innate and adaptive immunity, DC production methodologies limited by week-long incubation periods and supra-physiological cytokine concentrations are likely the opposite of natural DC generation. Alternatively, recent evidence suggests that rapid and direct conversion of abundant circulating monocytic precursors to DC (mDC) has been observed *in vivo* at the site of immune challenge, as well as *ex vivo*. *In vivo* the rapid production of inflammatory DC (infDC) from blood monocyte precursors was reported following pathogen exposure (Cheong et al., 2010; Segura and Amigorena, 2013), and *ex vivo* it is observed in the large-scale conversion of blood monocytes to functional DC associated with the FDA-approved cellular immunotherapy extracorporeal photopheresis (ECP) (Raval and Ratcliffe, 2018; Ventura et al., 2018). This indicates nature has already provided a DC generation pathway which is both rapid and cytokine-free, dependent only on physical forces and physiologic signals present at steady state in mammalian circulation.

2.1 DC generation via inflammatory and tissue signaling

The possibility of DC production from monocytes using physiologic approaches was first suggested by early seminal work of Mueller and colleagues, which showed that blood monocytes could be converted to DC rapidly and in the absence of cytokines in an *in vitro* model of endothelial translocation (Fig. 2A) (Randolph et al., 1998). These cells were then shown to be potent stimulators of allogenic T cells with *in vivo* activity (Randolph et al., 1999). In this model, monocytes spontaneously translocated across cultured endothelial cell barrier layers, were stimulated by microbial pathogen-associated molecular patterns (PAMPs) in the subendothelial space, and then "retro-translocated" and emerged in 24h or less with all the phenotypic and functional features of DC. This observation was later confirmed in multiple murine *in vivo* models of inflammatory DC

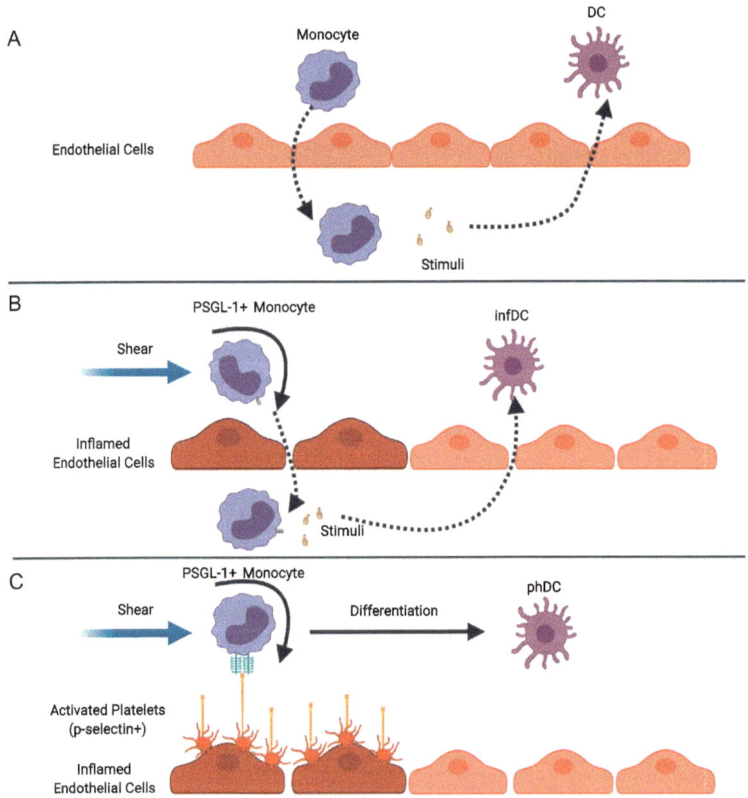

Fig. 2 Evolution of the understanding of physiological DC maturation (A) monocyte translocation to the subendothelial layer followed by stimulation and re-emergence with DC phenotype, (B) shear-mediated monocyte rolling *via* adhesion receptor/ligand tethering, (C) platelet-mediated monocyte maturation.

(infDC) production, reporting similar cellular players and stimulatory components, with monocyte-to-DC generation proceeding rapidly and without cytokine addition (Hespel and Moser, 2012). InfDCs generated in this manner are capable of stimulating potent Th1 and Th2-based immunity, confirming that physiologically-derived DC, particularly these arising from the "migratory" monocyte subtype populations expressing Ly6C (mouse) or $CD14^{high}$ (human), are an important component of the cellular armamentum against pathogens (Jakubzick et al., 2017). Additional data also point to the importance of migration through epithelium as a differentiating signal for human CD14+ monocytes, in work showing their conversion to IL-12p70-producing DC populations, capable of stimulating both Th1 and Th17 T cells, after translocating across the blood-brain barrier (BBB) inflamed epithelial cells (Ifergan et al., 2008).

The rapid differentiation of monocytes to "professional APC" following tissue transmigration, defined by criteria such as downstream ability to migrate via afferent lymphatics to lymph nodes, cross-present antigen, and stimulate naïve T cells in ways which mirror classical DC (cDC) in blood (Jakubzick et al., 2017), could be considered a specialized inflammatory extension of their better known fate as a blood-borne precursors of tissue macrophages. Monocytes, which circulate in the bloodstream as a major myeloid precursor population representing 10–15% of mononuclear leukocytes, can routinely be signaled to cross endothelial barriers via luminal surface markers, such as adhesion molecules *E*- and P-selectin, intercellular adhesion molecule (ICAM)-1, and vascular cell adhesion molecule (VCAM)-1, as well as tethered chemokines expressed by activated endothelial cells in response to inflammation or tissue damage (Fig. 2B). (Campbell et al., 1998; Hyduk and Cybulsky, 2009; León and Ardavín, 2008). Monocytes use their complementary ligands to interact with the activated endothelium and negotiate the high shear stress exerted by the blood flow in the vasculature. One such monocyte ligand, PSGL-1, is expressed at a significantly higher level by inflammatory $Ly6C^{hi}$ monocytes than by resident $Ly6C^{low}$ monocytes, pointing to that specific adhesion molecule and monocyte subtype as candidates for interactions leading to "infDC" instead of macrophage differentiation (Gerhardt and Ley, 2015). The majority of the necessary and sufficient signaling events at sites of infection or inflammation that facilitate monocyte differentiation to a DC rather than a macrophage pathway still remain to be elucidated. However, overwhelming evidence now points to the possibility that "physiologically-derived" DC may act as APC populations with similar capabilities and potency as cDC.

Recently, activated platelets have been identified as another important player in the signaling cascade to which monocytes are exposed during their transendothelial migration and potential DC conversion (Fig. 2C). Activated platelets have been shown to act as "bridges" and "pathfinders" during *in vivo* monocyte diapedesis through endothelial layers, effectively marking sites of monocyte extravasation (Kuckleburg et al., 2011; Zuchtriegel et al., 2016). This points to a previously unrecognized role of platelets as critical components of blood monocyte recruitment, homing, activation and differentiation to DC during extravasation to inflamed tissue. It was appreciated more than 20 years ago that activated platelets can form an integrin synapse with blood-borne monocytes, activating monocyte production of a myriad of monokines, including MCP-1, TNFa, IL-8 (Weyrich et al., 1995, 1996), MMP-9 (Galt et al., 2001) and CXCR-5 (Halvorsen et al., 2014). In addition, when artificial surfaces were coated with a platelet adhesion molecule P-selectin and allowed to interact with blood monocytes under either flow or static conditions, monocytes produced significant levels of these cytokines plus IL-1b, IL-6, IL-8, IL-12 and MIP-1b (Suzuki et al., 2013; Weyrich et al., 1995). These experiments identify the platelet P-selectin and its monocyte ligand PSGL-1 as a critical tethering and signaling junction. The importance of this axis as a principal site of both platelet/monocyte binding and signal transduction was also highlighted by additional data indicating that this selectin/selectin ligand interaction induces monocyte NF-kB and COX-2 signaling (Dixon et al., 2006) and preferential platelet binding to monocytes over other PBMC (Ahn et al., 2005).

If rapid APC generation at the site of pathogen infiltration or injury represents the type of immunological "emergency" response, platelets are uniquely situated both physiologically and spatially to be a critical player in such a reaction. As a testament to how little is presently known regarding the complexity and function of signaling molecules present in platelets, a recent proteomic study confirmed the presence of 827 proteins within platelet granules (Zufferey et al., 2014). A growing appreciation of the role of platelets beyond hemostasis has been driven by recent data confirming not only P-selectin-mediated leukocyte binding but also platelet GPIbα binding of leukocyte Mac-1 and LFA-1 and platelet CD40L binding monocyte CD40, leading to the conclusion by some authors that platelets should be re-classified as immune cells (Lam et al., 2011).

Taken together, these data indicate that under inflammatory conditions the physiological forces alone, likely involving stimulatory molecules provided by readily available blood components such as platelets, may offer a means to reprogram blood monocytes into functional APC. Such "physiological" DC could prove to be a rapid, inexpensive, and readily available alternative to the cytokine-derived DC, which have dominated the dendritic cell landscape for nearly two decades but to date have demonstrated limited clinical applicability.

2.2 DC generation via extracorporeal photopheresis (ECP)

Recent work from our group confirmed the importance of platelets not only as adhesion molecule sources for monocyte tethering underflow stress, but also as providers of critical signaling molecules driving monocyte-to-DC conversion. Serendipitously, this finding arose from studies aimed at uncovering the mechanistic basis of an empirically crated but highly clinically successful therapy, "Extracorporeal Photochemotherapy (ECP)," thus simultaneously contributing to the growing field of research into physiologic methods of DC differentiation, and demonstrating that an effective *ex vivo* method of generating such DCs and using them therapeutically already exists.

ECP was originally conceived as method to palliate therapeutically intractable leukemias, by exposing extracorporeally circulated blood to lethal doses of the photoactivatable DNA-crosslinking drug, 8-methoxypsoralen or 8-MOP (Ratcliffe et al., 2015). The goal was to markedly diminish the body burden of malignant cells while avoiding system toxicity. Surprisingly, return of the treated blood in the very first T cell lymphoma/leukemia patient, produced a selective anti-cancer immune reaction that eliminated the leukemic cells. An international clinical trial of ECP in leukemic CTCL was so successful that ECP was fast-tracked as the first FDA approved anti-cancer immunotherapy. Due to its empiric efficacy, and favorable safety profile, despite its long mysterious mechanism, the treatment gained widespread use in university medical centers. ECP is now regularly performed in more than 350 university centers in the USA and Europe, with more than 3 million treatments administered to date to more than 70,000 patients. The scope of the 30-year international quest for the cellular and molecular ingredients that produce ECP's immunotherapeutic responses is most succinctly highlighted by the more than 1200 peer reviewed publications focused on that journey. The motivation driving that search has been the expectation that the answer to that riddle could have immense clinical impact through tumor- and patient-specific treatment of a broad spectrum of immunogenic cancers.

As a therapy ECP is distinguished by three features: (1) efficacy as both an anti-cancer immunotherapy (cutaneous T cell lymphoma) and a tolerogenic therapy in organ transplant rejection and graft-vs-host disease; (2) specificity for the pathogenic clone(s), whether in CTCL or auto-aggressive disease; and (3) an unusually favorable safety profile (Edelson et al., 2018). From the outset, it is been clear that to have these features, ECP must somehow provide a partnership with the normal immune system, which itself is by far the most intricately potent immunotherapy ever designed (Berger et al., 2010; Kibbi et al., 2016). The clues were that, mimicking the physiologic immune system, ECP is bidirectionally balanced, effecting antigen-specific immunity and tolerance, through silent self-sorting of cell-derived antigens, whether they be extracted from invading microbes, malignant cells, or autologous tissues. The earliest evidence suggesting that ECP shared those principal features with the physiologic and homeostatic immune system was that, in the best responding CTCL patients, the malignant cells were immunologically eliminated while preserving overall patient immunocompetence, without leading to any immune deficiencies or opportunistic infections. Parallel antigen specificity and ability to control disease while preserving normal immune function was subsequently clinically established on the tolerogenic side of the immunologic equation, in organ transplant and GvHD settings.

The recognition that ECP must be tightly linked to the physiologic immune system itself presented investigative opportunities studded with hurdles. Since ECP's initial clinical accomplishments could not be explained by existing scientific paradigms, and yet were quite reproducibly impressive, the principles underlying them were presumed to be both novel and important, yet obscure. Uncovering those principles required reverse scientific engineering in relevant experimental animal systems, while accurately miniaturizing the clinical ECP device for application (Ventura et al., 2018, 2019). That challenge was met through development of an ECP device scalable from mouse to man (Ventura et al., 2018, 2019). Use of the miniaturized device in a mouse melanoma model permitted demonstration that: (1) experimental ECP could regularly induce immunity selective to the melanoma; (2) the immunity resulted from induction of monocyte-to-DC maturation; (3) activated platelets signaled the monocytes to enter the DC differentiation pathway; (4) cancer cells exposed to ultraviolet-activated 8-MOP underwent immunogenic cell death; (5) the ECP-induced DC readily internalized the apoptotic cancer cells and extracted and presented cancer-specific antigens from them; (6) CD4 and CD8 T cells collaborated to generate the therapeutic anti-cancer response. The pivotal causative role of the ECP-induced DC in the initiation of induced anti-cancer immunity has thus been experimentally established (Edelson, 2014).

The manner by which ECP-activated platelets extracorporeally trigger monocyte-to-DC differentiation has also been recently elucidated (Durazzo et al., 2014; Han et al., in press). During blood treatment in the ECP device, plasma fibrinogen adheres to the treatment chamber walls, creating a substrate for platelet adhesion. Platelets encountering the fibrinogen deposits are activated by recognition of platelet α2bβ3 and α5β1 integrins of fibrinogen RGD domains. Subsequently, blood monocytes underflow conditions not only roll, tether and interact directly with fibrinogen-immobilized activated platelets, but rapidly acquire the phenotypic (Durazzo et al., 2014) and functional (Han et al., in press) characteristics ofDC in a P-selectin/PSGL-1-dependent manner, very reminiscent of the PSGL-1-associated conversion of "migratory" monocytes linked to infDC formation described in the previous section of this review.

Combined, ECP's rich clinical record and its scientific underpinnings serve as a proof of the principle that physiologically-induced, antigen-loaded DC can stimulate potent clinical responses, even in the context of advanced immunologic diseases. Since DC are the master-switches of antigen-specific immunity and tolerance, their physiologic large-scale production by ECP raises the intriguing possibility that they can be clinically applied to any disorder for which the relevant antigens are available. These potential applications include all immunogenic cancers, anti-microbial vaccinations and the full range of auto-aggressive immunologic disorders (Mutyambizi et al., 2009; Shen et al., 2008).

3. Gene engineering tools

Clinical limitations exhibited by the current and dominant approaches in generating DCs, driven by use of function-skewing biomolecules such as cytokines, small molecules, inhibitors and proteins (i.e., PAMPS, DAMPS, antibodies, etc.), have generated interest in supplementary methods of generating clinically functional DCs. The ever-expanding arsenal of genetic engineering tools available for mammalian cell manipulation is inspiring innovative methods of intervening with creation, maturation and, ultimately, function of output professional APCs, with the goal of improving upon current methods of *ex vivo* DC generation. Such innovations can conceptually be applied to the modification and improvement of either the

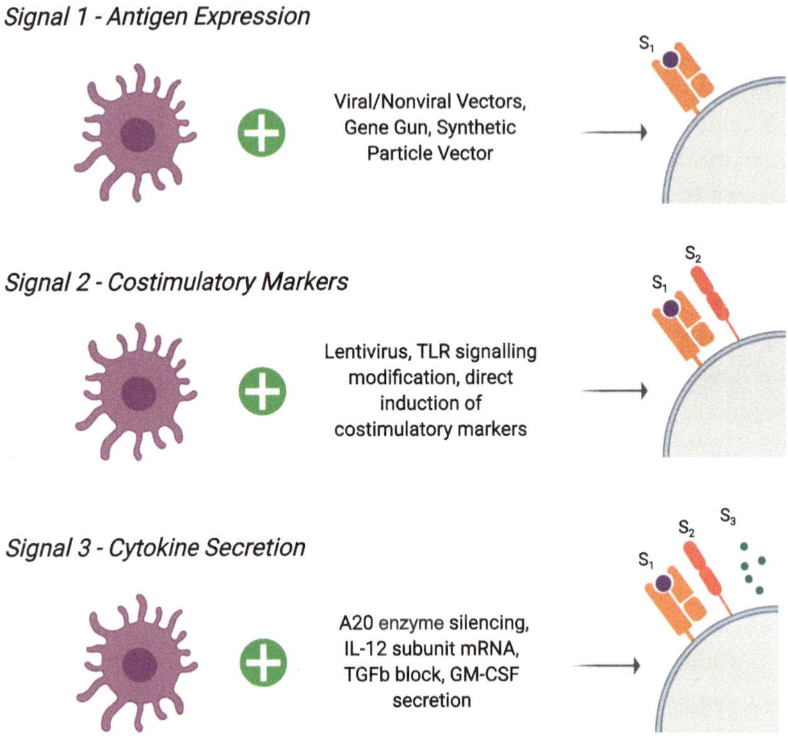

Fig. 3 Three canonical signals in DC activation and genetic engineering approaches to modulate signals.

cytokine-derived or the physiologically-derived DCs described in previous sections. This section will follow attempts to modulate the three canonical DC-mediated signals essential for robust T cell activation: (1) antigen presentation, (2) costimulatory signals, and (3) cytokine secretion. Fig. 3 summarizes the three canonical signals and the genetic engineering approaches to modulate these signals.

3.1 Antigen expression vectors (Signal 1)

The first requirement for DCs to mount an antigen-specific immune response is the delivery of antigen to the APC. In order to promote optimal antigen delivery followed by processing and presentation, developments in early genetic engineering-mediated immunization strategies rationally focused on antigen delivery. The goal of genetic immunization is to introduce the gene encoding a target antigen into the cytoplasm of a cell to induce endogenous production of antigen, with the presumption being that continued expression of antigen would be facilitated in its processing and access to MHC. The advantages of inducing expression of antigen in DCs is the specificity of antigen sequences delivered, opportunities to tailor MHC class I and II loading, and simultaneous delivery of diverse antigens ranging from defined tumor-specific peptides to uncharacterized whole tumor antigen mRNA for endogenous processing and sorting of epitopes (Ribas, 2005).

Genetic modification of dendritic cells *ex vivo* requires gene transfer to the isolated cells. To achieve such transduction of new genetic material, naked DNA or mRNA strands encoding target antigens can be delivered directly to DCs; however, transfection efficiencies can be greatly improved by utilizing tools such as lipid mediated transfection, electroporation, and others. Table 1 summarizes the methods used to modulate signal 1. Since antigen delivery to DCs was the earliest, and consequently the best-explored, genetic engineering-mediated strategy, these tools are described below in the context of genetic DC immunization.

3.1.1 Introduction of non-viral vectors into DC

Lipid-mediated transfection relies on cationic lipids, such as lipofectamine, which facilitate DNA and mRNA delivery into cells. The cationic lipids form a unilamellar liposomal structure with a positive surface charge and fuse with nucleic acids to form a transfection complex. By charged interactions

Table 1 Summary of genetic engineering approaches for signal 1.

Vector	Type	Characteristics
Non-viral vectors	Cationic lipids (lipofectamine)	Cationic liposomal structures interact with negatively charged cell walls
	Electroporation	High voltage shocks permeabilize cells
Viral vectors	Adenovirus, herpes simplex, retrovirus, lentivirus	Leverage viral infection mechanisms. Off-target effects are of concern
Synthetic particle vectors	Biodegradable polymeric nano/micro-particles	Tunable platform for encapsulation and delivery Low pH from degrading polymer potentially reduces transfection efficiency. Slow release kinetics
Gene guns	Particle microprojectiles	Gold microparticles wrapped with naked DNA delivered in high velocity at the intracellular level

between the cationic lipid headgroups with the negatively charged cell membrane, the transfection complex localizes to the cell surface for endocytosis and transfection. Strategies utilizing such lipid complexes have been successful in achieving desired DC transfection for expression of tumor antigen (Denis-Mize et al., 2000; Rughetti et al., 2000).

Electroporation is another popular gene delivery method applicable to most cell types. It uses high voltage electric shocks to temporarily permeabilize the cell membrane, enhancing introduction of inherently charged nucleic acid molecules into cells. Recent efforts with electroporation strategies have resulted in high transfection efficiencies of DC (Gilboa and Vieweg, 2004; Tuyaerts et al., 2003; Van Tendeloo et al., 2001).

Although transgene expression levels achievable with these non-viral gene delivery methods proved to be very low or undetectable, the high affinity of T cell receptors (TCRs) nevertheless enable antigen-specific T cells to recognize and mount an immune response against antigens presented by transduced DCs (Gilboa et al., 1998). The antigen expression and the resulting immunoprotective effects of these non-virally transfected DCs have been demonstrated in multiple animal models of antigenic tumor challenge (Kirk and Mulé, 2000; Ribas et al., 2000). However, the drawback of most physical methods of gene transfer is that they have been reported to confer some toxicity or loss of phenotype on recipient DCs (Wysocki et al., 2002).

3.1.2 Introduction of viral vectors into DC

To enhance the efficiency of DC transfection beyond the limits of physical methods and to reduce toxicity, viral vectors have become a preferred method of antigen expression in DCs (Kirk and Mulé, 2000; Tuyaerts et al., 2007a). Viral gene delivery methods aim to hijack existing mechanisms of viral infection to increase internalization and transfection rates of engineered genetic material of choice. Viruses used as vectors are attenuated or rendered replication-deficient by deletion of genes necessary for progeny virion generation, in order to prevent viral replication in mammalian cells.

Many viral vectors have been utilized for DC transduction, including adenovirus, adeno-associated virus, herpes simplex virus, vaccinia virus, retrovirus and lentivirus, each with their theoretical advantages and disadvantages. Retroviral vectors have size restrictions on the expressed polypeptides (Ulmer et al., 1993) and require cell division for transduction, necessitating early use in DC culture from bone marrow precursors. Human DCs are usually derived by differentiation from blood monocytes without cell division, therefore favoring the utilizations of lentivirus, poxvirus, herpes virus, and adeno-associated viruses that can infect resting cells (Ribas, 2005). However, viral vectors such as adenovirus vectors may lead to DC cytotoxicity, especially at high ratios of multiplicity of infection (MOI); cause unexpected vector-induced immunogenicity; and induce inflammation and oncogenesis *in vivo* due to the random insertion of viral genomic DNA into chromosomal DNA (Akita et al., 2010). Eventual effectiveness of vectors such as vaccinia for subsequent immunizations may be compromised by immune responses against the vectors themselves (Ulmer et al., 1993). Because of the variability in the genetic material that can be incorporated, stability of transgene expression, extent of viral titers and their toxicity to DCs, the viral vectors and their characteristics have been extensively studied and reviewed (Jenne et al., 2001; Kirk and Mulé, 2000; Morse and Lyerly, 2002; Wysocki et al., 2002).

3.1.3 Introduction of synthetic particle vectors into DC

A major enabling technology for genetic immunization of DCs is specific delivery of genetic material *to the desired target* cell at high efficiency by utilizing the inherent propensity of DCs to efficiently take up particulate matter. To this end, novel particle technology has been employed to improve upon the inefficiencies associated with naked DNA delivery while avoiding challenges in cytotoxicity of non-vector technologies and the adverse effects mediated by viral vectors. Synthetic, polymeric particulate vectors are often preferred due to their tunable design, well-established safety profile, and reduced susceptibility to non-specific anti-vector immune response or rejection.

A promising method for non-viral delivery of genetic vaccines is microparticulate DNA or RNA delivery systems formulated with a degradable polymer, such as poly lactic-*co*-glycolic acid (PLGA) (Hedley et al., 1998; McKeever et al., 2002; O'Hagan et al., 1998). Biodegradable polymer microspheres up to a diameter of 5 μm are readily internalized by phagocytic cells, and thus have been widely utilized as drug delivery platforms (Denis-Mize et al., 2000; Fahmy et al., 2005a, b). Because DCs utilize micropinocytosis, receptor-mediated internalization, or phagocytosis to uptake particulate antigens, base nanoparticle formulations have often been further modified with surface ligands for selective targeting (Bandyopadhyay et al., 2011; Hong et al., 2016; Saluja et al., 2014) and/or cloaking (Park et al., 2009; Sanders et al., 2002), or responsive reaction to external or physiologic cues (Little et al., 2005; Meyer and Wagner, 2006), widely diversifying the formulations and functionalities of these synthetic platforms.

Polymeric particle vectors have been successful in inducing antigen expression from delivered DNA for initiation of antigen-specific T cell-mediated responses. In an experiment where murine DCs were transfected by delivery of PLG and cationic surfactant, cetyltrimethylammonium bromide (CTAB), polymeric microparticle encapsulated with DNA encoding HIV p55 gag, expressed and processed antigen resulted in MHC-restricted gag-specific T cell response (Denis-Mize et al., 2000). In another use of the PLG-based cationic particle platform, PLGA/PBAE hybrid microparticles were able to transduce DCs *ex vivo* for expression and processing of a fusion protein containing an octapeptide epitope for presentation in MHC I. The addition of a pH-sensitive PBAE conferred comparative advantages against PLGA alone in intracellular delivery and transfection of DNA, as well as in *in vivo* vaccination using transduced DCs (Little et al., 2004).

Despite these advantages, even low-molecular-weight PLGA systems require up to 13 days to fully release encapsulated DNA after dendritic cell (DC) uptake *in vitro* (Walter et al., 2001a). Furthermore, biodegradation of PLGA can produce extremely low pH microenvironments (pH <3.5) after 3 days in an aqueous environment, leading to reduced plasmid DNA activity and transfection efficiency (Fu et al., 2000; Walter et al., 1999). Barriers ranging from initial uptake to confinement of particles in phagolysosomal compartments to endosomal degradation of polymeric vehicle have proved significant hurdles for successful delivery of DNA to the nucleus and, thus, detrimental for gene expression in DCs (Walter et al., 2001b). Improvements have been made to particle technology for DNA delivery with promising results, but novel delivery strategies and therapeutic targets need further development to enhance the potency of non-viral genetic vaccines (Little et al., 2005; Luo et al., 2003; Singh et al., 2000; Wang et al., 2004).

RNA interference (RNAi), a promising genetic engineering technology aimed at silencing disease-causing genes, has also been successfully applied to DC modification through the use of synthetic particle vectors. RNAi is particularly useful against "non-druggable" targets that can't be switched off by any other means. Although silencing of target genes after local delivery of siRNAs has been demonstrated, delivery of naked siRNA has proven challenging for RNAi therapies (Soutschek et al., 2004). In order to improve upon these limitations, synthetic nanocarrier vehicles have been introduced to realize the therapeutic promise of RNAi therapies. The requirement for siRNA to be recognized by a RNA-induced silencing complex (RISC) through endosomal escape presents a unique opportunity for synthetic nanocarrier delivery vehicles, as they concurrently and effectively achieve cellular uptake and endosomal escape (Basha et al., 2011; Gilleron et al., 2013). Utilization of siRNA in combination with synthetic nanocarrier vectors has yielded successes in a variety of disease models ranging from viral infections such as RSV and PIV (Bitko et al., 2005), HBV (Morrissey et al., 2005) and HSV (Palliser et al., 2006), to tumors in mice (Chono et al., 2008), to coronary artery diseases in non-human primates (Zimmermann et al., 2006).

Since APCs, particularly DCs, are known to readily uptake nanoparticles, nanocarrier-enabled RNAi therapeutics have focused on DCs are therapeutic targets. A widely utilized synthetic delivery platform for siRNA delivery is cationic lipid-based particles, conceptually similar to an evolved, advanced form of lipofectamine. Like the polymeric particles discussed above, a variety of lipid compositions, modifications and formulations have been developed for improving upon limitations associated with physiological barriers to uptake and delivery (Akita et al., 2010; Basha et al., 2011; Gilleron et al., 2013; Kedmi et al., 2018). One example of synthetic particle-delivered siRNA utilized for genetic modification of antigen presentation is the targeting of suppressor of cytokine signaling 1 (SOCS1), a gene target known for its negative-feedback regulation of antigen presentation pathways (Shen et al., 2004). In a study utilizing octaarginine-modified lipid nanoparticles, SOCS1 silencing in BMDC *ex vivo* demonstrated increased antigen-specific anti-tumor immunity (Akita et al., 2010).

Some of the nanoparticle delivery systems have also demonstrated *in vivo* efficacy, with improvements in nanoparticle homing, distribution, and enrichment in desired topology such as target organs (i.e., liver and tumor)—all factors conferring transfection stability during systemic administration of genetic material in lower doses and reduced dosing frequency and, thus, clinical viability. Synthetic gene vectors still face challenges to overcome for *in vivo* applications, such as poor mucosal penetration due to cationic charge (Sanders et al., 2009), and interference of cloaking polymers with later processes of endosomal destabilization and gene uptake (Meyer and Wagner, 2006). These deficits are continually improved upon with evolution of synthetic vector technologies (Hubbell et al., 2009). Although *in vivo* applications of these platforms have not yet been fully realized in clinic, technologies that successfully generate DCs have been attractive for *ex vivo* applications.

3.1.4 Gene guns

While viral and nanoparticle-mediated strategies of gene delivery to DCs *in vivo* have yet to fully realize their potential, it is worth mentioning an *in vivo* method that can successfully deliver antigen to tissue-resident DC. In this non-viral gene transfer method, particle microprojectiles (~1–3 μm in size, usually gold) wrapped with naked DNA are delivered at high velocity to the intracellular space, resulting in significant marker transgene expression. The "gene gun" method has been employed via intradermal or intramuscular injection (Yang and Sun, 1995). Since its development by McCabe (Yang et al., 1990) and colleagues and Sanford, Johnston et al. (Klein et al., 1987) have demonstrated that the gene gun technology can achieve efficient *in vivo* gene transfer using sub-microgram quantities of DNA per dose. In *in vivo* mouse models, gene guns were used to express transgene products such as luciferase (Williams et al., 1991), acetyl transferase, and b-galactosidases (Wolff et al., 1990). The protective capacity of these techniques have been demonstrated through monitoring immunization reactions against introduced foreign products, such as antibody titers against human growth hormone (hGH) (Tang et al., 1992), influenza (Fynan et al., 1993).

The mechanisms of genetic immunization via "gene gun" are not completely elucidated, as in this case gene transfers act non-specifically on muscular or dermal cells. However, evidence of cytotoxic T cell engagement against antigens introduced by intramuscular gene transfer of influenza viral protein (Ulmer et al., 1993) has further suggested that introduction of naked DNA can target APCs for expression of disease-associated proteins for antigen-specific activation of cell-mediated immunity. Condon et al. has demonstrated in mouse models that cutaneous genetic immunization can specifically target APCs, especially in DC-rich organs such as the skin, by eliciting protective immunity against lethal tumor challenges through cutaneous gene-gun transfection with OVA-encoding gene (Condon et al., 1996). The induction of potent antigen-specific CTL responses, combined with evidence skin-derived DC migration to the lymph nodes, establish non-viral "gene gun" gene delivery to be an efficient tool for genetic engineering of antigen presentation function in dendritic cells.

3.2 Adjuvant for post-differentiation maturation (Signal 2)

Dendritic cells integrate numerous external biomolecular stimuli to control adaptive immune responses. Since the direction and magnitude of such immune responses are determined not simply by the presentation of MHC I or MHC II-associated antigen on DCs, but also by the costimulatory molecules expressed in concert, efforts to generate DCs *ex vivo* for therapy have also focused on methods to appropriately modulate their maturation. The most common method has been to target the innately programmed pattern-recognition receptors (PRRs) on DCs that can sense endogenous or exogenous danger signals such as PAMPs, DAMPs, and their synthetic analogs (Macagno et al., 2007). Methods and genetic engineering tools are being developed to reach beyond antigen delivery toward directed maturation of DCs. Table 2 summarizes approached discussed in this section.

One strategy to induce DC maturation through genetic engineering is to exploit the DC stimulating capacity offered by the antigen delivery vehicle itself. In the case of lentiviral vectors, selective expression of antigen in DCs is the primary function for transduction; however, lentiviral vectors themselves could act as stimulating adjuvants. Lentiviral nucleic acids, genomic RNA and DNA, have been shown to stimulate DCs through TLRs (Pichlmair et al., 2007). Additionally, engagement of mitochondrial anti-viral signaling protein (MAVS) (Berg et al., 2012), cyclic guanosine monophosphate synthase (cGAS) (Lahaye et al., 2013), and stimulator of interferon genes (STING) pathways have been demonstrated to trigger DC activation

Table 2 Summary of genetic engineering approaches for signal 2.

Type	Vector	Method	Results
Vector-mediated activation	Virus	Lentiviral vector-mediated DC activation	Enhanced co-stimulation expression through TLR engagement and MAVS/cGAS/STING pathways
TLR signaling modification	Electroporation	mRNA encoding for caTLR4	Upregulation of DC maturation markers. Increased IL-12 and TNFa secretion
Direct manipulation	Virus	Ectopic expression of CD40L via adenovirus	Robust CD8 T cell response through upregulation of CD80, CD83, CD86
	Electroporation	GITRL encoding mRNA	Increased antigen-specific T cell stimulation
	siRNA	Silencing Notch ligands and DIgR2	Potent T cell proliferative responses. Off-target effects are of concern

(Kim et al., 2017). The lentiviral vector-mediated activation resulted in enhanced expression of immunogenic co-stimulation markers CD83 (Beignon et al., 2005) and CD86 (Kim et al., 2017).

Genetic tools have also been employed for directed maturation of DCs through modification of intracellular TLR signaling. Electroporating DCs with mRNA encoding constitutively active TLR4 (caTLR4) was demonstrated to enhance DC maturation, represented by upregulation of costimulatory molecules CD80, CD83, CD86 and secretion of proinflammatory cytokines IL-12 and TNFa (Cisco et al., 2004). Though the resulting mRNA transfection produced DCs similar in maturation state to LPS stimulation, they demonstrated the ability to selectively and specifically tap into DC maturation pathways for guided functional modulation.

Another strategy to control the functional phenotypes and maturation behaviors of DCs is through direct manipulation of costimulatory molecule expression. Engagement of CD40L, a costimulatory ligand on DCs, by activated CD4 T cells provides help in generating robust CD8 T cell responses. Further, CD40L ligation leads to maturation of DCs alongside upregulation of maturation markers as well as enhanced cytokine production. Thus, CD40L has been targeted for ectopic expression through DC transduction with adenovirus (Kikuchi et al., 2000), lentivirus, vaccinia virus (Feder-Mengus et al., 2005), and mRNA electroporation (Tcherepanova et al., 2008). The resulting modified DCs demonstrated potentiation of immunogenic responses, with enhanced expression of costimulatory markers CD80, CD83, and CD86 (Kikuchi et al., 2000; Koya et al., 2003), cytokine production (Feder-Mengus et al., 2005; Koya et al., 2003; Tcherepanova et al., 2008), activation of cytotoxic CD8 T cells (Koya et al., 2003), and functional anti-tumor capacity (Kikuchi et al., 2000).

Aside from CD40L, other elements in the tumor necrosis factor receptor (TNFR) superfamily have also been targeted to enhance DC maturation through ectopic expression. Glucocorticoid-induced TNF receptor-related protein ligand (GITRL), a known costimulator of naïve and activated T cells, was introduced in DCs through mRNA electroporation (Tuyaerts et al., 2007b) and showed increased antigen-specific CD8 stimulation by DCs. Similarly, co-transfection of mRNA coding for 4-1BBL in DCs, which positively regulates CD8 and CD4 mediated responses, along with tumor-associated antigen (TAA) HER-2/neu led to increased specific lysis of antigen-expressing target cells by *in vitro* induced CTL lines (Grünebach et al., 2005). Interestingly, 4-1BBL expression also led to CD40 and CD80 upregulation in transfected DCs, suggesting additional effects of RNA transfection in regulating DC co-stimulation.

Short interfering RNA (siRNA) is an emerging technology for silencing the activity of target genes. Efforts to utilize this technology for DC maturation have focused on silencing DC surface molecules that have suppressive effects on T cells, such as Notch ligands (Delta1, Jagged1, or Jagged2) and DC-derived immunoglobulin receptor 2 (DIgR2) (McKenzie et al., 2005). Knockdown of Notch ligands showed enhanced induction of CD4 T cell cytokines (Stallwood et al., 2006). DIgR2 silencing in DCs led to potent T cell proliferative responses that translated to enhanced antigen-specific *in vivo* immunization (Shi et al., 2006). However, an understanding of how these genetic tools affect DCs, particularly their off-target effects, is paramount in refining this technology for therapeutic use. In studying siRNA-mediated inhibition of TLR9 expression, Hornung et al. found that siRNA themselves could be immunostimulatory through inducing type I IFN in pDC. Surprisingly, liposome-complexed siRNA triggered TLR9-mediated detection and immunostimulatory responses in pDCs (Hornung et al., 2005), highlighting the need for further studies and investigations of the full capabilities of RNA-based therapeutics. Further optimization of RNAi therapeutics and their formulations are being explored to more effectively target DC populations for immunomodulation in antigen-specific immune activation or suppression (Basha et al., 2011).

3.3 Enhanced cytokine release/conditioning (Signal 3)

Paracrine release of cytokine soluble factors is a third and final signal in the 3-signal model for T cell activation by antigen-presenting cells. Cytokines and chemokines that are released by APCs condition the immune microenvironment and establish a context for polarization of effector cell function. Immunogenic T cell responses are characterized by Th1-polarization with cytokines such as type I IFN (IFN I), IFNg, and IL-12p70, for facilitating development and maintenance of CD8 T cells and Th1 CD4 T helper cells. Thus, controlling the cytokine secretory activity in *ex vivo* generated DCs, in addition to antigen expression and costimulatory molecule expression, is a critical element in the efforts to modulate the therapeutic *in vivo* immunological responses. Table 3 summarizes the approaches used to modulate signal 3.

One such target for manipulation of DC cytokine responses is A20, a ubiquitin-editing enzyme which negatively regulates the TLR and TNF receptor signaling pathways. As discussed in the previous section, TLR and TNF signaling play a crucial role in the determination of DC maturation. As such, silencing of A20 with a lentiviral vector expressing A20

Table 3 Summary of genetic engineering approaches for signal 3.

Type	Vector	Method	Results
Cytokine production via costimulatory upregulation	Lentivirus	A20 enzyme silencing siRNA	Enhanced costimulatory expression. Elevated IL-12, IL-6, and TNFa. Treg to Th1 phenotype skewing
Enhanced cytokine production	Adenovirus	IL-12 subunit mRNA	Upregulation of MHC I, CD40, and CD86. Enhanced IFNg secretion
Tolerance inhibition	Retrovirus	Transduction of dominant negative receptor for TGF-b (TbRIIDN)	Produced TGF-b-insensitive DCs. Increased IFNg and IL-12 production, potent tumor-specific CTL responses and tumor growth regression
Maturation enhancement	Adenovirus	Transduced GM-CSF cDNA with TAA gene	Anti-tumor responses and extended DC survival. Limited clinical efficacy

siRNA in *ex vivo* matured murine DCs has been reported to enhance costimulatory molecule expression and inhibited regulatory T cell-mediated (Treg) suppression (Song et al., 2008). Additionally, A20 silencing in siRNA lentivirus-transfected DCs led to elevated levels of IL-12p40, IL-6, and TNFa upon stimulation, and such increases in proinflammatory cytokine production was shown to effectively enhance tumor-infiltrating CTL and T helper responses in DC-based anti-cancer vaccine models. These results were paralleled in siRNA-transfected human monocyte-derived DCs, where A20 downregulation led to higher activation of NF-kB and activator protein-1, resulting in increased and sustained production of IL-6, IL-10, and IL-12p70 (Breckpot et al., 2009). The functional consequence of siRNA-induced A20 attenuation was DC-mediated skewing of naïve CD4 T cells toward IFNg-producing Th1 phenotype, leading to generation of enhanced DC vaccines potentiated for priming anti-cancer T cell responses.

IL-12p70 (IL-12), known for its wide-reaching relevance in bridging innate and adaptive immunity as well as supporting T cell, B-cell, NK cell development (Haddad et al., 2009), is a proinflammatory Th1-polarizing cytokine produced by DCs after stimulation. By inducing production of TNFa, IFNg, IL-2, and IL-18, IL-12 is also found to harbor potent clinical immunotherapeutic functions and anti-cancer immunomodulatory properties (Del Vecchio et al., 2007) leading to stimulation of antigen-specific immunity and dramatic anti-cancer effects in tumor models. However, excessive clinical toxicity and modest clinical responses to direct IL-12 cytokine therapy have pointed to the necessity of strategies that improve upon current methods of utilizing IL-12 for therapeutic use (Colombo and Trinchieri, 2002). To this end, modified DCs were created *ex vivo* for secretion of IL-12 by transduction with an adenovirus vector carrying IL-12 subunit mRNA, to provide a method for restricted IL-12 delivery to the immune microenvironment and precise modulation of Th1 responses (Okada et al., 2005). When co-transduced with antigen expression, these IL-12-producing DCs additionally demonstrated upregulation of MHC I, CD40, and CD86 as well as enhanced IFNg secretion and T cell stimulation capacity, ultimately resulting in an overall promotion of Th1-mediated therapeutic responses against murine tumor (melanoma and colon carcinoma) growth *in vivo* (Minkis et al., 2008; Ojima et al., 2006; Okada et al., 2005). Further, *ex vivo* transduction of DCs for production of Th1 cytokines downstream of IL-12, IL-2 and IL-18 have demonstrated similar enhancements in CTL proliferation, Th1-mediated IFN response, and tumor protection (Iinuma et al., 2006; Ogawa et al., 2004), suggesting that targeted modification of DC cytokine secretory function is a viable therapeutic strategy for reprogramming immunologic microenvironments and potentiating desired responses.

In many disease contexts, particularly cancer, proper mobilization of immune cells is hindered by the suppressive microenvironment the target creates in order to evade corrective immune intervention. For introduced DCs to carry out their preordained therapeutic functions in recruiting and stimulating disease-specific lymphocytes, it is paramount that they are sufficiently equipped to overcome disease-promoting obstacles such as regulatory cytokines transforming growth factor-b (TGF-b) and IL-10. Thus, a genetic modification solution has been proposed to render DCs less susceptible to such immunosuppressive factors, which preferentially recruit myeloid suppressor cells and Tregs, inhibit CTL, Th1 and NK responses, and cause apoptosis of cytotoxic T cells and DCs (Ghiringhelli et al., 2005; Luo et al., 2007). In an effort to evade the regulatory functions of TGF-b, retroviral transduction of a dominant-negative receptor for TGF-b, TGF-b type II receptor (TbRIIDN), has demonstrably produced transduced DCs which are TGF-b-insensitive (Wang et al., 2007). In *in vivo* vaccination models with antigen pulse, these *ex vivo* generated

TGF-b-insensitive DCs lead to increased IFNg and IL-12 production, potent tumor-specific CTL responses, as well as tumor growth regression. Another method of countering the detrimental effects of TGF-b has been to induce production of granulocyte-macrophage colony-stimulating factor (GM-CSF) within differentiated DCs to augment their survival and inhibit TGF-b-driven apoptosis. When DCs adenovirally transduced with GM-CSF cDNA were co-transduced with a TAA gene, they elicited potent therapeutic immunity in *in vivo* murine tumor models (Nakamura et al., 2002), which was paralleled by *in vitro* anti-tumor CTL responses in human DCs (Ojima et al., 2007).

As GM-CSF is a key cytokine utilized for *ex vivo* differentiation of DCs from stem cells or monocyte progenitors, it is not surprising that genetic modifications allowing for prolonged and continued GM-CSF exposure lead to extended DC survival. The continued presence of GM-CSF in the microenvironment of DCs may address the current clinical issues *ex vivo* generated DCs face, as representing an unstable form of DC which could revert their differentiation upon cytokine withdrawal (Ardavín et al., 2001; Romani et al., 1996), as well as their diminished capacity to generate robust clinical responses in the absence of GM-CSF in solid tumor microenvironments (Rosenberg et al., 2004; Timmerman et al., 2002). Given the critical drawback, however, that *ex vivo* generated DCs matured with GM-CSF have a poor track record of clinical responses with occasional instances of significant tumor regression and limited improvement in survival (O'Rourke et al., 2003; Palucka et al., 2006; Steinman, 2007; Steinman and Banchereau, 2007), compounded by findings that suggest GM-CSF-driven DC generation does not sufficiently represent DC development *in vivo* (Steinman, 2012; Vremec et al., 1997), alternative methods of generating and maturing DC populations which more closely resemble naturally-occurring DCs must be considered for therapeutic use.

3.4 Transcriptional (re)programming of DC

Ectopic expression of transcription factors has been classically utilized to reprogram cell fate in many different contexts, including the induction of pluripotent stem cells (Huangfu et al., 2008; Lowry et al., 2008), disease model generation (Weiser et al., 2018), and probing developmental processes (Aoyama et al., 1995). Extensive studies in dendritic cell development have identified transcription factors responsible for the cell fate choices of progenitor cells to DC lineages (Geissmann et al., 2010). For example, deficiencies of helix-loop-helix transcription factor E2-2 has been associated with loss of plasmacytoid DCs (pDCs) (Cisse et al., 2008) and transcription factor Batf3 with CD8a + classical DCs (cDCs). Additionally, in experiments demonstrating ectopic transcriptional expression for driving cell fate, PU.1 and RelB have been shown to induce monocytic differentiation into DCs (Auffray et al., 2009). However, the broad effects of the transcription factors in multiple cell types, dependence on expression levels and balance between multiple factors, and underdefined roles of each factor on specific progenitor populations have made it difficult to precisely elucidate the combinations of transcription factors required to reproduce *ex vivo* the *in vivo* development of DCs.

Utilizing lentiviral plasmids encoding transcription factors for ectopic expression, recently, mouse and human fibroblasts were demonstrated to be reprogrammed in cell fate for the creation of induced DCs (iDCs). A combination of PU.1, IRF8, and Batf3, transcription factors previously described as essential for generation of DCs, has been identified among 18 candidates through a screening process as the minimal transcription factor network necessary for inducing DC fate (Rosa et al., 2018). Although the combination of expressed transcription factors or their balance with antagonists or supplementary factors may not be optimized, these fibroblast-derived cells have been characterized to be genetically, morphologically, and functionally similar with cDC1 after 9 days after transduction in culture *ex vivo*.

Years of research had provided great insights into how DCs, as professional APC, promote the generation of antigen-specific immunity or tolerance through differential exposure of antigen, costimulatory molecules and conditioning cytokines to T cells. With growing understanding of genetic and transcription-driven mechanisms behind initiation and modulation of adaptive immune responses in DCs and how to control them, genetic tools have become an accepted and powerful approach for selective enhancement of DC functional properties. Having covered three *ex vivo* approaches in understanding DC differentiation and maturation, we dedicate the next section to an overview of clinical trials utilizing *ex vivo* generated DCs and their outcomes.

4. Clinical trials of DC-based vaccines

The use of DCs for immunotherapy applications has been a long-standing and tantalizing goal in the field of antigen presentation.

The clinical potential for therapeutic anti-cancer vaccines with cytokine-derived mDC generated a great deal of interest and excitement for more than a decade, beginning in the mid-late 1990s, when DC vaccination was utilized in hundreds of early-stage clinical trials targeting virtually every tumor type in which a defined tumor-associated Ag (TAA), whole tumor cells, or tumor lysates were available (Palucka and Banchereau, 2012; Steinman, 2012).

4.1 Early clinical trials with cytokine-derived mDCs

During this early period, DC vaccination was tested most often in patients with malignant melanoma, with more than 1250 patients treated, followed by prostate cancer (>750 patients treated), malignant glioma (>500 patients treated), and renal cell cancer (>250 patients treated) (Anguille et al., 2014). In those four malignancies only, the DC vaccine treatments reached Phase III clinical trials. Despite overall disappointing clinical results when analyzed across this broad range of malignancies (Galluzzi et al., 2012), indicating objective response rates in the 5–15% range (Butterfield, 2015), a large number of commercial interests attempted to bring DC-based therapies to the market. This effort culminated in the US FDA approval in 2010 of the DC-based therapy PROVENGE (Sipuleucel-T), a treatment based on the harvest of blood mononuclear cells from prostate cancer patients followed by incubation of the cells with a recombinant prostatic acid phosphatase (PAP) tumor-associated antigen in the form of a GM-CSF/PAP fusion protein (Hrkach et al., 2012; Kantoff et al., 2010). Despite it's relatively high cost ($100 k), limited survival benefit (4 months), and questions regarding it clinical trial design leading to the observed benefit (Huber et al., 2012), Provenge was utilized in 30,000 patients in the treatment of hormone-refractory prostate cancer (HRPC). However, large-scale adoption of the drug was hindered by its hefty price tag and uncertainty over insurance reimbursements. In 2014, parent company Dendreon filed for chapter 11 bankruptcy protection and since was sold several times, but has reorganized and is currently conducting a large-scale, placebo-controlled clinical trial (the "ProVent Trial") evaluating instead the effectiveness of PROVENGE in reducing disease progression in men with prostate cancer on active surveillance (AS) but no treatment (Rinde, 2019). Competing products in the prostate cancer space include DCVAC/PCa, a DC vaccine currently in Phase III trials consisting of autologous dendritic cells activated by *ex vivo* exposure to killed LNCaP prostate cancer cells

(Fucikova et al., 2018); rilimogene galvacirepvac (Prostvac), an off-the-shelf prostate cancer vaccine currently in Phase III trials consisting of a virus (vaccinia) as a vector to deliver the prostate specific antigen (PSA) along with other three co-stimulating molecules (LFA-3, ICAM-1 and B7.1) directly to cancer cells (Gulley et al., 2014); and aglatimagene besadenovec (ProstAtak), an oncolytic virus therapy (Grozescu and Popa, 2017).

Additional commercial entities sponsoring DC-based clinical trials include a vaccine loaded with autologous tumor lysate (DCVax-L, Northwest Therapeutics) in patients with newly diagnosed glioblastoma following surgery as add-on to the standard of care combining radiation and chemotherapy (Liau et al., 2018). APCEDEN® (APAC Biotech) is an mDC-based autologous vaccine which has been approved outside of the United States for four cancer indications- Prostate, Ovarian, Colorectal and Non-Small Cell Lung carcinoma (Kumar et al., 2017). Medigene Inc. also recently presented clinical data from the interim analysis of its ongoing Phase I/II clinical trial of its WT-1 and PRAME Ag-loaded "FastDC" vaccine for the treatment of acute myeloid leukemia (AML) (Fløisand et al., 2019). In addition, although not an injectable DC vaccine per se, the concept of targeting TAAs to DCs *in vivo* has recently been harnessed for the development of CDX1401 (TAA peptides + poly-ICLC) (Celldex Therapeutics, USA, NCT00045968) in the treatment of ovarian cancer and myelodysplastic syndromes (MDS) (Griffiths et al., 2018).

4.2 Naturally occurring DCs, neoantigens, and combination therapies

Despite challenges both clinically and commercially during the first wave of DC-based therapies, immunotherapy has once again come to the forefront of cancer therapeutics. Because of the clinical successes associated with checkpoint inhibition, CAR T cells, and other adoptive transfer strategies of anti-tumor T cells (Feigal et al., 2019), DC are enjoying a renaissance as both monotherapies and in combination with other immunotherapeutic or chemotherapeutic agents.

Therapeutic DC regimens under current clinical evaluation employ a variety of new cellular and antigenic components. These include rarer but naturally-occurring DC subsets, such as "natural" or conventional cDCs which can be isolated and purified rapidly and directly from blood without extended cytokine exposure; moDC cultured and matured by new and more potent methodologies, in combination with new antigenic sources including "neo antigen" loaded vaccines specific to the alterations present in the patients' unique tumor proteins or tumor-derived DNA/RNA; as well as new routes of administration including intranodal vaccinations and *in situ* DC targeting (Locy et al., 2018). Several excellent and exhaustive reviews of the clinical use of many of these next-generation DC-based vaccines have recently been published (Garg et al., 2017; Mastelic-Gavillet et al., 2019).

Perhaps the most promising of these recent developments is represented by vaccination with "natural," non-monocyte-derived DC populations, which can be loaded with more "personalized" Ag sources such as neoepitope-derived peptides. DC populations accessible directly from blood and tissue include cDC1 (CD141+) populations, which appear to be the most adept at cross-presentation and therefore efficiently prime CD8+ T cells (Haniffa et al., 2012); cDC2 (CD172a+) cells which are the predominant DC subset present in the blood and have been shown to be a potent CD4+ T cell activators including Th1, Th2 and Th17 responses (Leal Rojas et al., 2017); as well as plasmacytoid DC (pDC characterized by high type I IFN secretion) and skin-derived Langerhans cells (LC). Each of these DC subtypes can be rapidly isolated *ex vivo*, activated, Ag loaded, and injected back into the patients therapeutically (Patente et al., 2018; Saxena and Bhardwaj, 2018), giving such therapies significant advantages over mDC-based protocols.

Several examples of naturally-occurring DC subsets have already been associated with promising anti-tumor activity *in vivo*. Recent clinical studies utilizing intranodal administration of isolated and purified CD1c+ DCs, activated and loaded with HLA-A2.1-restricted tumor-derived peptides in patients with stage IV melanoma (NCT01690377), showed that cDC vaccines were capable of generating tumor-specific cytotoxic T lymphocyte (CTL) immunity that correlated with improved progression-free survival (PFS) (Schreibelt et al., 2016). Similar promise was shown by multiple vaccine trials associated with CD141+ DC as part of the European "Professional Cross-priming for Ovarian and Prostate Cancer" (PROCROP) initiative (www.procrop.eu). Overall, cDC2 subset vaccination has been shown to be safe, technically feasible, and effective in promoting prolonged progression-free survival (12–35 months) in 4 out of 14 patients (Locy et al., 2018). Similarly, a recent pDC vaccine has also been shown to be safe and feasible and resulted in the induction of antigen-specific CD4+ and CD8+ T cell responses and a systemic type I IFN signature (Tel et al., 2013). Efforts are also underway to combine cDC and pDC in combination internodal vaccination to see if this increases efficacy against prostate cancer (NCT02692976) and melanoma

(NCT02574377), respectively. In addition, LCs derived from CD34+ hematopoietic stem cells are currently being tested in two Phase I clinical trials in melanoma patients and patients with multiple myeloma (NCT01456104, NCT01995708).

In addition to these newer cDC and pDC reagents, more potent methodologies for the controlled differentiation of monocyte populations or myeloid precursors into DC are also currently under investigation in clinical trials, with different growth factors and cytokines capable of modulating DC differentiation and function. Besides the well-established culture in GM-CSF/IL-4, M-CSF, FLT3L, TGF-β, type I IFNs, TNF, IL-15 and TLR/PRR agonists such as nucleic acids (CpG), Imiquimod, LPS, monophosphoryl lipid A and BCG are being utilized to generate/mature a large variety of DCs with different functional abilities and potentially differing *in vivo* stimulatory capabilities (Bhardwaj et al., 2010; Castell-Rodríguez et al., 2017; Massa and Seliger, 2013). A comprehensive review of the varied DC-based vaccine reagents and their cellular precursor types which have successfully reached recent clinical trial is now available (Sprooten et al., 2019).

Besides efforts to optimize DC production, increased knowledge of the critical nature of the type of cell death necessary to render autologous tumor cells useful as potentially complex (and therefor resistant to tumor escape pathways) phagocytic substrates for DC internalization and cross-presentation has brought the field of immunogenic cell death (ICD) to the forefront of current tumor vaccination strategies. DCs can phagocyte dying tumor cells by means of a variety of specific receptors. In general, apoptotic cells are recognized by DC using their surface receptors integrin αVβ5, the CD36 molecule or by means of the phosphatidylserine receptor, while necrotic cells are recognized by molecules such as CD91, TLR-2, and TLR4 (Apetoh et al., 2004; Dhodapkar et al., 2008). But recent evidence points to the importance of a specific form of cell death, immunogenic cell death or ICD, which is generated by only a subset of chemotherapeutic agents or clinically-relevant anti-neoplastic treatments such as radiation therapy (RT), in rendering dying tumor cells ideal substrates for DC loading (Galluzzi et al., 2017; Vandenberk, et al. 2016; Wennerberg et al., 2017). Since a large proportion of current clinical trials utilize dying tumor cells or tumor cell lysates as vaccine Ag sources (Sprooten et al., 2019), the advantages of autologous or even allogenic tumor cells, namely, the complete set of tumor-relevant Ags (including neo antigens), independent of MHC haplotypes, is now well understood (Chiang et al., 2015; González et al., 2014).

Such dying and damaged tumor cells, if properly treated with ICD-inducing agents, contain associated immunological danger signals, including damage-associated molecular patterns (DAMPs) (Garg et al., 2012). ICD results in the cellular redistribution and extracellular release these DAMPs, which are capable of having significant adjuvant effects on DC and the potential to activate and restore tumor-targeted immune responses *in vivo* (Galluzzi et al., 2017; Rapoport and Anderson, 2019). The increased use of whole tumor/tumor lysates in DC-based therapy additionally offers the advantage of an antigenic profile including all potential MHC I and II epitopes, both defined and undefined, sorted naturally by the DC itself, with the simultaneous stimulation of both CD8 and CD4 anti-tumor T cells (Hanlon et al., 2019; Vermaelen, 2019). This may potentially alleviate some of the shortcomings attendant with the targeting of single or limited tumor-associated Ags (TAA) previously associated with DC-based vaccine failure in the past (Steinman and Mellman, 2004).

With this myriad of obstacles blunting the potential effectiveness of DC-based therapies, including existing tumor escape mechanisms such as tumor reduction of TAA expression and production of immunosuppressive cytokines/chemokines/cell surface checkpoint receptors, the protective influence of accessory suppressive cells such as Tregs and MDSC/TAM as well as general defects in the number and function of native DC (Fiedler and Hemann, 2019), more effective DC vaccines will likely need to share two important characteristics: (1) they will need be combined with additional immunotherapeutic measures to overcome tumor tolerance and (2) they will need to target multiple "personalized" antigens unique to that tumor, including neoantigens.

An excellent summary of the combination of DC-based vaccination with chemotherapy has recently been published (Van Gulijk et al., 2018), which highlights the aforementioned importance of chemotherapy-induced immunogenic cell death (ICD) in the induction of anti-tumor immunity (Kroemer et al., 2013). Besides standard chemotherapy, current combinations also include DC vaccines used concurrently with radiotherapy, immune checkpoint blockade (CPB), and a large group of immunomodulating agents including imiquimod, antibodies (anti-CD25, for example), cytokines (such as IL-2, IFNs and GM-CSF), and pathogen-derived TLR agonists, as well as adjuvant cell therapies such as co-delivery of cytokine-induced killer cells (CIK) or adoptively transferred T cells (ACT). Recent examples of promising combination therapies include the use of a DC-based vaccine with ipilimumab (an anti-CTLA-4 Ab) in melanoma, achieving an overall response rate of 38% (eight patients with complete responses and seven with partial responses) and a 6-month disease control rate of 51% (Wilgenhof et al., 2016), as well as chemotherapy combinations increasing OS in glioblastoma and mesothelioma (Cornelissen et al., 2016; Pellegatta et al., 2018).

Additional recent excitement and intense clinical interest surrounds the possibility of vaccinating with DC loaded with "personalized" neoantigens, since a common observation among patients given any immunotherapy is a high neoantigen load for their type of tumor being strongly associated with treatment responsiveness, often closely correlated with anti-tumor T cell responses (Schumacher and Schreiber, 2015). There are currently more than two dozen ongoing Phase 1 and Phase 2 trials targeting neoantigens utilizing different vaccine platforms such as DNA, RNA, synthetic long peptides, and dendritic cells (Hu et al., 2018; Hundal and Mardis, 2019), a direct result of the pioneering proof-of-principle neoepitope trials in melanoma and glioblastoma (Carreno et al., 2015; Keskin et al., 2019; Ott et al., 2017; Sahin et al., 2017). But real shortcoming in cost and availability cannot be ignored in current neoepitope identification strategies and scalability to most clinical settings, problems which may be partially solved with improved genomic and immune-peptidomic technology as well as identification of additional neoepitopes from genome-scale mutations. But even significant progress on these fronts will not change the reality that different malignancies display varying mutational loads, tumor heterogeneity will allow genetically unstable subclones to avoid neoepitope targeting, and current lack of MHC class II epitopes targeting may severely hamstring fulminant cytolytic T cell responses.

Taken together, these results offer both hope and challenges that next-generation DC-based vaccines and vaccine combinations will be more readily and affordably designed and tested, move more rapidly through Phase 3 trials and induce the durable responses and extensions of long-term survival in cancer patients that have always been their promise. To this end, improvements in the isolation and generation of potent DC populations, loaded with stimulatory and diverse tumor Ag cargoes, has begun to move the needle toward DC vaccination assuming it's place alongside other immunotherapies in the oncology armamentarium.

4.3 Prospective clinical applications of ECP-driven DC therapies

ECP is a dendritic cell-based immunotherapy with an outstanding clinical record of safety and efficacy in treating cutaneous T cell lymphoma (CTCL).

However, its FDA approval is limited to CTCL, and routine therapeutic uses only cover CTCL, heart and lung transplant, and GvHD. Because ECP's therapeutic effects reflect a tight partnership with the physiologic immune system, the treatment's mechanistic underpinnings are as complex as those of immunity itself. Yet, the key tunable variables which collectively govern ECP's linkage with the immune system are now largely identified, enabling significant enhancement of the treatment and tailoring to the clinical need. It is therefore worth considering the prospective uses of ECP in the wider therapeutic context.

The first, and most accessible, advantage learned from the recent scientific advances elaborated above, is that we now know how to polarize ECP to either immunizing or tolerizing mode. Since the current therapy delivers both the immunizing and tolerizing modalities simultaneously, in a "one size fits all" device, the capacity to polarize it promises to markedly augment the potency of the treatment by eliminating the "drag" of the counteracting mode. For example, if the goal is to immunize against a cancer, one can shut off the tolerizing vector by protecting incipient DCs from 8-MOP damage, while exposing the cancer cells to the drug to enhance their immunogenic cell death (Tatsuno et al., 2019). That can be accomplished by not exposing the extracorporeally passed monocytes to ultraviolet (UVA) light, thereby producing *only* immunizing DC. Alternatively, if the goal is to tolerize in the transplant or autoimmune setting, one can shut off the immunizing vector by ensuring that all of the passaged immune cells are heavily impacted by photoactivated 8-MOP, thereby producing *only* tolerizing DC. That can be accomplished in several ways, for instance by decreasing the hematocrit of the blood treated in the device, thus decreasing the light-shielding effect of the red blood cells and consequently ensuring greater UVA exposure of immune cells. In short, ECP's incorporation of a drug which is "turned on by light" presents a unique means of tuning this therapy to the desired mode.

Again because of ECP's partnership with the immune system, the full range of disorders that the immune system touches are conceptually amenable to management by polarized ECP. This breadth of potential therapeutic targets necessitates an organized hierarchy of clinical investigation. Since antigen specificity, which is ECP's central feature, is dependent on processing antigen by the ECP-induced DC, any clinical circumstance in which the key antigen source can be fed to the induced DC, is a candidate for tailored ECP therapy. Given the enormous length of that clinical menu, we have chosen to focus on two particularly attractive therapeutic targets first: personalized anti-microbial vaccination, and haplotype-matched stem cell transplantation.

ECP's efficient and rapid provision of physiologic DC, within a day, suggests an ease of personalized vaccination against the microbe of choice. In a preventive vaccine, as opposed to treatment of an existing disease, it is likely that relatively few antigen-loaded DC will be required to create a substantial effect. We envision a process involving drawing of about 25 cc of subject blood into a heparinized syringe, also containing appropriate microbial antigen source, such as killed virus. The anticoagulated blood, containing the antigen source, will then be passaged through the miniaturized ECP plate such as previously described (Ventura et al., 2018, 2019), to generate physiologic DC. The mix, containing the viral antigen source, will then be incubated overnight, before second day intravenous return to the patient as an anti-viral vaccine. Pilot studies of this approach are currently in the planning phase.

In tandem, ECP's antigen-specific tolerogenic effect, manifest both in treatment of otherwise therapeutically resistant post-stem cell allograft graft-vs-host disease, as well as in experimental animal histo-incompatible tissue transplantation, suggest an opportune target. Specifically, ECP will be used to facilitate stem cell allografts from mother-to-child, if the latter suffers from therapeutically resistant leukemia and is therefore a candidate for stem cell transplantation. The goal is to enable haplotype matched stem cells to reconstitute the depleted immune system of the afflicted child. ECP polarized to tolerization mode offers, from both clinical and experimental experience, optimism that accelerated graft-vs-host disease could be controlled, while leaving particularly potent graft-vs-leukemia effect. If a pilot study is encouraging, a larger multicenter trial will be initiated. This approach, while ambitious, offers two major advances: marked increase in intrafamilial donor availability, and potent anti-tumor immunity.

5. Concluding remarks

Since the initial discovery of DCs, years of research had provided great insights into how these professional APCs, through differential exposure of antigen, costimulatory molecules and conditioning cytokines, act as the master-switches of antigen-specific immunity or tolerance. Beginning with cytokine-generated monocyte-derived DCs, which initially provided a platform for immunobiological research, a variety of techniques have been developed to expand DCs *ex vivo* in large numbers with anticipated potentials for clinical translation as antigen-specific cellular immunotherapies. With a growing body of knowledge in DCs and the specific pathways through which they exert their superb immunomodulatory functions, the advent of recent advancements in genetic tools present an unprecedented opportunity for approaches in *ex vivo* generation of genetically tailored DCs. These methods represent the pinnacle of technological advancements in bringing design-driven approaches to generate *ex vivo* produced cells with deliberate and obvious advantages in clinical translational potentials as DC-based immunotherapies. Meanwhile, mechanisms of natural generation of DCs in normal physiology have largely remained underdefined. Pathways involving inflammatory stimulation and reactionary migration of monocytes have been suggested as determinant "physiologic" forces of DC differentiation, which have also been demonstrated in experiments with *ex vivo* platforms.

The continued excitement and technological advancements in *ex vivo* DC preparations, such as DC generation and antigen engineering, and combinatorial approaches exemplify persevering hope in the field of DC-based vaccines. Reviewing the current and historical clinical trials, however, we find a notable general lack of emphasis on generating DCs through methods that resemble DC biogenesis in normal physiology, from which our understanding of their critical immunologic functions derived. Contrastingly, the uncanny history of ECP's success, as a therapy shown to induce DC-driven therapeutic immune reactions in both cancers and autoimmune diseases, reflects promise that *ex vivo* DCs generated through physiologic cues and forces will meet the translational expectations once held for DCs. Perhaps it is not surprising that much of DC clinical trials thus far have yielded limited success, and it is in this context we speculate that a DCs generated in a more physiologically-relevant manner may also be more clinically-impactful.

Conflict of interest

O. S. is VP Immunology at Transimmune AG. R.L. E. has ownership interest (including patents) in, and is a consultant/advisory board member of, Transimmune, AG. No potential conflicts of interest were disclosed by the other authors.

References

Ahn, J.S., Agrawal, B., 2005. IL-4 is more effective than IL-13 for in vitro differentiation of dendritic cells from peripheral blood mononuclear cells. Int. Immunol. 17 (10), 1337–1346. Oxford University Press.

Ahn, K.C., Jun, A.J., Pawar, P., Jadhav, S., Napier, S., McCarty, O.J., Konstantopoulos, K., 2005. Preferential binding of platelets to monocytes over neutrophils under flow. Biochem. Biophys. Res. Commun. 329 (1), 345–355.

Akita, H., et al., 2010. Nanoparticles for ex vivo siRNA delivery to dendritic cells for cancer vaccines: programmed endosomal escape and dissociation. J. Control. Release 149 (1), 58–64. https://doi.org/10.1016/j.jconrel.2010.08.023. Elsevier B.V.

Amigorena, S., 2018. Dendritic cells on the way to glory. J. Immunol. 200 (3), 885–886. Am Assoc Immnol.

Anguille, S., et al., 2009. Short-term cultured, interleukin-15 differentiated dendritic cells have potent immunostimulatory properties. J. Transl. Med. 7 (1), 109. BioMed Central.

Anguille, S., Smits, E.L., Lion, E., van Tendeloo, V.F., Berneman, Z.N., 2014. Clinical use of dendritic cells for cancer therapy. Lancet Oncol. 15 (7), e257–e267.

Aoyama, T., Dong, C.H., Wu, Y., Carabelli, M., Sessa, G., Ruberti, I., Morelli, G., Chua, N.H., 1995. Ectopic expression of the arabidopsis transcriptional activator Athb-1 alters leaf cell fate in tobacco. Plant Cell 7 (11), 1773–1785.

Apetoh, L., Ghringheli, F., Tesnieres, A., Obeid, M., Ortiz, C., Criollo, A., et al., 2004. Toll like receptor 4 dependent contribution of the immune system to anticancer chemotherapy and radiotherapy. Nat. Med. 13, 1051–1060.

Ardavín, C., et al., 2001. Origin and differentiation of dendritic cells. Trends Immunol. 22 (12), 691–700. https://doi.org/10.1016/S1471-4906(01)02059-2. Elsevier.

Auffray, C., Sieweke, M.H., Geissmann, F., 2009. Blood monocytes: development, heterogeneity, and relationship with dendritic cells. Annu. Rev. Immunol. 27 (1), 669–692.

Balan, S., Kale, V.P., Limaye, L.S., 2009. A simple two-step culture system for the large-scale generation of mature and functional dendritic cells from umbilical cord blood CD34+ cells. Transfusion 49 (10), 2109–2121. Wiley Online Library.

Ballen, K.K., Gluckman, E., Broxmeyer, H.E., 2013. Umbilical cord blood transplantation: the first 25 years and beyond. Blood 122 (4), 491–498.

Banchereau, J., Steinman, R.M., 1998. Dendritic cells and the control of immunity. Nature 392 (6673), 245.

Bandyopadhyay, A., Fine, R.L., Demento, S., Bockenstedt, L.K., Fahmy, T.M., 2011. The impact of nanoparticle ligand density on dendritic-cell targeted vaccines. Biomaterials 32 (11), 3094–3105.

Basha, G., Novobrantseva, T.I., Rosin, N., Tam, Y.Y.C., Hafez, I.M., Wong, M.K., Sugo, T., Ruda, V.M., Qin, J., Klebanov, B., Ciufolini, M., 2011. Influence of cationic lipid composition on gene silencing properties of lipid nanoparticle formulations of siRNA in antigen-presenting cells. Mol. Ther. 19 (12), 2186–2200.

Beignon, A.S., McKenna, K., Skoberne, M., Manches, O., DaSilva, I., Kavanagh, D.G., Larsson, M., Gorelick, R.J., Lifson, J.D., Bhardwaj, N., 2005. Endocytosis of HIV-1 activates plasmacytoid dendritic cells via toll-like receptor-viral RNA interactions. J. Clin. Invest. 115 (11), 3265–3275.

Bell, G.M., Anderson, A.E., Diboll, J., Reece, R., Eltherington, O., Harry, R.A., Fouweather, T., MacDonald, C., Chadwick, T., McColl, E., Dunn, J., 2017. Fouweather, T., MacDonald, C., Chadwick, T., McColl, E. and Dunn, J., Autologous tolerogenic dendritic cells for rheumatoid and inflammatory arthritis. Ann. Rheum. Dis. 76 (1), 227–234.

Berg, R.K., Melchjorsen, J., Rintahaka, J., Diget, E., Søby, S., Horan, K.A., Gorelick, R.J., Matikainen, S., Larsen, C.S., Ostergaard, L., Paludan, S.R., 2012. Genomic HIV RNA induces innate immune responses through RIG-I-dependent sensing of secondary-structured RNA. PLoS One 7 (1), e29291.

Berger, C., et al., 2010. Rapid generation of maturationally synchronized human dendritic cells: contribution to the clinical efficacy of extracorporeal photochemotherapy. Blood 116 (23), 4838–4847. Am Soc Hematology.

Bernhard, H., et al., 1995. Generation of immunostimulatory dendritic cells from human CD34+ hematopoietic progenitor cells of the bone marrow and peripheral blood. Cancer Res. 55 (5), 1099–1104. AACR.

Bhardwaj, N., Gnjatic, S., Sawhney, N., 2010. TLR AGONISTS: are they good adjuvants? Cancer J. 16 (4), 382–391.

Bitko, V., Musiyenko, A., Shulyayeva, O., Barik, S., 2005. Inhibition of respiratory viruses by nasally administered siRNA. Nat. Med. 11 (1), 50–55.

Brandt, K., et al., 2003. Interleukin-21 inhibits dendritic cell activation and maturation. Blood 102 (12), 4090–4098. Am Soc Hematology.

Brasel, K., et al., 2000. Generation of murine dendritic cells from flt3-ligand–supplemented bone marrow cultures. Blood 96 (9), 3029–3039. Am Soc Hematology.

Breckpot, K., Aerts-Toegaert, C., Heirman, C., Peeters, U., Beyaert, R., Aerts, J.L., Thielemans, K., 2009. Attenuated expression of A20 markedly increases the efficacy of double-stranded RNA-activated dendritic cells as an anti-cancer vaccine. J. Immunol. 182 (2), 860–870.

Briseño, C.G., et al., 2016. Distinct transcriptional programs control cross-priming in classical and monocyte-derived dendritic cells. Cell Rep. 15 (11), 2462–2474. Elsevier.

Broxmeyer, H.E., et al., 1989. Human umbilical cord blood as a potential source of transplantable hematopoietic stem/progenitor cells. Proc. Natl. Acad. Sci. 86 (10), 3828–3832. National Acad Sciences.

Broxmeyer, H.E., et al., 1991. Umbilical cord blood hematopoietic stem and repopulating cells in human clinical transplantation. Blood Cells 17 (2), 313–329.

Buckland, M., Lombardi, G., 2009. Aspirin and the induction of tolerance by dendritic cells. In: Dendritic Cells. Springer, pp. 197–213.

Buckland, M., et al., 2006. Aspirin modified dendritic cells are potent inducers of allo-specific regulatory T-cells. Int. Immunopharmacol. 6 (13–14), 1895–1901. Elsevier.

Buelens, C., et al., 2002. Interleukin-3 and interferon β cooperate to induce differentiation of monocytes into dendritic cells with potent helper T-cell stimulatory properties. Blood 99 (3), 993–998. Am Soc Hematology.

Bürdek, M., et al., 2010. Three-day dendritic cells for vaccine development: antigen uptake, processing and presentation. J. Transl. Med. 8 (1), 90. BioMed Central.

Burgess, A.W., Metcalf, D., 1980. The nature and action of granulocyte-macrophage colony stimulating factors. Blood 56 (6), 947–958. Am Soc Hematology.

Burgess, A.W., Camakaris, J., Metcalf, D., 1977. Purification and properties of colony-stimulating factor from mouse lung-conditioned medium. J. Biol. Chem. 252 (6), 1998–2003. ASBMB.

Butterfield, L.H., 2015. Cancer vaccines. BMJ 350, h988. British Medical Journal Publishing Group.

Campbell, J.J., Hedrick, J., Zlotnik, A., Siani, M.A., Thompson, D.A., Butcher, E.C., 1998. Chemokines and the arrest of lymphocytes rolling under flow conditions. Science 279 (5349), 381–384.

Camporeale, A., et al., 2003. Critical impact of the kinetics of dendritic cells activation on the in vivo induction of tumor-specific T lymphocytes. Cancer Res. 63 (13), 3688 LP–3694.

Carreno, B.M., et al., 2015. A dendritic cell vaccine increases the breadth and diversity of melanoma neoantigen-specific T cells. Science 348 (6236), 803–808.

Castell-Rodríguez, A., Piñón-Zárate, G., Herrera-Enríquez, M., Jarquín-Yáñez, K., Medina-Solares, I., 2017. Dendritic cells: location, function, and clinical implications. In: Ghosh, A. (Ed.), Biology of Myelomonocytic Cells. IntechOpen. https://doi.org/10.5772/intechopen.68352. Available from: https://www.intechopen.com/books/biology-of-myelomonocytic-cells/dendritic-cells-location-function-and-clinical-implications.

Caux, C., et al., 1990. Tumor necrosis factor-alpha strongly potentiates interleukin-3 and granulocyte-macrophage colony-stimulating factor-induced proliferation of human CD34+ hematopoietic progenitor cells. Blood 75 (12), 2292–2298. Am Soc Hematology.

Caux, C., et al., 1996. CD34+ hematopoietic progenitors from human cord blood differentiate along two independent dendritic cell pathways in response to GM-CSF+ TNF alpha. J. Exp. Med. 184 (2), 695–706. Rockefeller University Press.

Cerboni, S., Gentili, M., Manel, N., 2013. Diversity of pathogen sensors in dendritic cells. In: Advances in Immunology. Elsevier, pp. 211–237.

Chiang, C.L., Coukos, G., Kandalaft, L.E., 2015. Whole tumor antigen vaccines: where are we? Vaccines 3 (2), 344–372.

Cheong, C., et al., 2010. Microbial stimulation fully differentiates monocytes to DC-SIGN/ CD209+ dendritic cells for immune T cell areas. Cell 143 (3), 416–429. Elsevier.

Chono, S., Li, S.D., Conwell, C.C., Huang, L., 2008. An efficient and low immunostimulatory nanoparticle formulation for systemic siRNA delivery to the tumor. J. Control. Release 131 (1), 64–69.

Chorny, A., et al., 2005. Vasoactive intestinal peptide induces regulatory dendritic cells with therapeutic effects on autoimmune disorders. Proc. Natl. Acad. Sci. 102 (38), 13562–13567. National Acad Sciences.

Cisco, R.M., Abdel-Wahab, Z., Dannull, J., Nair, S., Tyler, D.S., Gilboa, E., Vieweg, J., Daaka, Y., Pruitt, S.K., 2004. Induction of human dendritic cell maturation using transfection with RNA encoding a dominant positive toll-like receptor 4. J. Immunol. 172 (11), 7162–7168.

Cisse, B., Caton, M.L., Lehner, M., Maeda, T., Scheu, S., Locksley, R., Holmberg, D., Zweier, C., den Hollander, N.S., Kant, S.G., Holter, W., 2008. Transcription factor E2-2 is an essential and specific regulator of plasmacytoid dendritic cell development. Cell 135 (1), 37–48.

Collin, M., Bigley, V., 2018. Human dendritic cell subsets: an update. Immunology 154 (1), 3–20. Wiley Online Library.

Colombo, M.P., Trinchieri, G., 2002. Interleukin-12 in anti-tumor immunity and immunotherapy. Cytokine Growth Factor Rev. 13 (2), 155–168.

Condon, C., Watkins, S.C., Celluzzi, C.M., Thompson, K., Falo, L.D., 1996. DNA-based immunization by in vivo transfection of dendritic cells. Nat. Med. 2 (10), 1122–1128.

Cornelissen, R., et al., 2016. Extended tumor control after dendritic cell vaccination with low-dose cyclophosphamide as adjuvant treatment in patients with malignant pleural mesothelioma. Am. J. Respir. Crit. Care Med. 193 (9), 1023–1031. American Thoracic Society.

Dauer, M., et al., 2003. Mature dendritic cells derived from human monocytes within 48 hours: a novel strategy for dendritic cell differentiation from blood precursors. J. Immunol. 170 (8), 4069–4076. Am Assoc Immnol.

Dauer, M., et al., 2005. FastDC derived from human monocytes within 48 h effectively prime tumor antigen-specific cytotoxic T cells. J. Immunol. Methods 302 (1–2), 145–155. Elsevier.

de Vries, I.J.M., et al., 2003. Maturation of dendritic cells is a prerequisite for inducing immune responses in advanced melanoma patients. Clin. Cancer Res. 9 (14), 5091–5100. AACR.

Del Vecchio, M., Bajetta, E., Canova, S., Lotze, M.T., Wesa, A., Parmiani, G., Anichini, A., 2007. Interleukin-12: biological properties and clinical application. Clin. Cancer Res. 13 (16), 4677–4685.

Denis-Mize, K.S., Dupuis, M., MacKichan, M.L., Singh, M., Doe, B., O'Hagan, D., Ulmer, J.B., Donnelly, J.J., McDonald, D.M., Ott, G., 2000. Plasmid DNA adsorbed onto cationic microparticles mediates target gene expression and antigen presentation by dendritic cells. Gene Ther. 7 (24), 2105.

Dixon, D.A., et al., 2006. Expression of COX-2 in platelet-monocyte interactions occurs via combinatorial regulation involving adhesion and cytokine signaling. J. Clin. Invest. 116 (10), 2727–2738. Am Soc Clin Investig.

Dhodapkar, M.V., Dhodapkar, K.M., Palucka, A.K., 2008. Interactions of tumor cells with dendritic cells: balancing immunity and tolerance. Cell Death Differ. 15, 39–50.

Durazzo, T.S., Tigelaar, R.E., Filler, R., Hayday, A., Girardi, M., Edelson, R.L., 2014. Induction of monocyte-to-dendritic cell maturation by extracorporeal photochemotherapy: initiation via direct platelet signaling. Transfus. Apher. Sci. 50 (3), 370–378.

Ebner, S., et al., 2002. A novel role for IL-3: human monocytes cultured in the presence of IL-3 and IL-4 differentiate into dendritic cells that produce less IL-12 and shift Th cell responses toward a Th2 cytokine pattern. J. Immunol. 168 (12), 6199–6207. Am Assoc Immnol.

Edelson, R.L., 2014. Mechanistic insights into extracorporeal photochemotherapy: efficient induction of monocyte-to-dendritic cell maturation. Transfus. Apher. Sci. 50 (3), 322–329.

Edelson, R., Wu, Y., Schneiderman, J., 2018. American council on ECP (ACE): why now? J. Clin. Apher. 33 (4), 464–468. Wiley Online Library.

Fahmy, T.M., Samstein, R.M., et al., 2005a. Surface modification of biodegradable polyesters with fatty acid conjugates for improved drug targeting. Biomaterials 26 (28), 5727–5736.

Fahmy, T.M., Fong, P.M., et al., 2005b. Targeted for drug delivery. Mater. Today 8 (8), 18–26. Elsevier.

Farkas, Á., Kemény, L., 2011. Interferon-α in the generation of monocyte-derived dendritic cells: recent advances and implications for dermatology. Br. J. Dermatol. 165 (2), 247–254. Wiley Online Library.

Fearnley, D.B., Whyte, L.F., Carnoutsos, S.A., Cook, A.H., Hart, D.N.J., 1999. Monitoring human blood dendritic cell numbers in normal individuals and in stem cell transplantation. Blood 93 (2), 728–736.

Feder-Mengus, C., Schultz-Thater, E., Oertli, D., Marti, W.R., Heberer, M., Spagnoli, G.C., Zajac, P., 2005. Nonreplicating recombinant vaccinia virus expressing CD40 ligand enhances APC capacity to stimulate specific CD4+ and CD8+ T cell responses. Hum. Gene Ther. 16 (3), 348–360.

Fedoric, B., Krishnan, R., 2008. Rapamycin downregulates the inhibitory receptors ILT2, ILT3, ILT4 on human dendritic cells and yet induces T cell hyporesponsiveness independent of FoxP3 induction. Immunol. Lett. 120 (1–2), 49–56. Elsevier.

Feigal, E.G., et al., 2019. At the end of the beginning: immunotherapies as living drugs. Nat. Immunol. 20, 955–962. Nature Publishing Group.

Fiedler, E.C., Hemann, M.T., 2019. Aiding and abetting: how the tumor microenvironment protects cancer from chemotherapy. Ann. Rev. Cancer Biol. 3, 409–428.

Fløisand, Y., Bigalke, I., Josefsen, D., Geiger, C., Pinkernell, K., Kvalheim, G., 2019. Interim analysis of a Wt-1 and prame 'fast-DC'vaccine shows safety as active immmunotherapy for the prevention of AML relapse: PS1223. HemaSphere 3, 558.

Fogel-Petrovic, M., et al., 2007. Physiological concentrations of transforming growth factor β1 selectively inhibit human dendritic cell function. Int. Immunopharmacol. 7 (14), 1924–1933. Elsevier.

Fong, L., et al., 2001. Altered peptide ligand vaccination with Flt3 ligand expanded dendritic cells for tumor immunotherapy. Proc. Natl. Acad. Sci. 98 (15), 8809–8814. National Acad Sciences.

Fu, K., et al., 2000. Visual evidence of acidic environment within degrading poly(lactic-co-glycolic acid) (PLGA) microspheres. Pharm. Res. 17 (1), 100–106.

Fucikova, J., et al., 2018. Phase I/II trial of dendritic cell-based active cellular immunotherapy with DCVAC/PCa in patients with rising PSA after primary prostatectomy or salvage radiotherapy for the treatment of prostate cancer. Cancer Immunol. Immunother. 67 (1), 89–100. Springer.

Fynan, E.F., et al., 1993. DNA vaccines: protective immunizations by parenteral, mucosal, and gene-gun inoculations. Proc. Natl. Acad. Sci. U. S. A. 90 (24), 11478–11482.
Galluzzi, L., et al., 2012. Trial watch: dendritic cell-based interventions for cancer therapy. Oncoimmunology. 1 (7), 1111–1134. Taylor & Francis.
Galluzzi, L., et al., 2017. Immunogenic cell death in cancer and infectious disease. Nat. Rev. Immunol. 17 (2), 97. Nature Publishing Group.
Galt, S.W., Lindemann, S., Medd, D., Allen, L.L., Kraiss, L.W., Harris, E.S., Prescott, S.M., McIntyre, T.M., Weyrich, A.S., Zimmerman, G.A., 2001. Differential regulation of matrix metalloproteinase-9 by monocytes adherent to collagen and platelets. Circ. Res. 89 (6), 509–516.
Garcia-Gonzalez, P., et al., 2016. Tolerogenic dendritic cells for reprogramming of lymphocyte responses in autoimmune diseases. Autoimmun. Rev. 15 (11), 1071–1080. Elsevier.
Garg, A.D., Krysko, D.V., Verfaillie, T., Kaczmarek, A., Ferreira, G.B., Marysael, T., Rubio, N., Firczuk, M., Mathieu, C., Roebroek, A.J., et al., 2012. A novel pathway combining calreticulin exposure and ATP secretion in immunogenic cancer cell death. Embo J. 31, 1062–1079.
Garg, A.D., et al., 2017. Trial watch: immunogenic cell death induction by anticancer chemotherapeutics. Oncoimmunology 6 (12), e1386829. Taylor & Francis.
Geisel, J., et al., 2014. Sulforaphane protects from T cell–mediated autoimmune disease by inhibition of IL-23 and IL-12 in dendritic cells. J. Immunol. 192 (8), 3530–3539. Am Assoc Immnol.
Geissmann, F., Manz, M.G., Jung, S., Sieweke, M.H., Merad, M., Ley, K., 2010. Development of monocytes, macrophages, and dendritic cells. Science 327 (5966), 656–661.
Gerhardt, T., Ley, K., 2015. Monocyte trafficking across the vessel wall. Cardiovasc. Res. 107 (3), 321–330. Oxford University Press.
Ghiringhelli, F., Ménard, C., Terme, M., Flament, C., Taieb, J., Chaput, N., Puig, P.E., Novault, S., Escudier, B., Vivier, E., Lecesne, A., 2005. CD4+ CD25+ regulatory T cells inhibit natural killer cell functions in a transforming growth factor-β-dependent manner. J. Exp. Med. 202 (8), 1075–1085.
Gilboa, E., Vieweg, J., 2004. Cancer immunotherapy with mRNA-transfected dendritic cells. Immunol. Rev. 199, 251–263.
Gilboa, E., Nair, S.K., Lyerly, H.K., 1998. Immunotherapy of cancer with dendritic-cell-based vaccines. Cancer Immunol. Immunother. 46 (2), 82–87.
Gilleron, J., Querbes, W., Zeigerer, A., Borodovsky, A., Marsico, G., Schubert, U., Manygoats, K., Seifert, S., Andree, C., Stöter, M., Epstein-Barash, H., 2013. Image-based analysis of lipid nanoparticle-mediated siRNA delivery, intracellular trafficking and endosomal escape. Nat. Biotechnol. 31 (7), 638–646.
González, F.E., Gleisner, A., Falcón-Beas, F., Osorio, F., López, M.N., Salazar-Onfray, F., 2014. Tumor cell lysates as immunogenic sources for cancer vaccine design. Hum. Vaccin Immunother. 10 (11), 3261–3269.
Griffiths, E.A., et al., 2018. NY-ESO-1 vaccination in combination with decitabine induces antigen-specific T-lymphocyte responses in patients with myelodysplastic syndrome. Clin. Cancer Res. 24 (5), 1019–1029. AACR.
Grozescu, T., Popa, F., 2017. Immunotherapy and gene therapy in prostate cancer treatment. J. Med. Life 10 (1), 54. Carol Davila-University Press.
Grünebach, F., Kayser, K., Weck, M.M., Müller, M.R., Appel, S., Brossart, P., 2005. Cotransfection of dendritic cells with RNA coding for HER-2/neu and 4-1BBL increases the induction of tumor antigen specific cytotoxic T lymphocytes. Cancer Gene Ther. 12 (9), 749.
Gulley, J.L., et al., 2014. Immune impact induced by PROSTVAC (PSA-TRICOM), a therapeutic vaccine for prostate cancer. Cancer Immunol. Res. 2 (2), 133–141. AACR.

Hackstein, H., et al., 2003. Rapamycin inhibits IL-4—induced dendritic cell maturation in vitro and dendritic cell mobilization and function in vivo. Blood 101 (11), 4457–4463. Am Soc Hematology.

Haddad, E.A., Senger, L.K., Takei, F., 2009. An accessory role for B cells in the IL-12-induced activation of resting mouse NK cells. J. Immunol. 183 (6), 3608–3615.

Han P., Hanlon D., Filler R., Lee J.S., Robinson E., Arshad N., Sobolev O., Tatsuno K., Cote C., Rivera-Molina F., Yurter A., Fahmy T.M., Edelson R., Platelet P-selectin initiates cross-presentation and dendritic cell differentiation in monocytes. *Sci. Adv.* (in press).

Hanlon, D., Sobolev, O., Ventura, A., Vassall, A., Kibbi, N., Han, P., Yurter, A., Robinson, A., Filler, R., Tatsuno, K., Edelson, R.L., 2019. Rapid production of physiologic dendritic cells (phDC) for immunotherapy. Dendritic Cell Reprogramming. In: Methods in Molecular Biology. Springer. (in press).

Halvorsen, B., Smedbakken, L.M., Michelsen, A.E., Skjelland, M., Bjerkeli, V., Sagen, E.L., Taskén, K., Bendz, B., Gullestad, L., Holm, S., Biessen, E.A., 2014. Activated platelets promote increased monocyte expression of CXCR5 through prostaglandin E2-related mechanisms and enhance the anti-inflammatory effects of CXCL13. Atherosclerosis 234 (2), 352–359.

Haniffa, M., et al., 2012. Human tissues contain CD141hi cross-presenting dendritic cells with functional homology to mouse CD103+ nonlymphoid dendritic cells. Immunity 37 (1), 60–73. Elsevier.

Haniffa, M., Bigley, V., Collin, M., 2015. Human mononuclear phagocyte system reunited. In: Seminars in Cell & Developmental Biology. Elsevier, pp. 59–69.

Harris, K.M., 2011. Monocytes differentiated with GM-CSF and IL-15 initiate Th17 and Th1 responses that are contact-dependent and mediated by IL-15. J. Leukoc. Biol. 90 (4), 727–734. Wiley Online Library.

Harry, R.A., et al., 2010. Generation and characterisation of therapeutic tolerogenic dendritic cells for rheumatoid arthritis. Ann. Rheum. Dis. 69 (11), 2042–2050. BMJ Publishing Group Ltd.

Hedley, M.L., Curley, J., Urban, R., 1998. Microspheres Containing Plasmid-Encoded Antigens Elicit Cytotoxic T-Cell Responses. Nature Publishing Group.

Hegde, S., Pahne, J., Smola-Hess, S., 2004. Novel immunosuppressive properties of interleukin-6 in dendritic cells: inhibition of NF-κB binding activity and CCR7 expression. FASEB J. 18 (12), 1439–1441. Federation of American Societies for Experimental Biology.

Hespel, C., Moser, M., 2012. Role of inflammatory dendritic cells in innate and adaptive immunity. Eur. J. Immunol. 42 (10), 2535–2543.

Hong, E., et al., 2016. Configuration-dependent presentation of multivalent IL-15: IL-15Rα enhances the antigen-specific T cell response and anti-tumor immunity. J. Biol. Chem. 291 (17), 8931–8950. ASBMB.

Hornung, V., Guenthner-Biller, M., Bourquin, C., Ablasser, A., Schlee, M., Uematsu, S., Noronha, A., Manoharan, M., Akira, S., de Fougerolles, A., Endres, S., 2005. Sequence-specific potent induction of IFN-α by short interfering RNA in plasmacytoid dendritic cells through TLR7. Nat. Med. 11 (3), 263–270.

Hrkach, J., Von Hoff, D., Ali, M.M., Andrianova, E., Auer, J., Campbell, T., De Witt, D., Figa, M., Figueiredo, M., Horhota, A., Low, S., 2012. Preclinical development and clinical translation of a PSMA-targeted docetaxel nanoparticle with a differentiated pharmacological profile. Sci. Transl. Med. 4 (128), 128ra39.

Hu, Z., Ott, P.A., Wu, C.J., 2018. Towards personalized, tumour-specific, therapeutic vaccines for cancer. Nat. Rev. Immunol. 18 (3), 168.

Huangfu, D., Maehr, R., Guo, W., Eijkelenboom, A., Snitow, M., Chen, A.E., Melton, D.A., 2008. Induction of pluripotent stem cells by defined factors is greatly improved by small-molecule compounds. Nat. Biotechnol. 26 (7), 795–797.

Hubbell, J.A., Thomas, S.N., Swartz, M.A., 2009. Materials engineering for immunomodulation. Nature 462 (7272), 449–460.

Huber, M.L., et al., 2012. Interdisciplinary critique of sipuleucel-T as immunotherapy in castration-resistant prostate cancer. J. Natl. Cancer Inst. 104 (4), 273–279. Oxford University Press.

Hundal, J., Mardis, E.R., 2019. Personalized Cancer Vaccines in Clinical Trials. The Scientist.

Hyduk, S.J., Cybulsky, M.I., 2009. Role of α4β1 integrins in chemokine-induced monocyte arrest under conditions of shear stress. Microcirculation 16 (1), 17–30. Taylor & Francis.

Ifergan, I., Kébir, H., Bernard, M., Wosik, K., Dodelet-Devillers, A., Cayrol, R., Arbour, N., Prat, A., 2008. The blood-brain barrier induces differentiation of migrating monocytes into Th17-polarizing dendritic cells. Brain 131 (3), 785–799.

Iinuma, H., Okinaga, K., Fukushima, R., Inaba, T., Iwasaki, K., Okinaga, A., Takahashi, I., Kaneko, M., 2006. Superior protective and therapeutic effects of IL-12 and IL-18 gene-transduced dendritic neuroblastoma fusion cells on liver metastasis of murine neuroblastoma. J. Immunol. 176 (6), 3461–3469.

Inaba, K., et al., 1992. Generation of large numbers of dendritic cells from mouse bone marrow cultures supplemented with granulocyte/macrophage colony-stimulating factor. J. Exp. Med. 176 (6), 1693–1702. Rockefeller University Press.

Inaba, K., Inaba, M., Naito, M., et al., 1993a. Dendritic cell progenitors phagocytose particulates, including bacillus Calmette-Guerin organisms, and sensitize mice to mycobacterial antigens in vivo. J. Exp. Med. 178 (2), 479–488. Rockefeller University Press.

Inaba, K., Inaba, M., Deguchi, M., et al., 1993b. Granulocytes, macrophages, and dendritic cells arise from a common major histocompatibility complex class II-negative progenitor in mouse bone marrow. Proc. Natl. Acad. Sci. 90 (7), 3038–3042. National Acad Sciences.

Iruretagoyena, M.I., et al., 2006. Inhibition of nuclear factor-κB enhances the capacity of immature dendritic cells to induce antigen-specific tolerance in experimental autoimmune encephalomyelitis. J. Pharmacol. Exp. Ther. 318 (1), 59–67. ASPET.

Jakubzick, C.V., Randolph, G.J., Henson, P.M., 2017. Monocyte differentiation and antigen-presenting functions. Nat. Rev. Immunol. 17 (6), 349–362.

Jansen, J.H., et al., 1989. Inhibition of human macrophage colony formation by interleukin 4. J. Exp. Med. 170 (2), 577–582. Rockefeller University Press.

Jauregui-Amezaga, A., et al., 2015. Intraperitoneal administration of autologous tolerogenic dendritic cells for refractory Crohn's disease: a phase I study. J. Crohn's Colitis 9 (12), 1071–1078. Oxford University Press UK.

Jenne, L., Schuler, G., Steinkasserer, A., 2001. Viral vectors for dendritic cell-based immunotherapy. Trends Immunol. 22 (2), 102–107.

Jonuleit, H., et al., 1997. Pro-inflammatory cytokines and prostaglandins induce maturation of potent immunostimulatory dendritic cells under fetal calf serum-free conditions. Eur. J. Immunol. 27 (12), 3135–3142. Wiley Online Library.

Kalinski, P., Okada, H., 2010. Polarized dendritic cells as cancer vaccines: directing effector-type T cells to tumors. In: Seminars in Immunology. Elsevier, pp. 173–182.

Kaliński, P., et al., 1998. Prostaglandin E2 induces the final maturation of IL-12-deficient CD1a+ CD83+ dendritic cells: the levels of IL-12 are determined during the final dendritic cell maturation and are resistant to further modulation. J. Immunol. 161 (6), 2804–2809. Am Assoc Immnol.

Kaliński, P., et al., 2001. Prostaglandin E2 is a selective inducer of interleukin-12 p40 (IL-12p40) production and an inhibitor of bioactive IL-12p70 heterodimer. Blood 97 (11), 3466–3469. Am Soc Hematology.

Kalinski, P., et al., 2011. Dendritic cells in cancer immunotherapy: vaccines or autologous transplants? Immunol. Res. 50 (2–3), 235–247. Springer.

Kantoff, P.W., et al., 2010. Sipuleucel-T immunotherapy for castration-resistant prostate cancer. N. Engl. J. Med. 363 (5), 411–422. Mass Medical Soc.

Kazma, I., et al., 2013. Mycophenolic acid-treated dendritic cells generate regulatory CD4+ T cells that suppress CD8+ T cells' allocytotoxicity. Int. Immunol. 26 (3), 173–181. Oxford University Press UK.

Kedmi, R., Veiga, N., Ramishetti, S., Goldsmith, M., Rosenblum, D., Dammes, N., Hazan-Halevy, I., Nahary, L., Leviatan-Ben-Arye, S., Harlev, M., Behlke, M., 2018. A modular platform for targeted RNAi therapeutics. Nat. Nanotechnol. 13 (3), 214–219.

Keskin, D.B., et al., 2019. Neoantigen vaccine generates intratumoral T cell responses in phase Ib glioblastoma trial. Nature 565 (7738), 234.

Kibbi, N., et al., 2016. Induction of anti-tumor CD8 T cell responses by experimental ECP-induced human dendritic antigen presenting cells. Transfus. Apher. Sci. 55 (1), 146–152. https://doi.org/10.1016/j.transci.2016.06.001. Elsevier Ltd.

Kikuchi, T., Moore, M.A.S., Crystal, R.G., 2000. Dendritic cells modified to express CD40 ligand elicit therapeutic immunity against preexisting murine tumors. Blood 96 (1), 91–99. Am Soc Hematology.

Kim, J.T., Liu, Y., Kulkarni, R.P., Lee, K.K., Dai, B., Lovely, G., Ouyang, Y., Wang, P., Yang, L., Baltimore, D., 2017. Dendritic cell-targeted lentiviral vector immunization uses pseudotransduction and DNA-mediated STING and cGAS activation. Sci. Immunol. 2 (13). eaal1329. https://doi.org/10.1126/sciimmunol.aal1329.

Kirk, C.J., Mulé, J.J., 2000. Gene-modified dendritic cells for use in tumor vaccines. Hum. Gene Ther. 11 (6), 797–806.

Klein, T.M., Wolf, E.D., Wu, R., Sanford, J.C., 1987. High-velocity microprojectiles for delivering nucleic acids into living cells. Nature 327 (6117), 70–73.

Koya, R.C., Kasahara, N., Favaro, P.M., Lau, R., Ta, H.Q., Weber, J.S., Stripecke, R., 2003. Potent maturation of monocyte-derived dendritic cells after CD40L lentiviral gene delivery. J. Immunother. 26 (5), 451–460.

Kroemer, G., et al., 2013. Immunogenic cell death in cancer therapy. Annu. Rev. Immunol. 31, 51–72. Annual Reviews.

Kubo, S., et al., 2014. The JAK inhibitor, tofacitinib, reduces the T cell stimulatory capacity of human monocyte-derived dendritic cells. Ann. Rheum. Dis. 73 (12), 2192–2198. BMJ Publishing Group Ltd.

Kuckleburg, C.J., Yates, C.M., Kalia, N., Zhao, Y., Nash, G.B., Watson, S.P., Rainger, G.E., 2011. Endothelial cell-borne platelet bridges selectively recruit monocytes in human and mouse models of vascular inflammation. Cardiovasc. Res. 91 (1), 134–141.

Kumar, C., et al., 2017. A retrospective analysis comparing APCEDEN® dendritic cell immunotherapy with best supportive care in refractory cancer. Immunotherapy 9 (11), 889–897. Future Medicine.

Kvistborg, P., et al., 2009. Fast generation of dendritic cells. Cell. Immunol. 260 (1), 56–62. Elsevier.

Lahaye, X., Satoh, T., Gentili, M., Cerboni, S., Conrad, C., Hurbain, I., El Marjou, A., Lacabaratz, C., Lelièvre, J.D., Manel, N., 2013. The capsids of HIV-1 and HIV-2 determine immune detection of the viral cDNA by the innate sensor cGAS in dendritic cells. Immunity 39 (6), 1132–1142.

Lam, F.W., Vijayan, K.V., Rumbaut, R.E., 2011. Platelets and their interactions with other immune cells. Compr. Physiol. 5 (3), 1265–1280.

Langenkamp, A., Messi, M., Lanzavecchia, A., Sallusto, F., 2000. Kinetics of dendritic cell activation: impact on priming of TH1, TH2 and nonpolarized T cells. Nat. Immunol. 1 (4), 311–316.

Leal Rojas, I.M., et al., 2017. Human blood CD1c+ dendritic cells promote Th1 and Th17 effector function in memory CD4+ T cells. Front. Immunol. 8, 971. Frontiers.

Lee, J.-I., et al., 1999. Cyclosporine A inhibits the expression of costimulatory molecules on in vitro-generated dendritic cells: association with reduced nuclear translocation of nuclear factor kappa B1. Transplantation 68 (9), 1255–1263. LWW.
León, B., Ardavín, C., 2008. Monocyte migration to inflamed skin and lymph nodes is differentially controlled by L-selectin and PSGL-1. Blood 111 (6), 3126–3130. Am Soc Hematology.
Liau, L.M., et al., 2018. First results on survival from a large phase 3 clinical trial of an autologous dendritic cell vaccine in newly diagnosed glioblastoma. J. Transl. Med. 16 (1), 142. BioMed Central.
Little, S.R., Lynn, D.M., Ge, Q., Anderson, D.G., Puram, S.V., Chen, J., Eisen, H.N., Langer, R., 2004. Poly-β amino ester-containing microparticles enhance the activity of nonviral genetic vaccines. Proc. Natl. Acad. Sci. 101 (26), 9534–9539.
Little, S.R., Lynn, D.M., Puram, S.V., Langer, R., 2005. Formulation and characterization of poly (β amino ester) microparticles for genetic vaccine delivery. J. Control. Release 107 (3), 449–462. https://doi.org/10.1016/j.jconrel.2005.04.022. Elsevier.
Locy, H., et al., 2018. Dendritic cells: the tools for cancer treatment. In: Dendritic Cells. BoD–Books on Demand, p. 101.
Lowry, W.E., Richter, L., Yachechko, R., Pyle, A.D., Tchieu, J., Sridharan, R., Clark, A.T., Plath, K., 2008. Generation of human induced pluripotent stem cells from dermal fibroblasts. Proc. Natl. Acad. Sci. 105 (8), 2883–2888. https://doi.org/10.1073/pnas.0711983105.
Luo, Y., O'Hagan, D., Zhou, H., Singh, M., Ulmer, J., Reisfeld, R.A., Primus, F.J., Xiang, R., 2003. Plasmid DNA encoding human carcinoembryonic antigen (CEA) adsorbed onto cationic microparticles induces protective immunity against colon cancer in CEA-transgenic mice. Vaccine 21 (17–18), 1938–1947.
Luo, X., Tarbell, K.V., Yang, H., Pothoven, K., Bailey, S.L., Ding, R., Steinman, R.M., Suthanthiran, M., 2007. Dendritic cells with TGF-beta1 differentiate naive CD4 +CD25- T cells into islet-protective Foxp3+ regulatory T cells. Proc. Natl. Acad. Sci. 104 (8), 2821–2826.
Lutz, M.B., 2004. IL-3 in dendritic cell development and function: a comparison with GM-CSF and IL-4. Immunobiology 209 (1–2), 79–87. Elsevier.
Lutz, M.B., et al., 1999. An advanced culture method for generating large quantities of highly pure dendritic cells from mouse bone marrow. J. Immunol. Methods 223 (1), 77–92. Elsevier.
Lyman, S.D., Jacobsen, S.E.W., 1998. c-kit ligand and Flt3 ligand: stem/progenitor cell factors with overlapping yet distinct activities. Blood 91 (4), 1101–1134. Am Soc Hematology.
Macagno, A., Napolitani, G., Lanzavecchia, A., Sallusto, F., 2007. Duration, combination and timing: the signal integration model of dendritic cell activation. Trends Immunol. 28 (5), 227–233.
Mailliard, R.B., et al., 2004. α-type-1 polarized dendritic cells: a novel immunization tool with optimized CTL-inducing activity. Cancer Res. 64 (17), 5934–5937. AACR.
Maraskovsky, E., et al., 1996. Dramatic increase in the numbers of functionally mature dendritic cells in Flt3 ligand-treated mice: multiple dendritic cell subpopulations identified. J. Exp. Med. 184 (5), 1953–1962. Rockefeller University Press.
Maraskovsky, E., et al., 2000. In vivo generation of human dendritic cell subsets by Flt3 ligand. Blood 96 (3), 878–884. Am Soc Hematology.
Markowicz, S., Engleman, E.G., 1990. Granulocyte-macrophage colony-stimulating factor promotes differentiation and survival of human peripheral blood dendritic cells in vitro. J. Clin. Invest. 85 (3), 955–961. Am Soc Clin Investig.
Mascanfroni, I.D., et al., 2013. IL-27 acts on DCs to suppress the T cell response and autoimmunity by inducing expression of the immunoregulatory molecule CD39. Nat. Immunol. 14 (10), 1054. Nature Publishing Group.

Massa, C., Seliger, B., 2013. Fast dendritic cells stimulated with alternative maturation mixtures induce polyfunctional and long-lasting activation of innate and adaptive effector cells with tumor-killing capabilities. J. Immunol. 190, 3328–3337.

Mastelic-Gavillet, B., Balint, K., Boudousquié, C., Gannon, P.O., Kandalaft, L.E., 2019. Personalized dendritic cell vaccines—recent breakthroughs and encouraging clinical results. Front. Immunol. 10, 766.

Matsumoto, T., Hasegawa, H., Onishi, S., Ishizaki, J., Suemori, K., Yasukawa, M., 2013. Protein kinase C inhibitor generates stable human tolerogenic dendritic cells. J. Immunol. 191 (5), 2247–2257.

McKeever, U., et al., 2002. Protective immune responses elicited in mice by immunization with formulations of poly(lactide-co-glycolide) microparticles. Vaccine 20 (11–12), 1524–1531. https://doi.org/10.1016/S0264-410X(01)00509-6.

McKenzie, G.J., et al., 2005. Notch: a unique therapeutic target for immunomodulation. Expert Opin. Ther. Targets 9 (2), 395–410. https://doi.org/10.1517/14728222.9.2.395. Taylor & Francis.

Menges, M., et al., 2002. Repetitive injections of dendritic cells matured with tumor necrosis factor α induce antigen-specific protection of mice from autoimmunity. J. Exp. Med. 195 (1), 15–22. Rockefeller University Press.

Meyer, M., Wagner, E., 2006. pH-responsive shielding of non-viral gene vectors. Expert Opin. Drug Deliv. 3 (5), 563–571. https://doi.org/10.1517/17425247.3.5.563. Taylor & Francis.

Miller, J.C., et al., 2012. Deciphering the transcriptional network of the dendritic cell lineage. Nat. Immunol. 13 (9), 888. Nature Publishing Group.

Minkis, K., et al., 2008. Type 2 bias of T cells expanded from the blood of melanoma patients switched to type 1 by IL-12p70 mRNA–transfected dendritic cells. Cancer Res. 68 (22), 9441–9450. https://doi.org/10.1158/0008-5472.CAN-08-0900. AACR.

Mohty, M., et al., 2003. IFN-α skews monocyte differentiation into Toll-like receptor 7-expressing dendritic cells with potent functional activities. J. Immunol 171 (7), 3385–3393. Am Assoc Immnol.

Morrissey, D.V., et al., 2005. Potent and persistent in vivo anti-HBV activity of chemically modified siRNAs. Nat. Biotechnol. 23 (8), 1002–1007. https://doi.org/10.1038/nbt1122.

Morse, M.A., Lyerly, H.K., 2002. DNA and RNA modified dendritic cell vaccines. World J. Surg. 26 (7), 819–825. https://doi.org/10.1007/s00268-002-4058-0. Springer.

Morse, M.A., Lyerly, H.K., Li, Y., 1999. The role of IL-13 in the generation of dendritic cells in vitro. J. Immunother. 22 (6), 506–513.

Mosier, D.E., 1967. A requirement for two cell types for antibody formation in vitro. Science 158 (3808), 1573–1575. American Association for the Advancement of Science.

Mutyambizi, K., Berger, C.L., Edelson, R.L., 2009. The balance between immunity and tolerance: the role of Langerhans cells. Cell. Mol. Life Sci. 66 (5), 831–840. Springer.

Naik, S.H., Corcoran, L.M., Wu, L., 2005. Development of murine plasmacytoid dendritic cell subsets. Immunol. Cell Biol. 83 (5), 563–570. Wiley Online Library.

Nair, S., Archer, G.E., Tedder, T.F., 2012. Isolation and generation of human dendritic cells. Curr. Protoc. Immunol. 99 (1), 7–32. Wiley Online Library.

Nakamura, M., et al., 2002. Dendritic cells genetically engineered to simultaneously express endogenous tumor antigen and granulocyte macrophage colony-stimulating factor elicit potent therapeutic antitumor immunity. Clin. Cancer Res. 8 (8), 2742–2749. AACR.

Nicola, N.A., 1989. Hemopoietic cell growth factors and their receptors. Annu. Rev. Biochem. 58 (1), 45–77. Annual Reviews 4139 El Camino Way, PO Box 10139, Palo Alto, CA 94303-0139, USA.

O'Hagan, D.T., Singh, M., Gupta, R.K., 1998. Poly(lactide-co-glycolide) microparticles for the development of single-dose controlled-release vaccines. Adv. Drug Deliv. Rev 32 (3), 225–246.

O'Rourke, M.G.E., et al., 2003. Durable complete clinical responses in a phase I/II trial using an autologous melanoma cell/dendritic cell vaccine. Cancer Immunol. Immunother. 52 (6), 387–395. https://doi.org/10.1007/s00262-003-0375-x. Springer.

Ogawa, F., Iinuma, H., Okinaga, K., 2004. Dendritic cell vaccine therapy by immunization with fusion cells of interleukin-2 gene-transduced, spleen-derived dendritic cells and tumour cells. Scand. J. Immunol. 59 (5), 432–439. https://doi.org/10.1111/j.0300-9475.2004.01411.x.

Ojima, T., et al., 2006. The boosting effect of co-transduction with cytokine genes on cancer vaccine therapy using genetically modified dendritic cells expressing tumor-associated antigen. Int. J. Oncol. 28 (4), 947–953. Spandidos Publications.

Ojima, T., et al., 2007. Benefits of gene transduction of granulocyte macrophage colony-stimulating factor in cancer vaccine using genetically modified dendritic cells. Int. J. Oncol. 31 (4), 931–939. Spandidos Publications.

Okada, N., et al., 2005. Immunological properties and vaccine efficacy of murine dendritic cells simultaneously expressing melanoma-associated antigen and interleukin-12'. Cancer Gene Ther. 12 (1), 72–83. https://doi.org/10.1038/sj.cgt.7700772. Nature Publishing Group.

Osugi, Y., Vuckovic, S., Hart, D.N.J., 2002. Myeloid blood CD11c+ dendritic cells and monocyte-derived dendritic cells differ in their ability to stimulate T lymphocytes. Blood 100 (8), 2858–2866. Am Soc Hematology.

Ott, P.A., et al., 2017. An immunogenic personal neoantigen vaccine for patients with melanoma. Nature 547 (7662), 217.

Palliser, D., et al., 2006. An siRNA-based microbicide protects mice from lethal herpes simplex virus 2 infection. Nature 439 (7072), 89–94. https://doi.org/10.1038/nature04263.

Palucka, K., Banchereau, J., 2012. Cancer immunotherapy via dendritic cells. Nat. Rev. Cancer 12 (4), 265–277. https://doi.org/10.1038/nrc3258.

Palucka, K., Banchereau, J., 2013. Dendritic-cell-based therapeutic cancer vaccines. Immunity 39 (1), 38–48. Elsevier.

Palucka, A.K., et al., 2006. Dendritic cells loaded with killed allogeneic melanoma cells can induce objective clinical responses and MART-1 specific CD8+ T-cell immunity. J. Immunother. 29 (5), 545–557. https://doi.org/10.1097/01.cji.0000211309.90621.8b. LWW.

Palucka, K., et al., 2011. Dendritic cells and immunity against cancer. J. Intern. Med. 269 (1), 64–73Wiley Online Library.

Park, S.-J., et al., 2004. IL-6 regulates in vivo dendritic cell differentiation through STAT3 activation. J. Immunol. 173 (6), 3844–3854. Am Assoc Immnol.

Park, J., et al., 2009. PEGylated PLGA nanoparticles for the improved delivery of doxorubicin. Nanomedicine 5 (4), 410–418. https://doi.org/10.1016/j.nano.2009.02.002. Elsevier Inc.

Patente, T.A., et al., 2018. Human dendritic cells: their heterogeneity and clinical application potential in cancer immunotherapy. Front. Immunol. 9, 1–18.

Pellegatta, S., et al., 2018. Survival gain in glioblastoma patients treated with dendritic cell immunotherapy is associated with increased NK but not CD8+ T cell activation in the presence of adjuvant temozolomide. Oncoimmunology 7 (4), e1412901. Taylor & Francis.

Peng, J.C., Thomas, R., Nielsen, L.K., 2005. Generation and maturation of dendritic cells for clinical application under serum-free conditions. J. Immunother. 28 (6), 599–609. LWW.

Penna, G., Adorini, L., 2000. 1α, 25-Dihydroxyvitamin D3 inhibits differentiation, maturation, activation, and survival of dendritic cells leading to impaired alloreactive T cell activation. J. Immunol. 164 (5), 2405–2411. Am Assoc Immnol.
Pichlmair, A., et al., 2007. Tubulovesicular structures within vesicular stomatitis virus G protein-pseudotyped lentiviral vector preparations carry DNA and stimulate antiviral responses via toll-like receptor 9. J. Virol. 81 (2), 539–547. https://doi.org/10.1128/jvi.01818-06. Am Soc Microbiol.
Piemonti, L., et al., 1995. IL-13 supports differentiation of dendritic cells from circulating precursors in concert with GM-CSF. Eur. Cytokine Netw. 6 (4), 245–252.
Randolph, G.J., et al., 1998. Differentiation of monocytes into dendritic cells in a model of transendothelial trafficking. Science 282, 480–484. https://doi.org/10.1126/science.287.5452.495. October.
Randolph, G.J., Inaba, K., Robbiani, D.F., Steinman, R.M., Muller, W.A., 1999. Differentiation of phagocytic monocytes into lymph node dendritic cells in vivo. Immunity 11 (6), 753–761.
Rapoport, B.L., Anderson, R., 2019. Realizing the clinical potential of immunogenic cell death in cancer chemotherapy and radiotherapy. Int. J. Mol. Sci. 20 (4), E959.
Ratcliffe, N., et al., 2015. National institutes of health state of the science symposium in therapeutic apheresis: scientific opportunities in extracorporeal photopheresis. Transfus. Med. Rev. 29 (1), 62–70. Elsevier.
Ratta, M., et al., 2002. Dendritic cells are functionally defective in multiple myeloma: the role of interleukin-6. Blood 100 (1), 230–237. Am Soc Hematology.
Raval, J.S., Ratcliffe, N.R., 2018. Extracorporeal photopheresis and personalized medicine in the 21st century: the future's so bright!. J. Clin. Apher. 33 (4), 461–463. https://doi.org/10.1002/jca.21633.
Reid, C.D., et al., 1992. Interactions of tumor necrosis factor with granulocyte-macrophage colony-stimulating factor and other cytokines in the regulation of dendritic cell growth in vitro from early bipotent CD34+ progenitors in human bone marrow. J. Immunol. 149 (8), 2681–2688. Am Assoc Immnol.
Reis e Sousa, C., 2006. Dendritic cells in a mature age. Nat. Rev. Immunol. 6 (6), 476. Nature Publishing Group.
Ribas, A., 2005. Genetically modified dendritic cells for cancer immunotherapy. Curr. Gene Ther. 5 (6), 619–628.
Ribas, A., et al., 2000. Immune deviation and Fas-mediated deletion limit antitumor activity after multiple dendritic cell vaccinations in mice. Cancer Res. 60 (8), 2218–2224. AACR.
Rinde, M., 2019. Sipuleucel-T Shows Potential With New Trial Data, But Questions Regarding Clinical Relevance Remain, Targeted Oncology. Available at: https://www.targetedonc.com/publications/targeted-therapy-news/2019/april2-2019/sipuleucelt-shows-potential-with-new-trial-data-but-questions-regarding-clinical-relevance-remain.
Rogers, N.M., Kireta, S., Coates, P.T.H., 2010. Curcumin induces maturation-arrested dendritic cells that expand regulatory T cells in vitro and in vivo. Clin. Exp. Immunol. 162 (3), 460–473. Wiley Online Library.
Romani, N., et al., 1996. Generation of mature dendritic cells from human blood. An improved method with special regard to clinical applicability. J. Immunol. Methods 196 (2), 137–151. Elsevier.
Rosa, F.F., et al., 2018. Direct reprogramming of fibroblasts into antigen-presenting dendritic cells. Sci. Immunol. 3 (30), eaau4292. https://doi.org/10.1126/sciimmunol.aau4292.
Rosenberg, S.A., Yang, J.C., Restifo, N.P., 2004. Cancer immunotherapy: moving beyond current vaccines. Nat. Med. 10 (9), 909–915. https://doi.org/10.1038/nm1100.
Rughetti, A., et al., 2000. Transfected human dendritic cells to induce antitumor immunity. Gene Ther. 7 (17), 1458. https://doi.org/10.1038/sj.gt.3301266. Nature Publishing Group.

Sahin, U., et al., 2017. Personalized RNA mutanome vaccines mobilize poly-specific therapeutic immunity against cancer. Nature 547 (7662), 222.
Sakakibara, M., et al., 2006. Quick generation of fully mature dendritic cells from monocytes with OK432, low-dose prostanoid, and interferon-α as potent immune enhancers. J. Immunother. 29 (1), 67–77. LWW.
Sallusto, F., Lanzavecchia, A., 1994. Efficient presentation of soluble antigen by cultured human dendritic cells is maintained by granulocyte/macrophage colony-stimulating factor plus interleukin 4 and downregulated by tumor necrosis factor alpha. J. Exp. Med. 179 (4), 1109–1118. Rockefeller University Press.
Saluja, S.S., et al., 2014. Targeting human dendritic cells via DEC-205 using PLGA nanoparticles leads to enhanced cross-presentation of a melanoma-associated antigen. Int. J. Nanomedicine 9, 5231. Dove Press.
Sanders, N.N., et al., 2002. Pegylated GL67 lipoplexes retain their gene transfection activity after exposure to components of CF mucus. Gene Ther. 9 (6), 363. https://doi.org/10.1038/sj/gt/3301663. Nature Publishing Group.
Sanders, N., et al., 2009. Extracellular barriers in respiratory gene therapy. Adv. Drug Deliv. Rev. 61 (2), 115–127. https://doi.org/10.1016/j.addr.2008.09.011. Elsevier.
Sato, K., et al., 1999. Interleukin-13 is involved in functional maturation of human peripheral blood monocyte-derived dendritic cells. Exp. Hematol. 27 (2), 326–336. Elsevier.
Saxena, M., Bhardwaj, N., 2018. Re-emergence of dendritic cell vaccines for cancer treatment. Trends Cancer. 4 (2), 119–137. Elsevier.
Schadendorf, D., et al., 2006. Dacarbazine (DTIC) versus vaccination with autologous peptide-pulsed dendritic cells (DC) in first-line treatment of patients with metastatic melanoma: a randomized phase III trial of the DC study group of the DeCOG. Ann. Oncol. 17 (4), 563–570. Oxford University Press.
Schaefer, L., 2014. Complexity of danger: the diverse nature of damage-associated molecular patterns. J. Biol. Chem. 289 (51), 35237–35245. ASBMB.
Schaller, T.H., Sampson, J.H., 2017. Advances and challenges: dendritic cell vaccination strategies for glioblastoma. Expert Rev. Vaccines 16 (1), 27–36. Taylor & Francis.
Schreibelt, G., et al., 2016. Effective clinical responses in metastatic melanoma patients after vaccination with primary myeloid dendritic cells. Clin. Cancer Res. 22 (9), 2155–2166. AACR.
Schüler, T., Blankenstein, T., 2003. Cutting edge: CD8+ effector T cells reject tumors by direct antigen recognition but indirect action on host cells. J. Immunol. 170 (9), 4427–4431. https://doi.org/10.4049/jimmunol.170.9.4427. *(Baltimore, Md.: 1950).*
Schuler, G., Steinman, R.M., 1985. Murine epidermal Langerhans cells mature into potent immunostimulatory dendritic cells in vitro. J. Exp. Med. 161 (3), 526–546. Rockefeller University Press.
Schuler, G., Romani, N., Steinman, R.M., 1985. A comparison of murine epidermal Langerhans cells with spleen dendritic cells. J. Invest. Dermatol. 85, 99s–106s.
Schumacher, T.N., Schreiber, R.D., 2015. Neoantigens in cancer immunotherapy. Science 348 (6230), 69–74.
Segura, E., Amigorena, S., 2013. Inflammatory dendritic cells in mice and humans. Trends Immunol. 34 (9), 440–445. Elsevier.
Shen, L., et al., 2004. Silencing of SOCS1 enhances antigen presentation by dendritic cells and antigen-specific anti-tumor immunity. Nat. Biotechnol. 22 (12), 1546–1553. https://doi.org/10.1038/nbt1035.
Shen, X., et al., 2008. Development of immunogenic tumor-loaded dendritic cells through physical perturbation and apoptotic cell loading. Immunol. Invest. 37 (8), 798–821. Taylor & Francis.
Shi, L., et al., 2006. DIgR2, dendritic cell-derived immunoglobulin receptor 2, is one representative of a family of IgSF inhibitory receptors and mediates negative regulation of

dendritic cell-initiated antigen-specific T-cell responses. Blood 108 (8), 2678–2686. https://doi.org/10.1182/blood-2006-04-015404. Am Soc Hematology.

Siena, S., et al., 1995. Massive ex vivo generation of functional dendritic cells from mobilized CD34+ blood progenitors for anticancer therapy. Exp. Hematol. 23 (14), 1463–1471.

Singh, M., et al., 2000. Cationic microparticles: a potent delivery system for DNA vaccines. Proc. Natl. Acad. Sci. 97 (2), 811–816. National Acad Sciences.

Song, X.-T., et al., 2008. A20 is an antigen presentation attenuator, and its inhibition overcomes regulatory T cell mediated suppression. Nat. Med. 14 (3), 258. https://doi.org/10.1038/nm1721. Nature Publishing Group.

Song, S.-S., et al., 2014. TGF-β favors bone marrow-derived dendritic cells to acquire tolerogenic properties. Immunol. Invest. 43 (4), 360–369. Taylor & Francis.

Soutschek, J., et al., 2004. Therapeutic silencing of an endogenous gene by systemic administration of modified siRNAs. Nature 432 (7014), 173. Nature Publishing Group.

Sprooten, J., Ceusters, J., Coosemans, A., Agostinis, A., Vleeschouwer, S., Zitvogel, L., Kroemer, G., Galluzzi, L., Garg, A., 2019. Trial watch: dendritic cell vaccination for cancer immunotherapy. OncoImmunology 8, 11.

Stallwood, Y., et al., 2006. Small interfering RNA-mediated knockdown of notch ligands in primary CD4+ T cells and dendritic cells enhances cytokine production. J. Immunol. 177 (2), 885–895. https://doi.org/10.4049/jimmunol.177.2.885. Am Assoc Immnol.

Steinbrink, K., et al., 1997. Induction of tolerance by IL-10-treated dendritic cells. J. Immunol. 159 (10), 4772–4780. Am Assoc Immnol.

Steiner, G., et al., 1985. Epidermal cells as accessory cells in the generation of allo-reactive and hapten-specific cytotoxic T lymphocyte (CTL) responses. J. Immunol. 134 (2), 736–741. Am Assoc Immnol.

Steinman, R.M., 2007. Dendritic cells: understanding immunogenicity. Eur. J. Immunol. 37 (Suppl. 1), 53–60. https://doi.org/10.1002/eji.200737400.

Steinman, R.M., 2012. Decisions about dendritic cells: past, present, and future. Annu. Rev. Immunol. 30 (1), 1–22. https://doi.org/10.1146/annurev-immunol-100311-102839.

Steinman, R.M., Banchereau, J., 2007. Taking dendritic cells into medicine. Nature 449 (7161), 419–426. https://doi.org/10.1038/nature06175.

Steinman, R.M., Cohn, Z.A., 1973. Identification of a novel cell type in peripheral lymphoid organs of mice: I. Morphology, quantitation, tissue distribution. J. Exp. Med. 137 (5), 1142–1162. Rockefeller University Press.

Steinman, R.M., Cohn, Z.A., 1974. Identification of a novel cell type in peripheral lymphoid organs of mice: II. Functional properties in vitro. J. Exp. Med. 139 (2), 380–397. Rockefeller University Press.

Steinman, R.M., Mellman, I., 2004. Immunotherapy: bewitched, bothered, and bewildered no more. Science 305 (5681), 197–200. https://doi.org/10.1126/science.1099688.

Steinman, R.M., Hawiger, D., Nussenzweig, M.C., 2003. Tolerogenic dendritic cells. Annu. Rev. Immunol. 21 (1), 685–711. Annual Reviews 4139 El Camino Way, PO Box 10139, Palo Alto, CA 94303-0139, USA.

Sun, C.-M., et al., 2007. Small intestine lamina propria dendritic cells promote de novo generation of Foxp3 T reg cells via retinoic acid. J. Exp. Med. 204 (8), 1775–1785. Rockefeller University Press.

Suzuki, J., Hamada, E., Shodai, T., Kamoshida, G., Kudo, S., Itoh, S., Koike, J., Nagata, K., Irimura, T., Tsuji, T., 2013. Cytokine secretion from human monocytes potentiated by P-selectin-mediated cell adhesion. Int. Arch. Allergy Immunol. 160 (2), 152–160.

Švajger, U., Obermajer, N., Jeras, M., 2010. Dendritic cells treated with resveratrol during differentiation from monocytes gain substantial tolerogenic properties upon activation. Immunology 129 (4), 525–535. Wiley Online Library.

Szabolcs, P., Moore, M.A., Young, J.W., 1995. Expansion of immunostimulatory dendritic cells among the myeloid progeny of human CD34+ bone marrow precursors cultured

with c-kit ligand, granulocyte-macrophage colony-stimulating factor, and TNF-alpha. J. Immunol. 154 (11), 5851–5861. Am Assoc Immnol.
Tacken, P.J., et al., 2007. Dendritic-cell immunotherapy: from ex vivo loading to in vivo targeting. Nat. Rev. Immunol. 7 (10), 790. Nature Publishing Group.
Takenaka, M.C., Quintana, F.J., 2017. Tolerogenic dendritic cells. In: Seminars in Immunopathology. Springer, pp. 113–120.
Tang, D.C., Devit, M., Johnston, S.A., 1992. Genetic immunization is a simple method for eliciting an immune response. Nature 356 (6365), 152–154. https://doi.org/10.1038/356152a0.
Tatsuno, K., et al., 2019. Extracorporeal photochemotherapy induces bona fide immunogenic cell death. Cell Death Dis. 10 (8), 578. Nature Publishing Group.
Tcherepanova, I.Y., et al., 2008. Ectopic expression of a truncated CD40L protein from synthetic post-transcriptionally capped RNA in dendritic cells induces high levels of IL-12 secretion. BMC Mol. Biol. 9 (1), 1–13. https://doi.org/10.1186/1471-2199-9-90. BioMed Central.
Tel, J., et al., 2013. Natural human plasmacytoid dendritic cells induce antigen-specific T-cell responses in melanoma patients. Cancer Res. 73 (3), 1063–1075. AACR.
Ten Brinke, A., et al., 2007. The clinical grade maturation cocktail monophosphoryl lipid A plus IFNγ generates monocyte-derived dendritic cells with the capacity to migrate and induce Th1 polarization. Vaccine 25 (41), 7145–7152. Elsevier.
Thurner, B., et al., 1999. Generation of large numbers of fully mature and stable dendritic cells from leukapheresis products for clinical application. J. Immunol. Methods 223 (1), 1–15. Elsevier.
Timmerman, J.M., et al., 2002. Idiotype-pulsed dendritic cell vaccination for B-cell lymphoma: clinical and immune responses in 35 patients. Blood 99 (5), 1517–1526. https://doi.org/10.1182/blood.V99.5.1517. Am Soc Hematology.
Tuyaerts, S., et al., 2003. Induction of Influenza Matrix Protein 1 and MelanA-specific T lymphocytes in vitro using mRNA-electroporated dendritic cells. Cancer Gene Ther. 10 (9), 696. https://doi.org/10.1038/sj.cgt.7700622. Nature Publishing Group.
Tuyaerts, S., Aerts, J.L., et al., 2007a. Current approaches in dendritic cell generation and future implications for cancer immunotherapy. Cancer Immunol. Immunother. 56 (10), 1513–1537. https://doi.org/10.1007/s00262-007-0334-z. Springer.
Tuyaerts, S., Van Meirvenne, S., et al., 2007b. Expression of human GITRL on myeloid dendritic cells enhances their immunostimulatory function but does not abrogate the suppressive effect of CD4+CD25+ regulatory T cells. J. Leukoc. Biol. 82 (1), 93–105. https://doi.org/10.1189/jlb.0906568. Wiley Online Library.
Ulmer, J.B., et al., 1993. Heterologous protection against influenza by injection of DNA encoding a viral protein. Science 259 (5102), 1745–1749. American Association for the Advancement of Science.
Van Gulijk, M., Dammeijer, F., Aerts, J.G., Vroman, H., 2018. Combination strategies to optimize efficacy of dendritic cell-based immunotherapy. Front. Immunol. 9, 2759.
Vandenberk, L., Belmans, J., Van Woensel, M., Riva, M., Van Gool, SW., 2016. Exploiting the immunogenic potential of cancer cells for improved dendritic cell vaccines. Front. Immunol 6, 663.
Van Tendeloo, V.F.I., et al., 2001. Highly efficient gene delivery by mRNA electroporation in human hematopoietic cells: superiority to lipofection and passive pulsing of mRNA and to electroporation of plasmid cDNA for tumor antigen loading of dendritic cells. Blood 98 (1), 49–56. Am Soc Hematology.
Ventura, A., et al., 2018. Extracorporeal photochemotherapy drives monocyte-to-dendritic cell maturation to induce anticancer immunity. Cancer Res. 78 (14), 4045–4058. https://doi.org/10.1158/0008-5472.CAN-18-0171.

Ventura, A., Vassall, A., Yurter, A., Robinson, E., Filler, R., Hanlon, D., Meeth, K., Ezaldein, H., Girardi, M., Sobolev, O., Bosenberg, M.W., 2019. Novel protocol for generating physiologic immunogenic dendritic cells. J. Vis. Exp. (147), e59370.
Vermaelen, K., 2019. Vaccine strategies to improve anti-cancer cellular immune responses. Front. Immunol. 10, 8.
Vieira, P.L., et al., 2000. Development of Th1-inducing capacity in myeloid dendritic cells requires environmental instruction. J. Immunol. 164 (9), 4507–4512. Am Assoc Immnol.
Vilella, R., et al., 2004. Pilot study of treatment of biochemotherapy-refractory stage IV melanoma patients with autologous dendritic cells pulsed with a heterologous melanoma cell line lysate. Cancer Immunol. Immunother. 53 (7), 651–658. Springer.
Villani, A.-C., et al., 2017. Single-cell RNA-seq reveals new types of human blood dendritic cells, monocytes, and progenitors. Science 356 (6335), eaah4573. https://doi.org/10.1126/science.aah4573.
Vopenkova, K., et al., 2012. Complex evaluation of human monocyte-derived dendritic cells for cancer immunotherapy. J. Cell. Mol. Med. 16 (11), 2827–2837. Wiley Online Library.
Vremec, D., et al., 1997. The influence of granulocyte/macrophage colony-stimulating factor on dendritic cell levels in mouse lymphoid organs. Eur. J. Immunol. 27 (1), 40–44. https://doi.org/10.1002/eji.1830270107. Wiley Online Library.
Walter, E., et al., 1999. Microencapsulation of DNA using poly(DL-lactide-co-glycolide): stability issues and release characteristics. J. Control. Release 61 (3), 361–374. https://doi.org/10.1016/S0168-3659(99)00151-0.
Walter, E., Dreher, D., Kok, M., Thiele, L., Kiama, S.G., Gehr, P., Merkle, H.P., 2001a. Hydrophilic poly(DL-lactide-co-glycolide) microspheres for the delivery of DNA to human-derived macrophages and dendritic cells. J. Control. Release 76 (1–2), 149–168.
Walter, E., Thiele, L., Merkle, H.P., 2001b. Gene delivery systems to phagocytic antigen-presenting cells. S.T.P. Pharma Sciences 11 (1), 45–56. Editions de santé.
Wang, C., et al., 2004. Molecularly engineered poly (ortho ester) microspheres for enhanced delivery of DNA vaccines. Nat. Mater. 3 (3), 190. Nature Publishing Group.
Wang, F.L., Qin, W.J., Wen, W.H., Tian, F., Song, B., Zhang, Q., Lee, C., Guo, Y.L., Wang, H., 2007. TGF-β insensitive dendritic cells: an efficient vaccine for murine prostate cancer. Cancer Immunol. Immunother. 56 (11), 1785–1793.
Watchmaker, P.B., et al., 2010. Independent regulation of chemokine responsiveness and cytolytic function versus CD8+ T cell expansion by dendritic cells. J. Immunol. 184 (2), 591–597. Am Assoc Immnol.
Weiser, C., Petkova, M.V., Rengstl, B., Döring, C., von Laer, D., Hartmann, S., Küppers, R., Hansmann, M.L., Newrzela, S., 2018. Ectopic expression of transcription factor BATF3 induces B-cell lymphomas in a murine B-cell transplantation model. Oncotarget 9 (22), 15942–15951.
Wennerberg, E., Vanpouille-Box, C., Bornstein, S., Yamazaki, T., Demaria, S., Galluzzi, L., 2017. Immune recognition of irradiated cancer cells. Immunol. Rev. 280 (1), 220–230.
Weyrich, A.S., McIntyre, T.M., McEver, R.P., Prescott, S.M., Zimmerman, G.A., 1995. Monocyte tethering by P-selectin regulates monocyte chemotactic protein-1 and tumor necrosis factor-alpha secretion. Signal integration and NF-kappa B translocation. J. Clin. Invest. 95 (5), 2297–2303.
Weyrich, A.S., Elstad, M.R., McEver, R.P., McIntyre, T.M., Moore, K.L., Morrissey, J.H., Prescott, S.M., Zimmerman, G.A., 1996. Activated platelets signal chemokine synthesis by human monocytes. J. Clin. Investig. 97 (6), 1525–1534.
Wilgenhof, S., et al., 2016. Phase II study of autologous monocyte-derived mRNA electroporated dendritic cells (TriMixDC-MEL) plus ipilimumab in patients with pretreated advanced melanoma'. J. Clin. Oncol. 34 (12), 1330–1338. American Society of Clinical Oncology.

Williams, R.S., Johnston, S.A., Riedy, M., DeVit, M.J., McElligott, S.G., Sanford, J.C., 1991. Introduction of foreign genes into tissues of living mice by DNA-coated microprojectiles. Proc. Natl. Acad. Sci. 88 (7), 2726–2730.

Witmer-Pack, M.D., et al., 1987. Granulocyte/macrophage colony-stimulating factor is essential for the viability and function of cultured murine epidermal Langerhans cells. J. Exp. Med. 166 (5), 1484–1498. Rockefeller University Press.

Wolff, J.A., Malone, R.W., Williams, P., Chong, W., Acsadi, G., Jani, A., Felgner, P.L., 1990. Direct gene transfer into mouse muscle in vivo. Science 247 (4949), 1465–1468. https://doi.org/10.1126/science.1690918.

Wysocki, P.J., Grabarczyk, P., Mackiewicz-Wysocka, M., Kowalczyk, D.W., Mackiewicz, A., 2002. Genetically modified dendritic cells—a new, promising cancer treatment strategy? Expert Opin. Biol. Ther. 2 (8), 835–845.

Xia, C., et al., 2005. Dexamethasone induces IL-10-producing monocyte-derived dendritic cells with durable immaturity. Scand. J. Immunol. 62 (1), 45–54. Wiley Online Library.

Xie, J., Qian, J., Yang, J., Wang, S., Freeman III, M.E., Yi, Q., 2005. Critical roles of Raf/MEK/ERK and PI3K/AKT signaling and inactivation of p38 MAP kinase in the differentiation and survival of monocyte-derived immature dendritic cells. Exp. Hematol. 33 (5), 564–572.

Xu, Y., et al., 2007. Differential development of murine dendritic cells by GM-CSF versus Flt3 ligand has implications for inflammation and trafficking. J. Immunol. 179 (11), 7577–7584. Am Assoc Immnol.

Yang, N.-S., Sun, W.H., 1995. Gene gun and other non-viral approaches for cancer gene therapy. Nat. Publ. Group 13, 5–9. October.

Yang, N.S., et al., 1990. In vivo and in vitro gene transfer to mammalian somatic cells by particle bombardment. Proc. Natl. Acad. Sci. U. S. A. 87 (24), 9568–9572.

Zhang, Y., et al., 1997. Induction of dendritic cell differentiation by granulocyte-macrophage colony-stimulating factor, stem cell factor, and tumor necrosis factor α in vitro from lineage phenotypes-negative c-kit+ murine hematopoietic progenitor cells. Blood 90 (12), 4842–4853. Am Soc Hematology

Zimmermann, T.S., et al., 2006. RNAi-mediated gene silencing in non-human primates. Nature 441 (7089), 111. Nature Publishing Group.

Zobywalski, A., et al., 2007. Generation of clinical grade dendritic cells with capacity to produce biologically active IL-12p70. J. Transl. Med. 5 (1), 18.

Zuchtriegel, G., Uhl, B., Puhr-Westerheide, D., Pörnbacher, M., Lauber, K., Krombach, F., Reichel, C.A., 2016. Platelets guide leukocytes to their sites of extravasation. PLoS Biol. 14 (5), 1–28. https://doi.org/10.1371/journal.pbio.1002459.

Zufferey, A., Schvartz, D., Nolli, S., Reny, J.L., Sanchez, J.C., Fontana, P., 2014. Characterization of the platelet granule proteome: evidence of the presence of MHC1 in alpha-granules. J. Proteomics 101, 130–140. https://doi.org/10.1016/j.jprot.2014.02.008. Elsevier B.V.

FSC
www.fsc.org
MIX
Papier | Fördert gute Waldnutzung
FSC® C083411

Kolibri 360 GmbH
Ferdinand-Jühlke-Straße 7
99095 Erfurt, Deutschland
produktsicherheit@kolibri360.de